VOLUME ONE HUNDRED AND THIRTY SIX

ADVANCES IN
CANCER RESEARCH

Selenium and Selenoproteins in Cancer

VOLUME ONE HUNDRED AND THIRTY SIX

ADVANCES IN CANCER RESEARCH

Selenium and Selenoproteins in Cancer

Edited by

KENNETH D. TEW
*Medical University of South Carolina,
Charleston, SC, United States*

FRANCESCO GALLI
*Laboratory of Clinical Biochemistry and
Human Nutrition,
Department of Pharmaceutical Sciences,
University of Perugia,
Perugia, Italy*

ELSEVIER

ACADEMIC PRESS
An imprint of Elsevier

Academic Press is an imprint of Elsevier
50 Hampshire Street, 5th Floor, Cambridge, MA 02139, United States
525 B Street, Suite 1800, San Diego, CA 92101-4495, United States
The Boulevard, Langford Lane, Kidlington, Oxford OX5 1GB, United Kingdom
125 London Wall, London, EC2Y 5AS, United Kingdom

First edition 2017

ISBN: 978-0-12-812016-3
ISSN: 0065-230X

For information on all Academic Press publications
visit our website at https://www.elsevier.com/books-and-journals

Publisher: Zoe Kruze
Acquisition Editor: Zoe Kruze
Editorial Project Manager: Fenton Coulthurst
Production Project Manager: James Selvam
Cover Designer: Mark Rogers

Typeset by SPi Global, India

CONTENTS

CONTRIBUTORS

Elias S.J. Arnér
Karolinska Institutet, Stockholm, Sweden

Desirée Bartolini
University of Perugia, Perugia, Italy

Silvia Cilloni
Environmental, Genetic and Nutritional Epidemiology Research Center (CREAGEN), University of Modena and Reggio Emilia, Modena, Italy

Catherine M. Crespi
Jonsson Comprehensive Cancer Center, UCLA Fielding School of Public Health, University of California, Los Angeles, CA, United States

Bastihalli T. Diwakar
Center for Molecular Immunology and Infectious Disease and Center for Molecular Toxicology and Carcinogenesis, The Pennsylvania State University, University Park, PA, United States

Andreza Fabro de Bem
Center of Biological Sciences (CCB), Federal University of Santa Catarina (UFSC), Florianópolis, SC; Institute of Biological Sciences, University of Brasilia, Brasilia, Brazil

Tommaso Filippini
Environmental, Genetic and Nutritional Epidemiology Research Center (CREAGEN), University of Modena and Reggio Emilia, Modena, Italy

Camile C. Fontelles
Faculty of Pharmaceutical Sciences, University of São Paulo, São Paulo-SP, Brazil

Francesco Galli
University of Perugia, Perugia, Italy

Peter R. Hoffmann
John A. Burns School of Medicine, University of Hawaii, Honolulu, HI, United States

Ewa Jabłońska
Nofer Institute of Occupational Medicine, Lodz, Poland

Anna P. Kipp
Institute of Nutrition, Friedrich Schiller University Jena, Jena, Germany

Arvind M. Korwar
Center for Molecular Immunology and Infectious Disease and Center for Molecular Toxicology and Carcinogenesis, The Pennsylvania State University, University Park, PA, United States

Michael P. Marciel
John A. Burns School of Medicine, University of Hawaii, Honolulu, HI, United States

Giuseppe Murdolo
Section of Internal Medicine, Endocrine and Metabolic Sciences, University of Perugia, Perugia, Italy

Thomas P. Ong
Faculty of Pharmaceutical Sciences, University of São Paulo; Food Research Center (FoRC), São Paulo-SP, Brazil

Robert F. Paulson
Center for Molecular Immunology and Infectious Disease and Center for Molecular Toxicology and Carcinogenesis, The Pennsylvania State University, University Park, PA, United States

Marta Piroddi
University of Perugia, Perugia, Italy

K. Sandeep Prabhu
Center for Molecular Immunology and Infectious Disease and Center for Molecular Toxicology and Carcinogenesis, The Pennsylvania State University, University Park, PA, United States

Rafael Radi
Center for Free Radical and Biomedical Research (CEINBIO), Universidad de la República, Montevideo, Uruguay

Edyta Reszka
Nofer Institute of Occupational Medicine, Lodz, Poland

Luca Sancineto
University of Perugia, Perugia, Italy

Claudio Santi
University of Perugia, Perugia, Italy

Sarah P. Short
Vanderbilt University Medical Center, Nashville, TN, United States

Kenneth D. Tew
Medical University of South Carolina, Charleston, SC, United States

Pierangelo Toquato
University of Perugia, Perugia, Italy

Cristina Tortoioli
Section of Internal Medicine, Endocrine and Metabolic Sciences, University of Perugia, Perugia, Italy

Marco Vinceti
Environmental, Genetic and Nutritional Epidemiology Research Center (CREAGEN), University of Modena and Reggio Emilia, Modena, Italy; Boston University School of Public Health, Boston, MA, United States

Christopher S. Williams
Vanderbilt University School of Medicine, Vanderbilt University; Vanderbilt Ingram Cancer Center, Vanderbilt University Medical Center; Veterans Affairs Tennessee Valley HealthCare System, Nashville, TN, United States

PREFACE

This volume of *Advances in Cancer Research* collects 10 chapters contributed by a group of scholars and opinion leaders in the field of selenium and selenoproteins in the different aspects of cancer biology and therapy. This group of distinguished authors discussed this subject starting from nutritional, cellular, and molecular aspects of Se in normal and cancer cells, to possible pharmacological and therapeutic applications of Se compounds in different types of malignancies.

Epidemiology and clinical data on nutritional, chemoprevention, and chemotherapy properties of Se are critically examined in Chapter 1. An adequate intake of Se is demonstrated to help protecting cells and to prevent carcinogenesis in its different phases; this level of evidence and an impressive number of promising in vitro and preclinical reports supported the investigation of Se compounds as anticancer agents in some clinical trials carried out on different types of malignancies (inorganic and organic forms of these compounds used in these studies have been reviewed in other chapters and particularly in Chapter 10). Rather frustrating or even negative conclusions were drawn from such trials that have been the subject of recent meta-analyses, and even adverse effects and potential health risks were reported for certain therapy protocols. Such disappointing findings and the apparent conflict between promising preclinical data and negative clinical results clearly demonstrate the need for further research to better understand biochemistry and pharmacology aspects of Se before planning for further investigation either in cancer prevention or in therapy protocols.

Although the biochemistry and molecular biology of Se have extensively been investigated in the last years with most sophisticated laboratory approaches and omics technology, we still have much to know on how this elemental species can affect cancer in its different and complex mechanisms and phases of progression, and how the biology of Se in its various aspects may keep normal and cancer cells in the tumor environment as separate targets of therapy, hence resulting in specific, efficient, and well-tolerated treatments.

Once introduced in the human organism as food micronutrient, Se is incorporated into selenoproteins, which are described in the context of tumorigenesis and cancer progression in Chapter 2, and metastasization in

Chapter 3. Selenoproteins promote essential homeostatic functions, such as the detoxification of cellular and extracellular oxidants, membrane lipid metabolism and signaling, intermediary metabolism, redox regulation of protein thiols with effects on signal transduction and transcriptional pathways, stress response, cell cycle regulation, differentiation and death programs, as well as iodine and thyroid hormone metabolism. The family of glutathione peroxidases described in Chapter 4 and that of thioredoxin reductase discussed in Chapter 5 encompass two main groups of Se proteins involved in several of these functions, the role of which in cancer prevention was recognized since the beginning of their identification.

Selenium is present in these proteins essentially under the form of Se cysteine (Sec) and rarely as Se methionine (Sem). This is the exclusive way for Se to be integrated and to exert its physiological functions in human tissues; indeed, under the free form, for instance as free Sec/Sem or inorganic Se, it cannot produce the same functions of the organized protein form; rather, it harms cells and tissue components being responsible for the generation of reactive species and redox perturbation and depletion of protein Cys residues and free thiols. If these aspects of Se toxicology provide the actual anticancer mechanism of Se compounds, it is thus difficult to imagine straightforward applications of these compounds in cancer therapy without side effects; at the same time, it is difficult to consider the anticancer potential of Se compounds unchained from their processing into the cellular metabolism from either the incorporation in Se proteins or transformation to activated metabolites through detoxification pathways and spontaneous reaction processes that are discussed in Chapter 10. These aspects of cellular biochemistry and metabolism of Se make pharmacokinetics and dynamics of Se compounds quite difficult to decipher and complicate to follow at the clinical level with speciation analysis and other laboratory protocols. On the other hand, they represent a double-edged sword in designing cancer-specific strategies of intervention being the tumor microenvironment (discussed in Chapter 6) composed of several entities with completely different settings in terms of role, gene expression patterns, and functional control of cellular processes, such as cell cycle progression, energy metabolism and redox homeostasis, detoxification/drug metabolism and production of mediators of inflammation and growth factors, repression of immune surveillance and angiogenesis.

In this respect, the expression and activity of Se proteins, such as glutathione peroxidase or thioredoxin reductase isoforms (discussed in Chapter 4 and 5, respectively), vary in the cellular components of the tumor microenvironment and during cancer progression. Both transcriptional and functional regulation mechanisms are responsible for these changes, which appear to affect the carcinogenic process in all its components. Their characterization is still in progress in different types of human cancers also with the aim of identify cancer-specific therapies that include the targeting of these Se proteins with Se compounds.

These and other aspects of Se metabolism and function are examined in detail in Chapter 7 as far as breast cancer is concerned. This type of cancer represents the first and second cause of death by cancer in women from less developed and more developed regions, respectively.

Epigenetics of cancer and the role of Se in this context are described in detail in Chapter 8; this is an emerging subject in cancer and selenoprotein gene regulation, which holds the potential to identify cancer-specific therapies.

Authors in Chapter 9 have reviewed the available information recently appeared in the literature on the role that Se may have in the different aspects of cancer stem cell biology, also including early, but suggestive, evidence on therapy protocols with Se compounds in this selected population of cancer cells to counteract their self-renewal, differentiation, and diffusion properties. The role of cancer stem cells in hematological cancers, and particulalry in leukemias, is also discussed in Chapter 6, focusing on the impact that Se compounds may have on the interaction between leukemia stem cells, other tumor cells, and the immune components of this type of cancer.

The negative findings of the past generation of clinical trials on Se compounds have led to reconsider drug development and intervention strategies. Chapter 10 provides an overview on the series of inorganic and organic forms of Se that have been investigated as chemoprevention and chemotherapy agents in different types of solid tumors and hematological cancers. The design and synthesis of a series of Se-organic compounds in the last years are providing new clinical perspectives pointing to a generation of molecules with increased efficacy and specificity of action already assessed at the preclinical level (also described in Chapter 10). Now, which is the chance for the new generations of Se-organic compounds to enter clinical trials is premature to say, but some attempts are in progress and others can be anticipated by the available information in the literature.

What emerges from this thematic collection is that the interest in the scientific community on Se and selenoproteins in cancer is keeping high as it is for the quality of scientific production. In this context, the possibility to develop successful therapy protocols based on this elemental molecule and selenoprotein manipulation continues to represent promising and fascinating aspects of cancer research.

The editors would like to express their gratitude to all the authors who participated to such editorial initiative that we are sure will be of great interest for the readers of the *Advances in Cancer Research* series.

Kenneth D. Tew
Medical University of South Carolina,
Charleston, SC, United States

Francesco Galli
Laboratory of Clinical Biochemistry and Human Nutrition
Department of Pharmaceutical Sciences
University of Perugia, Perugia, Italy

CHAPTER ONE

The Epidemiology of Selenium and Human Cancer

Marco Vinceti[*,†,1], **Tommaso Filippini**[*], **Silvia Cilloni**[*], **Catherine M. Crespi**[‡]

[*]Environmental, Genetic and Nutritional Epidemiology Research Center (CREAGEN), University of Modena and Reggio Emilia, Modena, Italy
[†]Boston University School of Public Health, Boston, MA, United States
[‡]Jonsson Comprehensive Cancer Center, UCLA Fielding School of Public Health, University of California, Los Angeles, CA, United States
[1]Corresponding author: e-mail address: marco.vinceti@unimore.it

Contents

Abstract

The relation between selenium and cancer has been one of the most hotly debated topics in human health over the last decades. Early observational studies reported an inverse relation between selenium exposure and cancer risk. Subsequently, randomized controlled trials showed that selenium supplementation does not reduce the risk of cancer and may even increase it for some types, including advanced prostate cancer and skin cancer. An increased risk of diabetes has also been reported. These findings have been consistent in the most methodologically sound trials, suggesting that the early observational studies were misleading. Other studies have investigated selenium compounds as adjuvant therapy for cancer. Though there is currently insufficient evidence regarding the utility and safety of selenium compounds for such treatments, this issue is worthy of further investigation. The study of selenium and cancer is complicated by the existence of a diverse array of organic and inorganic selenium compounds, each with

Advances in Cancer Research, Volume 136
ISSN 0065-230X
http://dx.doi.org/10.1016/bs.acr.2017.07.001

distinct biological properties, and this must be taken into consideration in the interpretation of both observational and experimental human studies.

1. INTRODUCTION

Selenium is a naturally occurring chemical element with both nutritional and toxicological effects that have received considerable attention in medicine and biology (Hatfield, Tsuji, Carlson, & Gladyshev, 2014; Jablonska & Vinceti, 2015). Trace amounts of selenium are necessary for cellular function and for the synthesis of several selenoproteins with antioxidant properties, which appear to be essential for life in mammals including humans (Arner, 2016; Labunskyy, Hatfield, & Gladyshev, 2014) and have been associated with protection against diseases linked to free radical damage, including infectious disease, cancer, and cardiovascular disease (Rayman, 2012). However, excessive selenium intake can lead to selenosis, a serious and sometimes fatal illness (Vinceti, Wei, Malagoli, Bergomi, & Vivoli, 2001), and more subtle overexposure may induce adverse health effects even at low levels (Vinceti, Maraldi, Bergomi, & Malagoli, 2009) through several mechanisms, including the pro-oxidant properties of some selenium species, which contrasts sharply with the antioxidant activity of selenoproteins (Brigelius-Flohe & Flohe, 2017; Hatfield et al., 2014; Labunskyy et al., 2014; Lazard, Dauplais, Blanquet, & Plateau, 2015; Letavayova, Vlasakova, Spallholz, Brozmanova, & Chovanec, 2008; Misra, Boylan, Selvam, Spallholz, & Bjornstedt, 2015; Porto, Moreira, Haddad-Ribeiro, Belo, & Neves, 2016; Spallholz, 1997; Touat-Hamici, Legrain, Bulteau, & Chavatte, 2014).

Selenium exists in several inorganic and organic chemical forms, and the biological properties and health effects of the various species can be quite different (Borella, Bargellini, & Medici, 1996; Fan & Vinceti, 2015; Fernandes & Gandin, 2015; Gammelgaard, Rasmussen, Gabel-Jensen, & Steffansen, 2012; Marschall, Bornhorst, Kuehnelt, & Schwerdtle, 2016; Vinceti, Solovyev, et al., 2013; Weekley & Harris, 2013). Human exposure to selenium in its various chemical species can occur through food, water, and air, with diet being the major source of human selenium exposure (Fairweather-Tait et al., 2011; Vinceti, Crespi, Malagoli, Del Giovane, & Krogh, 2013; Vinceti, Dennert, et al., 2014; Vinceti et al., 2009). Foods mainly contain organic selenium species, while in drinking water and air selenium is found mainly in its inorganic forms (Fan & Vinceti, 2015).

The assessment of selenium status in humans is a complex and controversial issue, with important implications for risk assessment and the design and interpretation of epidemiologic studies (Fairweather-Tait et al., 2011; Vinceti, Burlingame, et al., 2016; Vinceti, Crespi, et al., 2013). Three main approaches have been used for exposure assessment: dietary assessment methods, measurement of tissue-specific selenium levels, and proteomic endpoints such as the expression of selenoproteins in selected tissues. Each of these approaches has advantages and limitations.

Dietary assessment methods, based on short-term dietary recall or more often on dietary questionnaires, have the advantage of being broadly acceptable to study participants and therefore less prone to induce selection bias, in addition to being less expensive and less organizationally complex. Dietary assessment also has the advantage of allowing, when chemical analyses of selenium speciation in foodstuffs had been used, the determination of the intake of single selenium species before any possible metabolic conversion within the organism. On the downside, the accuracy of dietary assessment methods can suffer due to the variability of selenium content in certain food groups or specific foodstuffs, which can vary substantially even within a small geographic area. In the case of questionnaires, they share the measurement error inherent in retrospective dietary assessment methods investigating food intake.

Biomarkers of selenium exposure involve the measurement of overall selenium content in blood (serum, plasma, and less frequently whole blood and erythrocytes), urine, hair, (toe)nail, and rarely cerebrospinal fluid (CSF) or other highly specific body tissues (Ashton et al., 2009; Filippini et al., 2017; Mandrioli et al., 2017; Vinceti et al., 2012). Though the use of biomarkers has the advantage over dietary measurements of revealing the true level of selenium in specific body compartments, it has frequently overlooked limitations. First, the overall amount of selenium in the body tissues depends on the chemical forms of the element to which the organism has been exposed, with the organic species being more strongly retained in the body despite their lower toxicity and biological activity (Panter et al., 1996; Vinceti, Grill, et al., 2015; Vinceti, Solovyev, et al., 2013). In addition, the distribution of ingested selenium in different body compartments is influenced by its specific chemical species as well as by the intake of dietary factors, such as methionine (Salbe & Levander, 1990; Vinceti, Grill, et al., 2015). Moreover, some toxic elements such as cadmium or other heavy metals may alter the retention of ingested selenium in the body by increasing its excretion as well as modify its bioavailability, further weakening the value of biomarkers for assessing selenium intake or biological activity (Vinceti, Grill, et al., 2015). It should also be noted that some biomarkers such as blood and urine selenium levels appear to reflect short-term amount of exposure, while long-term exposure (around 6–12 months) is better reflected by hair or nail selenium levels. Another methodological issue affecting biomarkers is the limited degree of correlation between the selenium levels in different body compartments, which also depends on the chemical species of the element, thus raising additional risk of exposure misclassification. CSF selenium levels, for example, tend to correlate with blood levels only for some of the organic forms of the element, while for the more toxic, inorganic species such as selenite or selenate there is little or no correlation (Vinceti, Solovyev, et al., 2013). This raises an issue of exposure misclassification for studies aiming to investigate selenium status in the central nervous system through peripheral indicators of exposure (Vinceti, Mandrioli, et al., 2014).

Finally, the amount and activity of antioxidant selenoproteins, particularly glutathione peroxidases and plasma selenoprotein P, have been frequently used as indirect indicators of selenium intake, and their low levels have been frequently interpreted as a consequence and a biomarker of selenium deficiency (Fairweather-Tait et al., 2011; Jablonska & Vinceti, 2015; Labunskyy et al., 2014; Vinceti et al., 2009). However, this approach overlooks the fact that selenoprotein levels and activity increase as a compensatory response to oxidative stress, and therefore, several stressors with prooxidant activity, including high amounts of some selenium species, may upregulate selenoproteins (as well as other antioxidant enzymes) thanks to this mechanism (Hafeman, Sunde, & Hoekstra, 1974; Jablonska & Vinceti, 2015; Touat-Hamici et al., 2014; Vinceti et al., 2009). In addition, no clear relation between selenoprotein activity per se and health outcomes has been established so far in epidemiologic studies. These complex relations between selenium exposure and selenoprotein activity may explain why some regulatory agencies have not adopted a proteomic approach based on maximal upregulation of selenoprotein levels by ingested selenium when setting the dietary reference values for this element (Jablonska & Vinceti, 2015; Vinceti et al., 2017).

The human health effects of selenium, aside from overt toxicity, have been debated for many years. Health effects deriving from both its deficiency and overexposure are potentially broad and can include cancer and cardiovascular, metabolic, endocrine, and neurological disease. Among these outcomes, the link between selenium intake and cancer risk has been the most investigated in literature, and that which attracted the largest interest also from the general population.

Due to the uncertainties in the health effects of selenium and in the assessment of selenium status and biological activities, it is no surprise that consensus is lacking about the exact amount of selenium exposure that may be associated with adverse health effects, with reference to both deficiency and overexposure (Jablonska & Vinceti, 2015; Vinceti et al., 2017). These uncertainties are reflected in the widely varying environmental standards and dietary recommendations for this element worldwide (Jablonska & Vinceti, 2015; Vinceti et al., 2017).

With specific reference to cancer, interest in a possible etiological role of selenium was first sparked by laboratory animal studies, suggesting the ability of selenium to enhance cancer growth (Nelson, Fitzhugh, & Calvery, 1943; Schroeder & Mitchener, 1972; Seifter, Ehrich, Hudyma, & Mueller, 1946) and later to counteract it (Clayton & Baumann, 1949; Shamberger & Rudolph, 1966). Similarly, ambiguous effects have been observed in more recent studies, which have reported that forms of selenium may enhance carcinogenesis (Birt et al., 1988; Chen et al., 2000; National Toxicology Program, 2011; Novoselov et al., 2005; Su, Tang, Tang, & Gao, 2005; Woutersen, Appel, & Van Garderen-Hoetmer, 1999), reduce it (Guo, Hsia, Hsiung, & Chen, 2015; INCHEM, 1987; Nakahara et al., 2013; Steinbrenner, Speckmann, & Sies, 2013; Wang, Sun, Tan, Wu, & Zhang, 2014; Wrobel, Wolff, Xiao, Power, & Toborek, 2016; Yang, Jia, Chen, Yang, & Li, 2012), or exert ambivalent effects through the selenoproteins (Brigelius-Flohe & Kipp, 2016; Hatfield, Yoo, Carlson, & Gladyshev, 2009; Kasaikina et al., 2013; Varlamova & Cheremushkina, 2017).

In the 1970s, the first epidemiologic studies, using ecologic designs, were conducted. These studies were followed by a large number of observational studies including case–control and cohort investigations, and later by randomized controlled trials (RCTs). More recently, an additional topic has been added to the debate, namely, the possibility that selenium administration may be beneficial in cancer patients (Fernandes & Gandin, 2015; Misra et al., 2015; Wallenberg, Misra, & Bjornstedt, 2014).

For decades, controversies regarding selenium have abounded, with claims of health benefits met with counterclaims of harm. An atmosphere of *selenophobia*, or fear of the health effects of selenium, in the early 1970s (Frost, 1972) gave way to a culture of *selenophilia* focused on the health benefits of selenium in the late 1980s (Casey, 1988). Driving the *selenophilia* side of the debate has been the fascinating possibility that changes of just a few μg/day in dietary intake of selenium may modify the risk of cancer.

Here we briefly review the literature, published through February 28, 2017, concerning the relation between selenium and cancer in humans, starting with observational studies and then focusing on RCTs. We also summarize current evidence about the potential use of selenium compounds in cancer therapy. Our purpose is not to give an exhaustive review of the vast literature on selenium and cancer, but rather to provide a historical perspective and highlight important studies and developments that have shaped current scientific thinking about this curious element.

2. SELENIUM AND CANCER: SCOPE OF RESEARCH

To identify articles for this review and to provide an updated overview of the research on selenium and cancer, we conducted PubMed searches, which are summarized in Fig. 1. To identify epidemiologic studies about selenium and cancer, we conducted a PubMed search using the MeSH terms "selenium," "humans," "epidemiology" OR "epidemiologic methods", and "neoplasms" on February 28, 2017. To determine the number and percentage of these studies that pertained to clinical trials, we further added "Clinical Trial" as Publication Type to the PubMed search string.

This search identified a total of 650 epidemiologic studies regarding selenium and cancer, with 15% (96/650) pertaining to clinical trials. As shown in Fig. 1, overall the most commonly studied cancers have been prostate, breast, lung, and colorectal. Experimental studies have most commonly focused on all cancer, prostate cancer, skin cancer, gastrointestinal cancer, and bladder cancer. Prostate cancer, in particular, has been extensively studied with regard to an etiologic role of selenium intake, which has led to a preponderance of male subjects in cohort studies and particularly in RCTs.

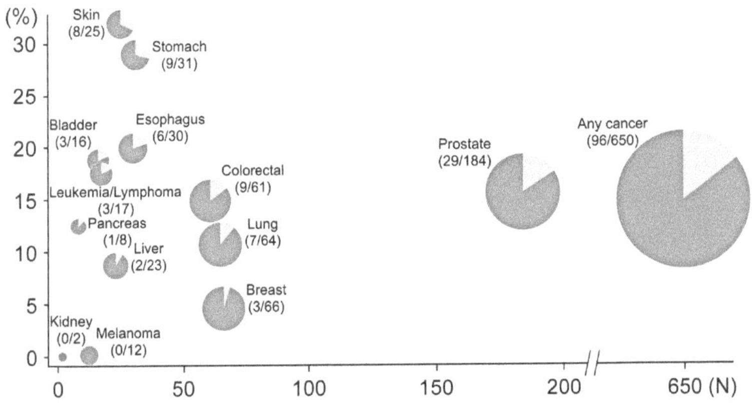

Fig. 1 Search hits for PubMed search performed in February 28, 2017. String [all MeSH Terms]: "selenium" AND "humans" AND ("epidemiology" OR "epidemiologic methods") AND "neoplasms" (or the individual type of cancer). Search for clinical trials was performed by adding "clinical trial" [Publication Type] to the search string. In *parentheses* the number of clinical trials and of the total studies for each MeSH descriptor is reported. The percentage of clinical trials on the total hits is reported on the vertical axis.

3. OBSERVATIONAL STUDIES

The first epidemiologic study conducted on selenium and cancer appears to have been reported in the 1969 publication by Shamberger and Frost (1969). Using an ecologic study design, the authors identified an inverse correlation of the cancer mortality rates in the United States and Canada with several indicators of selenium exposure, such as selenium in crops and plants and blood selenium levels, thus suggesting a potential cancer-preventive effect of higher selenium intake. This study was subsequently replicated by other ecologic studies carried out in various areas of the world. A few of these studies confirmed the original findings of Shamberger and Frost, but others found a null or direct association (Chen, Geissler, Parpia, Li, & Campbell, 1992; Clark, Cantor, & Allaway, 1991; Guo, Chow, Zheng, Li, & Blot, 1994; Guo, Hsing, et al., 1994; Keshavarzi, Moore, Najmeddin, & Rahmani, 2012; Kneller et al., 1992; Schrauzer, White, & Schneider, 1977). These studies suffered from the well-known limitations of ecologic studies, particularly inadequate assessment of selenium exposure status and the inability to take into account confounders at the individual level. These studies led to further investigations using stronger methodological designs.

Subsequent to the ecologic studies, a large number of investigations with cross-sectional, case–control, and cohort designs were conducted to evaluate the possible relation between selenium and cancer risk (Vinceti, Burlingame, et al., 2016; Vinceti, Crespi, et al., 2013; Vinceti, Dennert, et al., 2014; Vinceti, Rovesti, Bergomi, & Vivoli, 2000). Despite the strengths that these individual-based studies offered compared to ecologic designs, they still suffered from many important limitations. We turn our attention to these limitations next.

A key limitation of observational studies on selenium and cancer risk concerns exposure assessment. Exposure classification in these studies used biomarkers such as blood, serum, plasma, urine, hair or toenail selenium levels, or dietary assessment methods based on food frequency questionnaires. All these methods may have substantial limitations (Fairweather-Tait et al., 2011; Haldimann, Venner, & Zimmerli, 1996; Vinceti, Crespi, et al., 2013). For example, the use of blood levels of selenium to assess its status is limited by variations in the retention and the tissue-specific distribution of selenium not only according to its amount of exposure but also to its chemical forms and the concomitant exposure to other substances, such as methionine and heavy metals (Vinceti, Grill, et al., 2015; Vinceti, Grioni, et al., 2015). In addition, peripheral circulating levels of selenium such as blood content may not

GPx1, it too appears to have an antiinflammatory role and can reduce expression of proinflammatory cytokines (Banning, Florian, et al., 2008). Furthermore, GPx1 and GPx2 appear to synergize, as mice that lack both GPx1 and GPx2 develop ileocolitis despite little phenotype being observed with loss of either GPx individually (Esworthy et al., 2001). Interestingly, these mice also develop spontaneous tumorigenesis, which was dependent on inflammation driven by interactions with pathogenic microflora, as germ-free mice remained tumor free (Chu et al., 2004). Placing GPx2-null mice on an inflammatory carcinogenesis protocol also determined that GPx2-null mice had more severe inflammation as well as higher tumor incidence in the colon (Krehl et al., 2012), overall indicating a protective role for GPx2 in colitis and colitis-associated tumorigenesis. However, the effect of GPx2 in noninflammatory-driven cancers appears to be the opposite. GPx2 is often found to be overexpressed in sporadic (i.e. nonhereditary and noninflammatory) colon cancers as well as cancers from the esophagus and liver, and it appears to promote carcinogenesis in this context (Florian et al., 2001; Lei, Tian, Zhang, Zhao, & Su, 2016; Suzuki et al., 2013). GPx2-null mice had decreased ACF formation after AOM treatment, and its knockdown has been shown to decrease growth of HT-29 colon cancer xenografts, patient-derived colon tumor spheroids, and castration-resistant prostate cancer cells (Banning, Kipp, et al., 2008; Emmink et al., 2014; Muller et al., 2013; Naiki et al., 2014).

While the roles of GPx1 and GPx2 in tumorigenesis appear to be tissue or context specific, perhaps the GPx which most clearly functions as a tumor suppressor is the extracellular family member, GPx3. This GPx is most highly expressed in the kidneys and secreted extracellularly into the plasma, although expression can also be detected in other tissue types (Avissar et al., 1994; Whitin, Bhamre, Tham, & Cohen, 2002). In tumor cells, GPx3 is a frequent target of hypermethylation, and its downregulation and/or methylation is associated with poorer prognosis and chemotherapeutic resistance in a wide range of cancer types, a topic discussed in depth in chapter "Selenium and epigenetics in cancer: Focus on DNA methylation" by Jabłońska and Reszka (Falck et al., 2010; Pelosof et al., 2017; Yu et al., 2007). Functionally, GPx3 overexpression has been reported to decrease anchorage-independent growth, xenograft size, and metastasis of prostate cancer cells (Yu et al., 2007). Interestingly, these authors also report downregulation of c-Met, a tyrosine kinase receptor which mediates signaling downstream of hepatic growth factor signaling. Other studies have also determined that GPx3's antitumorigenic function may be through direct interaction with p53-induced gene 3 (PIG3), a p53 target gene which exposure misclassification and act as a confounder. Cadmium, for example, is a known carcinogenic element which may confound the association between other environmental or dietary factors and cancer risk, but it may also alter selenium status by increasing its excretion and lowering its blood levels, thus inducing exposure misclassification (Vinceti, Dennert, et al., 2014; Vinceti, Rovesti, et al., 2000).

Finally, an additional source of bias in case–control and cross-sectional studies is the ability of clinical or preclinical disease to affect selenium status through disease-induced changes in dietary habits or through selective uptake of selenium by cancer cells. Such processes can induce spurious inverse associations between selenium status and disease risk and thus greatly reduce the reliability of case–control and cross-sectional studies. This phenomenon may explain why most case–control studies have found lower selenium body tissue content in cancer patients compared with controls, or in patients with advanced disease compared with early-stage disease.

Given these limitations of observational studies, it is not surprising that even longitudinal studies, generally regarded as the most reliable observational study design, have yielded conflicting results. Most longitudinal studies have had cohort-nested case–control designs and have assessed exposure through determination of baseline serum, toenail, or dietary selenium (Vinceti, Dennert, et al., 2014), and far less frequently through baseline selenoprotein status (Epplein, Burk, Cai, Hargreaves, & Blot, 2014; Hughes et al., 2016, 2015; Li, Peng, Mao, & Qian, 2015; Penney et al., 2013, 2010; Takata et al., 2012). The majority of these studies have suggested an inverse association between selenium exposure at baseline and subsequent cancer incidence, but many have found null or even direct relations (Vinceti, Dennert, et al., 2014).

Lung cancer incidence, for example, has been found to be associated in prospective studies with both low (Kabuto et al., 1994; Knekt, Marniemi, Teppo, Heliovaara, & Aromaa, 1998; van den Brandt et al., 1993) and high (Menkes et al., 1986; Suadicani, Hein, & Gyntelberg, 2012) baseline selenium status, as well as to have no association (Goodman, Schaffer, Bankson, Hughes, & Omenn, 2001; Ratnasinghe et al., 2000) or to have different associations depending on genetic background (Jablonska, Gromadzinska, Sobala, Reszka, & Wasowicz, 2008). Most of these studies had a limited range of baseline selenium intake. In addition, no dose–response relation between selenium exposure and cancer risk emerged, as shown by an analysis of the results according to the study participants' selenium status (Vinceti, Dennert, et al., 2014).

Other cancer types for which conflicting results have been found include breast cancer, for which associations range from imprecisely inverse (Knekt et al., 1990) to null and direct (Coates, Weiss, Daling, Morris, & Labbe, 1988; Garland et al., 1995; van Noord, Maas, van der Tweel, & Collette, 1993), and liver cancer, where as recently as 2016 studies reported both inverse (Hughes et al., 2016) and null associations (Ma et al., 2017). Interestingly, the study which found a null association between baseline selenium exposure and liver cancer incidence (Ma et al., 2017) was characterized by a lower background selenium intake than the study finding an inverse association (Hughes et al., 2016), therefore suggesting that these different results were not directly linked to selenium deficiency.

Another interesting observation yielded by the longitudinal studies is that the relation between baseline selenium exposure and subsequent disease status can differ according to the indicator of exposure used. For example, prostate cancer risk has been frequently reported to be inversely associated with baseline selenium status when the latter was assessed through toenail or blood selenium content, but no association emerged when selenium status was determined through dietary assessment (Geybels et al., 2014; Graff et al., 2017; Park et al., 2015; Vinceti, Dennert, et al., 2014). In other studies, self-administration of selenium supplements and baseline plasma selenium were found to be associated with excess risk of overall or aggressive prostate cancer or its recurrence (Chan et al., 2009; Dalkin et al., 2001; Kenfield, Van Blarigan, DuPre, & Stampfer, 2015; Lawson et al., 2007; Zhang, Coogan, Palmer, Strom, & Rosenberg, 2009). Since subsequent experimental studies have found no association between selenium exposure and prostate cancer (Vinceti, Crespi, et al., 2013; Vinceti, Dennert, et al., 2014), we may infer a posteriori that bias, due to exposure misclassification and/or confounding, occurred when blood or toenail selenium was used to assess exposure, but did not occur when dietary assessment methods were used. This is unexpected, since dietary assessment methods have been frequently claimed to be less reliable than biomarkers of selenium status to estimate usual selenium intake, due mainly to the high variability in selenium content in foods even within small geographical areas (Fairweather-Tait et al., 2011; Vinceti, Crespi, et al., 2013; Vinceti, Dennert, et al., 2014). A similar issue has been found for lung cancer, where assessment of baseline exposure status using blood levels of selenium yielded results that differed from the results subsequently found by randomized trials (no effect or a statistically imprecise increase (Karp et al., 2013; Lippman et al., 2009)). Again, the RCT results were more consistent with the observational studies based on direct assessments of selenium dietary intake than those based on other methods of exposure assessment such as blood selenium levels (Vinceti, Dennert, et al., 2014).

Overall, given the many limitations and sources of potential bias, it appears that observational studies have little potential in elucidating the relation between selenium and cancer risk, as frequently occurring in nutritional epidemiology differently from other epidemiologic areas, such as environmental and pharmacological epidemiology (Rothman, 2014; Sorensen, Lash, & Rothman, 2006; Vinceti & Rothman, 2016). Further studies of this type will have limited value, despite any potential improvement in statistical power, exposure assessment (i.e., determining the various selenium species), control of potential confounders, or assessment of genetic background. Accordingly, we next turn our attention to RCTs.

4. RANDOMIZED CONTROLLED TRIALS

The identification of selenium-containing enzymes with antioxidant properties (Flohe, Gunzler, & Schock, 1973; Rotruck et al., 1973), the numerous laboratory studies suggesting that selenium counteracts cancer cell growth (INCHEM, 1987), and the results of the first cohort studies (Lockitch, 1989; Vinceti, Crespi, et al., 2013) suggested the possibility of an inverse relation between selenium status and cancer risk in the human. It is therefore not surprising that randomized trials evaluating whether selenium supplementation would reduce human cancer risk were designed and conducted in the 1990s–2000s. Thus a paradoxical situation occurred, in that a substance with established toxicity became the object of human experiments focusing on its possible chemoprevention activity. For these trials, to ensure safety from the potential toxic effects of selenium, an upper intake level (no observed adverse effect level divided by 2) of 400 μg/day (Institute of Medicine Food and Nutrition Board, 2000) was established based on early environmental studies carried out in China (Yang, Wang, Zhou, & Sun, 1983).

In Table 1, we report key details of the randomized trials of selenium administration that included primary tumor incidence detection as a primary or secondary endpoint and whose results were published before February 28, 2017. The table summarizes the type of study population including its selenium status at baseline (measured) and as achieved after supplementation (estimated), follow-up details, assessment of potential bias, and the main results including both primary and secondary outcomes. Table 2 summarizes the occurrence of adverse effects in the selenium arms of these studies, ascertained through follow-up surveillance. We next discuss the first such trial.

Table 1 List of Randomized Placebo Controlled Trials Using Selenium Supplementation for Chemoprevention of Cancer

Trial and Year	Study Population[a]	Follow-Up and/or Duration	Baseline Se Level (μg/L)	Type of Se Supplement	Achieved Intake of Se Group[b]	Risk of Bias[c]	Main Results	RR/HR (95% CI)	References
NPC: Nutritional Prevention of Cancer study 1996	1250 (621/629) subjects aged 18–80 (years) with history of NMSC	Mean follow-up 7.9 years. Trial duration up to 13 years (end of blinded period)	114.2	200 μg/day high selenium yeast	276.1	High risk	Any cancer	HR 0.75 (0.58–0.97)	Clark et al. (1996); Duffield-Lillico et al. (2002); Duffield-Lillico, Slate, et al. (2003)
							Bladder cancer	HR 1.28 (0.50–3.25)	
							Breast cancer	HR 1.89 (0.69–5.14)	
							Colorectal cancer	HR 0.46 (0.21–1.02)	
							Lung cancer	HR 0.74 (0.44–1.24)	
							Melanoma	HR 1.18 (0.49–2.85)	
							NMSC	HR 1.17 (1.02–1.34)	
							Prostate cancer	HR 0.48 (0.28–0.80)	
NPC: Nutritional Prevention of Cancer substudy 1996	423 (210/213) subjects aged 18–80 with history of NMSC	Trial duration up to 6.3 years (end of blinded period)	116.5	400 μg/day high selenium yeast	477.7	High risk	Any cancer	RR 1.10 (0.57–2.17)	Reid et al. (2008)
							NMSC	HR 0.91 (0.69–1.20)	

OTR: Organ Transplant Recipients 2007	184 (91/93) organ graft recipients aged 18–65	Trial duration up to 5 years (3 of treatment and 2 only of follow-up)	Not reported	200 μg/day selenium-enriched yeast	—	High risk	Skin keratosis Skin cancer	RR[d] 1.09 (0.73–1.62) RR[d] 3.07 (0.64–14.80)	Dreno, Euvrard, Frances, Moyse, and Nandeuil (2007)
SELECT: Selenium and Vitamin E Cancer Prevention Trial 2009	17,748 (8752/8696) relatively healthy men aged ≥50	Median follow-up 5.46 years	136.3	200 μg/day selenized yeast/ L-selenomethionine	290.9	Low risk	Any cancer Bladder cancer Colorectal cancer Colorectal adenoma Lung cancer Prostate cancer	HR 1.01 (0.92–1.11) HR 1.13 (0.83–1.50) HR 1.05 (0.74–1.49) RR 0.96 (0.90–1.02) HR 1.12 (0.81–1.55) HR 1.04 (0.91–1.19)	Klein et al. (2011); Lance et al. (2017); Lippman et al. (2009); Lotan et al. (2012)
BRCA: Chemoprevention BRCA1 + women 2011	1135 women BRCA1 carriers	Median follow-up 2.92 years	Not reported	250 μg/day inorganic selenite	—	No information	Any cancer Breast cancer Ovarian cancer	HR 1.4 (0.9–2.0) HR 1.3 (0.7–2.5) HR 1.3 (0.6–2.7)	Lubinski et al. (2011)

Continued

Table 1 List of Randomized Placebo Controlled Trials Using Selenium Supplementation for Chemoprevention of Cancer—cont'd

Trial and Year	Study Population	Follow-Up and/or Duration	Baseline Se Level (μg/L)	Type of Se Supplement	Achieved Intake of Se Group	Risk of Bias	Main Results	RR/HR (95% CI)	References
SWOG: Southwest Oncology Group Trial S9917 2011	423 (212/211) men aged ≥40 with HGPIN but prostate cancer free	Duration up to 3 years	136.6	200 μg/day selenium as selenomethionine	291.1	Low risk	Prostate cancer	RR 0.97 (0.68–1.39)	Marshall et al. (2011)
NBT: Negative Biopsy Trial 2013	699 (467/232) men at high risk for prostate cancer aged <80	Median follow-up 3 years. Duration up to 5 years	126.1	234 with 200 μg/day and 233 with 400 μg/day selenized yeast	284.1 and 484.1	Low risk	200 μg/day Any cancer Colon cancer Melanoma NMSC Prostate cancer 400 μg/day Any cancer Colon cancer Melanoma	200 μg/day RR^d 1.04 (0.69–1.60) RR^d 0.99 (0.06–15.76) RR^d 1.49 (0.25–8.82) RR^d 2.64 (0.71–9.84) RR^d 0.94 (0.52–1.70) 400 μg/day RR^d 0.86 (0.56–1.33) RR^d 1.00 (0.06–15.82) RR^d 0.66 (0.11–3.94)	Algotar et al. (2013)

							NMSC	RR[d] 0.88 (0.52–1.50)	
							Prostate cancer	RR[d] 0.90 (0.48–1.70)	
							All subjects	All subjects	
							Any cancer	RR[d] 0.93 (0.64–1.34)	
							Colon cancer	RR[d] 0.99 (0.09–10.90)	
							Melanoma	RR[d] 0.83 (0.20–3.43)	
							NMSC	RR[d] 0.80 (0.49–1.31)	
							Prostate cancer	RR[d] 0.90 (0.57–1.41)	
ECOG: Eastern Cooperative Oncology Group Trial 5597 2013	1561 (1040/521) subjects with history of resection of non–small-cell lung cancer aged ≥18	Duration up to 4 years	Not reported	200 μg/day selenized yeast	—	Low risk	Any cancer	RR[b] 1.02 (0.80–1.30)	Karp et al. (2013)
							Bladder cancer	RR[b] 0.75 (0.27–2.10)	
							Breast cancer	RR[b] 2.04 (0.44–9.55)	
							Colorectal cancer	RR[b] 0.50 (0.13–2.00)	
							Lung cancer	RR[b] 1.25 (0.82–1.92)	
							Melanoma	RR[b] 1.25 (0.24–6.43)	
							NMSC	RR[b] 0.66 (0.37–1.19)	
							Prostate cancer	RR[b] 0.87 (0.39–1.95)	

Continued

Table 1 List of Randomized Placebo Controlled Trials Using Selenium Supplementation for Chemoprevention of Cancer—cont'd

Trial and Year	Study Population	Follow-Up and/or Duration	Baseline Se Level (μg/L)	Type of Se Supplement	Achieved Intake of Se Group	Risk of Bias	Main Results	RR/HR (95% CI)	References
SELCEL: Selenium and Celecoxib Trial 2016	1374 (685/689) subjects with history of colorectal adenomas aged 40–80	Median follow-up 2.96 years	135.4	200 μg/day selenized yeast	290.3	Low risk	Any adenoma	RR 1.03 (0.91–1-16)	Thompson et al. (2016)
							Advanced adenoma	RR 1.02 (0.74–1.43)	
							Multiple adenoma	RR 1.47 (1.08–2.02)	
							NMSC	RR 1.34 (0.76–2.37)	

[a]Number of total subjects in the follow-up (treatment/placebo) and study description.
[b]Estimated baseline intake dividing serum/plasma selenium by 1.5 (Haldimann et al., 1996).
[c]According to Cochrane Handbook guidelines (Chapter 8).
[d]Computed using the *csi* routine of STATA 14.2 (Stata Corp., College Station, TX).
Abbreviations: *HGPIN*, high-grade prostatic intraepithelial neoplasia; *NMSC*, nonmelanoma skin cancer.

Table 2 Adverse Effects in Preventive Randomized Controlled Trials of Selenium Supplementation: Where Available, We Report the Number of Events in the Treated and Control Groups and the Relative Risk (RR), as Reported in the Study or Calculated From Raw Data[a]

Trial and Year	Dermatologic Effects (i.e., Alopecia, Dermatitis, and Nail Changes)	Nonmelanoma Skin Cancer	Diabetes	Other Effects	References
NPC 1996	—	Cases not reported RR 1.17 (95% CI 1.02–1.34)	58/39 RR 1.49 (95% CI 1.01–2.20)	—	Clark et al. (1996); Duffield-Lillico et al. (2002); Duffield-Lillico, Slate, et al. (2003); Stranges et al. (2007)
NPC 1996 substudy	—	98/108 RR 0.91 (95% CI 0.69–1.20)	—	—	Reid et al. (2008)
OTR 2007	—	6/2 RR 3.07 (95% CI 0.64–14.80)	—	—	Dreno et al. (2007)
SELECT 2009	Alopecia: 265/206 RR 1.28 (95% CI 1.07–1.53) Dermatitis: 619/524 RR 1.17 (95% CI 1.05–1.31) Nail changes: 1087/1035 RR 1.04 (95% CI 0.96–1.13)	—	724/669 RR 1.08 (95% CI 0.97–1.19)	Cardiovascular events: 1080/1050 RR 1.02 (95% CI 0.94–1.11) Fatigue: 666/610 RR 1.08 (95% CI 0.98–1.21) Halitosis: 503/427 RR 1.17 (95% CI 1.03–1.33) Nausea: 253/212 RR 1.19 (95% CI 0.99–1.42)	Klein et al. (2011); Lance et al. (2017); Lippman et al. (2009); Lotan et al. (2012)
BRCA 2011	—	—	—	—	Lubinski et al. (2011)

Continued

Table 2 Adverse Effects in Preventive Randomized Controlled Trials of Selenium Supplementation: Where Available, We Report the Number of Events in the Treated and Control Groups and the Relative Risk (RR), as Reported in the Study or Calculated From Raw Data—cont'd

Trial and Year	Dermatologic Effects (i.e., Alopecia, Dermatitis, and Nail Changes)	Nonmelanoma Skin Cancer	Diabetes	Other Effects	References
SWOG S9917 2011[b]	Dermatologic events: 1/0	—	—	Cardiovascular events: 0/1 Gastrointestinal effects: 0/1	Marshall et al. (2011)
NBT 2013 (200 μg/die)	Brittle hairs and nails: 24/26 RR 0.92 (95% CI 0.54–1.55)	8/3 RR 2.64 (95% CI 0.71–9.84)	12/7 RR 1.70 (95% CI 0.68–4.24)	Cataract/glaucoma: 14/13 RR 1.07 (95% CI 0.51–2.22) Garlic breath, kidney/liver abnormalities: 13/14 RR 0.92 (95% CI 0.44–1.92)	Algotar et al. (2013)
NBT 2013 (400 μg/die)	Brittle hairs and nails: 20/26 RR 0.77 (95% CI 0.44–1.33)	3/3 RR 1.00 (95% CI 0.20–4.88)	12/7 RR 1.71 (95% CI 0.68–4.26)	Cataract/glaucoma: 15/13 RR 1.15 (95% CI 0.56–2.36) Garlic breath, kidney/liver abnormalities: 11/14 RR 0.78 (95% CI 0.36–1.69)	Algotar et al. (2013)
ECOG 5597 2013[c]	Alopecia: 4%/5% RR 0.80 (95% CI 0.48–1.34) Dermatitis: 4%/2% RR 1.59 (95% CI 0.75–3.37) Nail changes: 5%/7% RR 0.72 (95% CI 0.46–1.12)	25/19 RR 0.80 (95% CI 0.44–1.45)	26/12 RR 1.19 (95% CI 0.61–2.35)	Fatigue: 7%/7% RR 1.02 (95% CI 0.68–1.53)	Karp et al. (2013)
SELCEL 2016	Brittle hair and nails: 30/35 RR 0.86 (95% CI 0.53–1.39)	Squamous cell carcinoma: 27/21 RR 1.34 (95% CI 0.76–2.37)	31/25 RR 1.25 (95% CI 0.74–2.11)	—	Thompson et al. (2016)

[a]Computed using the *csi* routine of STATA 14.2 (Stata Corp., College Station, TX).
[b]Only events of high grade reported here.
[c]Percentages of adverse events in selenium and placebo groups are reported if numbers are not available.

5. THE NUTRITIONAL PREVENTION OF CANCER TRIAL

The first selenium trial and one of the two most famous ones was the Nutritional Prevention of Cancer (NPC) trial, conducted by investigators at the Arizona Cancer Center (Clark et al., 1996; Duffield-Lillico, Dalkin, et al., 2003; Duffield-Lillico et al., 2002; Duffield-Lillico, Slate, et al., 2003; Reid et al., 2002, 2004). This trial was designed to test the hypothesis that dietary supplementation with selenium at 200 μg/day would decrease the risk of incident nonmelanoma skin cancer in a cohort of 1312 patients with recent diagnosis of basal or squamous cell skin cancer, life expectancy of not less than 5 years, and no treatment for internal cancers in the last 5 years. The subjects, recruited from 1983 through 1991, were randomized to either selenized yeast containing 200 μg/day organic selenium (mostly as selenomethionine and other selenocompounds (Reid et al., 2002)) or placebo, and followed until February 1, 1996 (the end of the blinded phase of patient treatment and follow-up), with an overall number of 9301 person-years in the entire period of follow-up (Duffield-Lillico et al., 2002).

The primary outcome of the NPC trial was the incidence of nonmelanoma skin cancer, basal cell, and squamous cell skin cancer. Several secondary outcomes were added to the study protocol after the study began, including incidence and mortality for all cancers and several major site-specific cancers. The final results of NPC were published in 2002–2003 (Duffield-Lillico, Dalkin, et al., 2003; Duffield-Lillico et al., 2002; Duffield-Lillico, Slate, et al., 2003). However, in 1996, a preliminary (ad interim) analysis was published based on follow-up data through December 1, 1993, based on the recommendation of the trial's Safety Monitoring and Advisory Committee to unblind the study results (Clark et al., 1996). This preliminary report has always attracted more interest than the final reports, as indicated by a February 2017 Google Scholar search conducted by our group which showed over 3000 hits for the preliminary 1996 report, 532 hits for the 2002 report, and even fewer (272) for the final report addressing the primary outcome. The report documented a sharp decrease in overall cancer incidence (−39%) and mortality (−50%) as well as in incidence of lung, colorectal, and prostate cancer (Clark et al., 1996). Prostate cancer had a decreased incidence of 65% compared to the placebo group, yielding a hazard ratio (HR) of 0.35 (95% CI 0.18–0.65). In contrast, there was little evidence for any effect of selenium supplementation on the primary outcome of nonmelanoma skin cancer incidence.

In contrast to this preliminary report, the final study results published in 2002–2003 were mixed and raised some important concerns (Table 1). On the positive side, there continued to be evidence of decreased prostate and colorectal cancer risk, and evidence of decreased risk for all cancers and lung cancer (Duffield-Lillico et al., 2002; Reid et al., 2002). However, there was an excess risk, though statistically imprecise, for breast cancer and for melanoma, bladder, and lymphatic malignancies. Further analyses concerning the prostate cancer endpoint showed a weak inverse relation between baseline plasma selenium levels in placebo-treated participants, but an unexpected direct association between baseline plasma selenium levels and subsequent prostate cancer incidence in selenium-treated subjects (Dalkin et al., 2001).

Furthermore, the primary outcome of the trial, the incidence of non-melanoma skin cancer, showed an excess incidence of the disease in selenium-treated subjects (Duffield-Lillico, Slate, et al., 2003), contradicting the original hypothesis that prompted the trial. An excess risk of glaucoma was also found among selenium-treated subjects, though the results were not published in a peer-reviewed article (Marshall, Reid, & Lillico, 2002). This prompted an observational study that also found this association (Bruhn, Stamer, Herrygers, Levine, & Noecker, 2009).

Another major source of concern was raised by Stranges et al. in their secondary analysis of the NPC cohort regarding a condition thus far not associated with selenium intake, type 2 diabetes (Stranges et al., 2007). In this study, published a decade after completion of the trial, the investigators found an increased risk of diabetes in selenium-treated subjects with suggestion of a dose–response relationship according to baseline selenium status.

In addition to these sources of concern for adverse effects of selenium, a serious methodological issue was discovered shortly after the publication of the final results of the trial. The authors revealed in their 2003 report (Duffield-Lillico, Dalkin, et al., 2003) that there had been a severe detection bias, with a lower tendency to be biopsied in the selenium arm compared to the placebo arm (14% vs 35%). This bias greatly undermines the reliability of the trial results at least with regard to the prostate cancer results. It is unclear if and how it affected other cancer endpoints. Taking into account a recently published study on factors influencing prostate biopsy rates in selenium and vitamin E cancer prevention trial (SELECT) and another trial (Tangen et al., 2016), it might be hypothesized that also the increase in diabetes incidence associated with selenium supplementation in NPC may have induced a lower biopsy rate in these subjects, thus contributing to the occurrence of this detection bias.

6. THE SELENIUM AND VITAMIN E CANCER PREVENTION TRIAL

The SELECT was conducted as a response to the apparent prostate cancer risk reduction findings of the NPC trial and of a trial of vitamin E (The Alpha-Tocopherol Beta Carotene Cancer Prevention Study Group, 1994), and findings from other trials examining a combination of vitamin E and selenium and other antioxidant substances (Lippman et al., 2009). This trial, a major effort by groups of oncology investigators and supported by the US National Cancer Institute, is for its size, cost (over 114 million US $), and soundness of scientific design not only the most important trial for selenium but also one of the most important trials ever implemented in the field of nutritional epidemiology and cancer chemoprevention (Goodman et al., 2013; Lippman et al., 2005).

Subjects participating in SELECT were randomized to four arms, selenomethionine (200 μg selenium/day) only, vitamin E only, both selenomethionine and vitamin E, or placebo. The trial enrolled over 35,000 men from several centers in the United States, Canada, and Puerto Rico. Given our interest in selenium, we focus our attention on the selenium only and placebo arms, which had 8752 and 8696 subjects, respectively. The primary outcome was incidence of prostate cancer. Eligibility criteria included age 50 and above for black men or 55 and above for the other races, a prostate-specific antigen value not exceeding 4.0 ng/L, and no suspicion of prostate cancer as detected through a digital rectal examination. Recruitment and consequent follow-up lasted from August 2001 through June 2004. The trial included several secondary outcomes such as other major cancers, cardiovascular diseases, adverse effects such as life-threatening events, minor side effects (alopecia, dermatitis, fatigue, halitosis, nail changes, and nausea), and eventually type 2 diabetes incidence, which was included as soon as its excess risk associated with selenium treatment emerged in the NPC trial.

The supplementation period of the SELECT was interrupted on October 23, 2008 when the trial Data Safety Monitoring Committee recommended discontinuation of the study for lack of efficacy and two other major concerns, an excess prostate cancer incidence detected in vitamin E-treated subjects and excess diabetes incidence among selenium-supplemented subjects (Klein et al., 2011; Lippman et al., 2009).

The final findings of the SELECT comparing selenium supplementation to placebo are summarized in Table 1 (Albanes et al., 2014; Kristal et al., 2014;

Lance et al., 2017; Lippman et al., 2009; Lotan et al., 2012; Martinez et al., 2014). There was no difference in incidence of all cancers, colorectal, lung, or overall prostate cancer. A slight increase in bladder cancer incidence occurred (HR 1.13), though this estimate was statistically unstable, with 99% CI 0.78–1.63 (Lotan et al., 2012). However, the most concerning results were related to some subtypes of prostate cancer and the adverse effects of the selenium treatment. Risk of aggressive prostate cancer (Gleason score $\geq$ 7) was increased according to extended follow-up results updated to July 5, 2011 in the selenium arm (HR 1.21, 99% CI 0.90–1.63) (Klein et al., 2011), and selenium supplementation was associated with excess advanced prostate cancer risk among subjects in the two highest quintiles of baseline toenail selenium concentration (HR 1.62, 95% CI 0.95–2.77), with a dose–response effect in these two upper categories of baseline selenium status (Kristal et al., 2014). Further investigation revealed that selenium administration increased cancer risk in subjects carrying a specific genotype, the CC at rs11781886, which was associated with decreased expression of the androgen-regulated prostate tumor suppressor protein NKX3.1 (Martinez et al., 2014).

Another report from SELECT found that selenium supplementation was associated with increased prostate cancer risk in subjects in the highest quintile of baseline plasma α-tocopherol. The increase was much greater for high-grade disease (Gleason score 7–10, HR 2.58, 95% CI 1.27–5.26) than for less advanced disease (HR 1.10, 95% CI 0.67–1.79), and for the first period of follow-up (<3 years, HR 2.09, 95% CI 1.15–3.79) than for the subsequent period (HR 1.14, 95% CI 0.69–1.86) (Albanes et al., 2014). Finally, there was an increased risk of type 2 diabetes in the selenium arm (724 new cases) compared with placebo (669), with a relative risk (RR) of 1.07 (99% CI 0.94–1.22) (Table 2). This excess diabetes incidence was a source of some concern for the trial Data and Safety Monitoring Committee, along with the increased prostate cancer incidence in the vitamin E arm observed in the interim data as of August 1, 2008, where *P*-values for such excess were 0.08 and 0.09, respectively (Lippman et al., 2009). Interestingly, this excess incidence of diabetes in the selenium arm decreased over time after the end of the supplementation period, further supporting a causal link with selenium (Klein et al., 2011).

As shown in Table 2, the selenium arm also experienced some additional adverse effects which had originally been included in the trial population surveillance because they were known to be associated with selenium overexposure (Vinceti et al., 2001): alopecia, dermatitis (both grades 1–2 and 3–4), fatigue (grades 1–2 but not 3–4), and nausea (grades 1–2 but not 3–4) (Lippman et al., 2009).

The unexpected discontinuation of the SELECT and its null or concerning results have played a major role in informing the selenium and cancer debate and have had a sobering effect on expectations of selenium as a game-changing chemopreventive substance. However, a question remains: why were the SELECT results so different from the NPC results, particularly for prostate cancer? Some investigators have suggested the lower baseline selenium status in NPC compared with SELECT as a possible explanation (Lu et al., 2016; Rayman, Combs, & Waters, 2009). However, the difference in baseline circulating selenium status between the two populations was very low, just 21 μg/day comparing plasma selenium in NPC with serum selenium of SELECT (Vinceti, Burlingame, et al., 2016), as shown in Fig. 2 (which also reports the results of another more

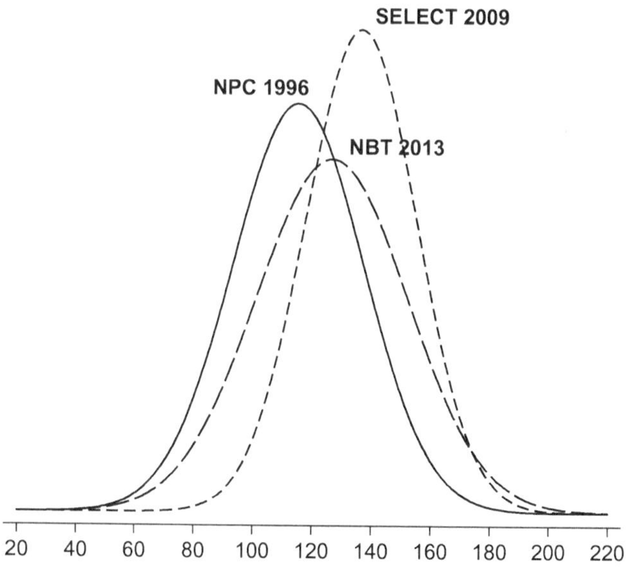

Fig. 2 Baseline circulating selenium levels in the NPC trial as plasma selenium (Duffield-Lillico, Dalkin, et al., 2003), NBT (Algotar et al., 2013), and SELECT (Lippman et al., 2009) as serum selenium—data computed from median and interquartile values with the equation found in http://handbook.cochrane.org/chapter_7/7_7_3_5_mediansand_inter quartile_ranges.htm (Higgins & Green, 2011).

recent trial, NBT (Algotar et al., 2013)). This corresponds to a dietary intake difference of about 14 μg/day, which is a small amount when compared with the daily intake of 77 μg/day in NPC and 91 μg/day in SELECT (computed using the 1:1.5 ratio between intake and plasma levels suggested by Haldimann et al. (1996). In addition, both intakes were above the nutritional requirements issued by several organizations (Vinceti et al., 2017), indicating that no nutritional selenium deficiency was affecting the populations of these two trials. Finally, NPC administered selenized yeast and SELECT administered selenomethionine. However, both forms were organic and the NPC selenium supplement included selenomethionine as a major source of the element (Jablonska & Vinceti, 2015; Vinceti, Burlingame, et al., 2016; Waters et al., 2012).

If we rule out these potential differences, two major ones still remain that may explain the different results for prostate cancer as well as for colorectal and lung cancer: the effect of the detection bias that occurred in SELECT and the peculiarity of the NPC population, which was composed of a highly specific population, patients with a history of nonmelanoma skin cancer. The first issue is certainly a major one for prostate cancer and may well explain the different results, but as previously mentioned, it is unclear if and how it may have affected the remaining NPC results reported in 2002–2003 (Duffield-Lillico et al., 2002; Duffield-Lillico, Slate, et al., 2003).

It is important to note that some findings from the two trials were consistent; in particular, an excess diabetes risk was observed in both trials. Unfortunately, nonmelanoma skin cancer was not included among the endpoints in SELECT and therefore the excess risk found in selenium-supplemented NPC subjects could not be further verified in this much larger trial.

7. THE OTHER TRIALS

After the NPC trial and during SELECT, many other trials concerning selective selenium administration were designed and implemented, adding to some early trials performed in China before NPC (Li, 1992; Li et al., 2000; Yu, Zhu, & Li, 1997; Yu et al., 1991), which however were excluded from subsequent meta-analyses due to their low methodological quality (Fairweather-Tait et al., 2011; Vinceti, Crespi, et al., 2013; Vinceti, Dennert, et al., 2014; Vinceti et al., 2009). Though none of these RCTs had the size and statistical power of SELECT, most of them had sound study design and have added relevant information about the effect of selenium on cancer risk. We focus on summarizing studies published from 2007 through February 28, 2017.

The main characteristics and results of these trials are reported in Table 1, and adverse effects of selenium supplementation are summarized in Table 2. Table 1 also reports the risk of bias of the various trials as assessed by our group using the recently published *RoB 2.0* tool developed by the Cochrane Collaboration (Higgins et al., 2016), available at https://sites.google.com/site/riskofbiastool/welcome/rob-2-0-tool.

Overall, the results of these trials showed remarkable consistency and have delineated a clear pattern of the relation between selenium exposure and cancer risk, therefore helping to correct the confusing pattern generated by the observational studies. The trials have consistently shown that selenium administration, in the form of 200 μg/day of organic selenium exclusively or predominantly as selenomethionine, with the exception of 250 μg/day of inorganic selenium in the Polish trial, does not reduce the risk of overall cancer or of major cancer types (Table 1). This conclusion is also reached when performing summary meta-analyses of these trials, as reported in Fig. 3. To perform these meta-analyses, we computed a summary RR of all cancer or selected cancers along with its 95% confidence limits by inputting the RR or HR of cancer in the selenium-supplemented arm compared to the placebo arm from each RCT into a random effects model using the *metan* STATA 14.2 routine (Stata Corp., College Station, TX, 2017). The study-specific RR or HR used in the models were the most adjusted RR or HR available from these studies, a method we have previously described (Vinceti, Filippini, et al., 2016). The figures also report the I^2 parameter of heterogeneity (Higgins, Thompson, Deeks, & Altman, 2003).

As shown in Fig. 3, summary estimates did not show evidence of any beneficial effect of selenium supplementation on cancer risk when looking at all cancers (Fig. 3A) and major cancer types such as colorectal cancer, melanoma and nonmelanoma skin cancer, and lung, breast, bladder, and prostate cancers (Fig. 3B–H). Limiting the analysis to the RCTs with the best methodological quality, these results were confirmed and suggested if anything a potential excess risk for breast and lung cancer and for melanoma. In a single study in patients with a history of resected lung cancer (Karp et al., 2013), selenium supplementation was also unexpectedly associated with a statistically unstable increase of some cancers such as hepatobiliary cancer, though the small number of cases made it difficult to evaluate the results.

Despite the comparability of the selenium doses used in these RCTs and the fact that they were all carried out in Western populations, there is heterogeneity in the study populations as well as in their baseline selenium exposure, as shown in Table 1. These studies enrolled men at high risk of prostate

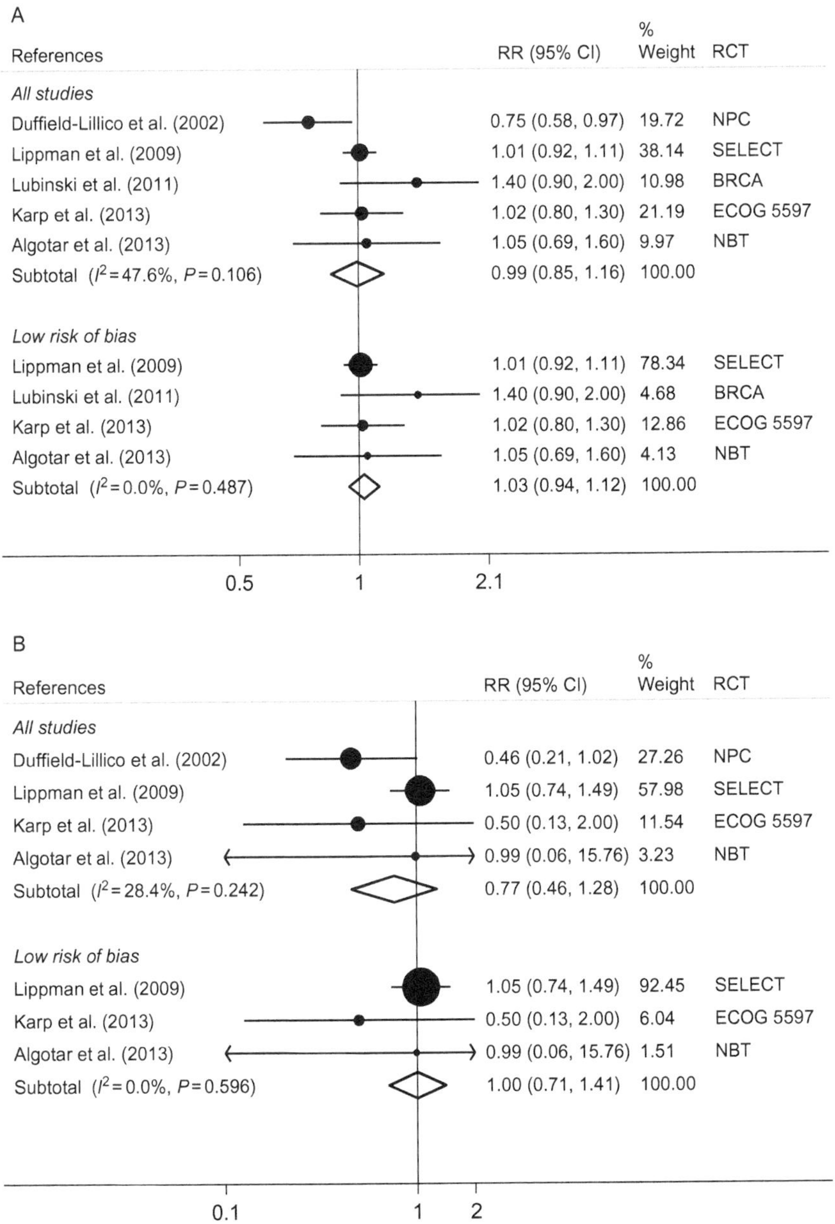

Fig. 3 Summary relative risks (RRs) with 95% confidence interval (CI) of cancer for all randomized controlled trials and for trials with low risk of bias. Participants received 200 μg/day of organic selenium supplementation (except for Lubinski et al., who administered 250 μg/day of inorganic selenium) or placebo. Based on the most adjusted RR from published results whenever available, and on 95% CI (when unavailable) computed using the standard error (SE) recovered from the 99% CI by computing SE = (LOG (upper 99% CI)—LOG (lower 99% CI))/5.15, then EXP(LOG(RR) = ±1.96*SE). (A) Any cancer. (B) Colorectal cancer. (C) Melanoma. (D) Nonmelanoma skin cancer. (E) Lung cancer. (F) Breast cancer. (G) Bladder cancer. (H) Prostate cancer.

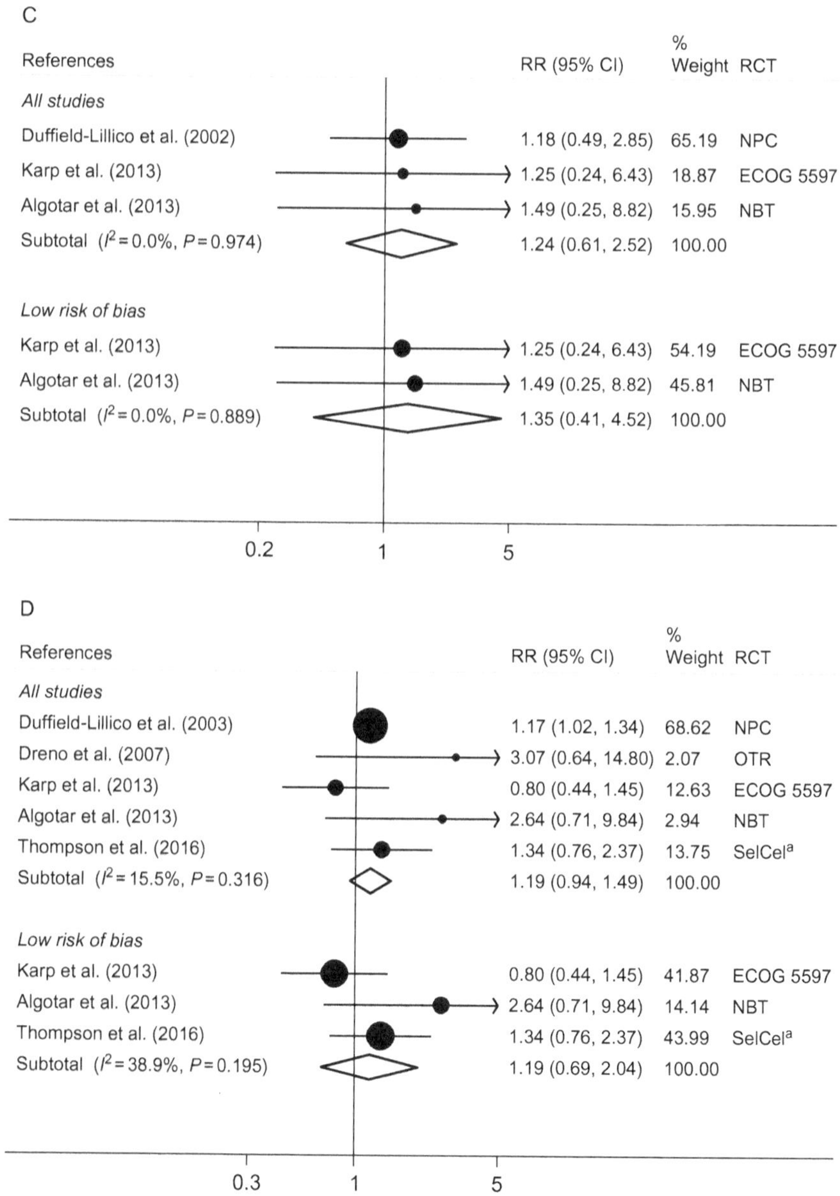

[a]Only squamous cell carcinoma.

Fig. 3—Cont'd

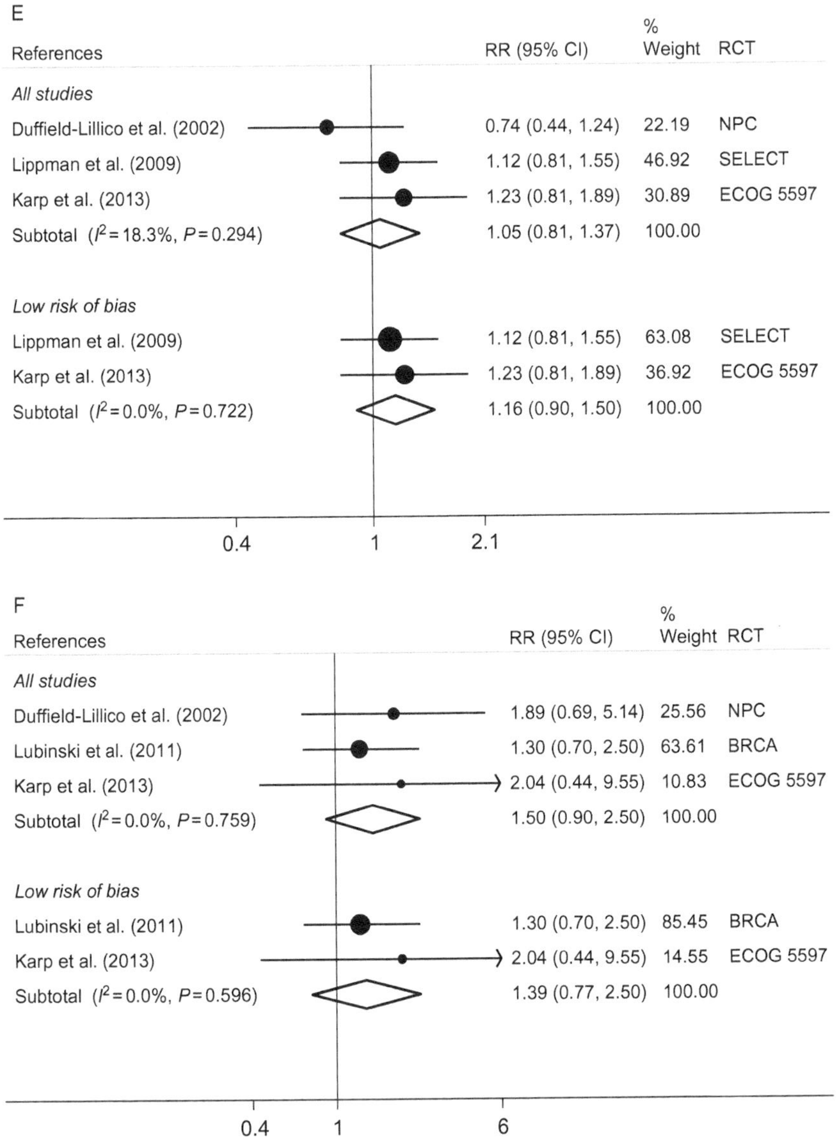
E
References
RR (95% CI)
% Weight
RCT
All studies
Duffield-Lillico et al. (2002)
0.74 (0.44, 1.24)
22.19
NPC
Lippman et al. (2009)
1.12 (0.81, 1.55)
46.92
SELECT
Karp et al. (2013)
1.23 (0.81, 1.89)
30.89
ECOG 5597
Subtotal (I^2 = 18.3%, P = 0.294)
1.05 (0.81, 1.37)
100.00
Low risk of bias
Lippman et al. (2009)
1.12 (0.81, 1.55)
63.08
SELECT
Karp et al. (2013)
1.23 (0.81, 1.89)
36.92
ECOG 5597
Subtotal (I^2 = 0.0%, P = 0.722)
1.16 (0.90, 1.50)
100.00
0.4
1
2.1
F
References
RR (95% CI)
% Weight
RCT
All studies
Duffield-Lillico et al. (2002)
1.89 (0.69, 5.14)
25.56
NPC
Lubinski et al. (2011)
1.30 (0.70, 2.50)
63.61
BRCA
Karp et al. (2013)
2.04 (0.44, 9.55)
10.83
ECOG 5597
Subtotal (I^2 = 0.0%, P = 0.759)
1.50 (0.90, 2.50)
100.00
Low risk of bias
Lubinski et al. (2011)
1.30 (0.70, 2.50)
85.45
BRCA
Karp et al. (2013)
2.04 (0.44, 9.55)
14.55
ECOG 5597
Subtotal (I^2 = 0.0%, P = 0.596)
1.39 (0.77, 2.50)
100.00
0.4
1
6

Fig. 3—Cont'd

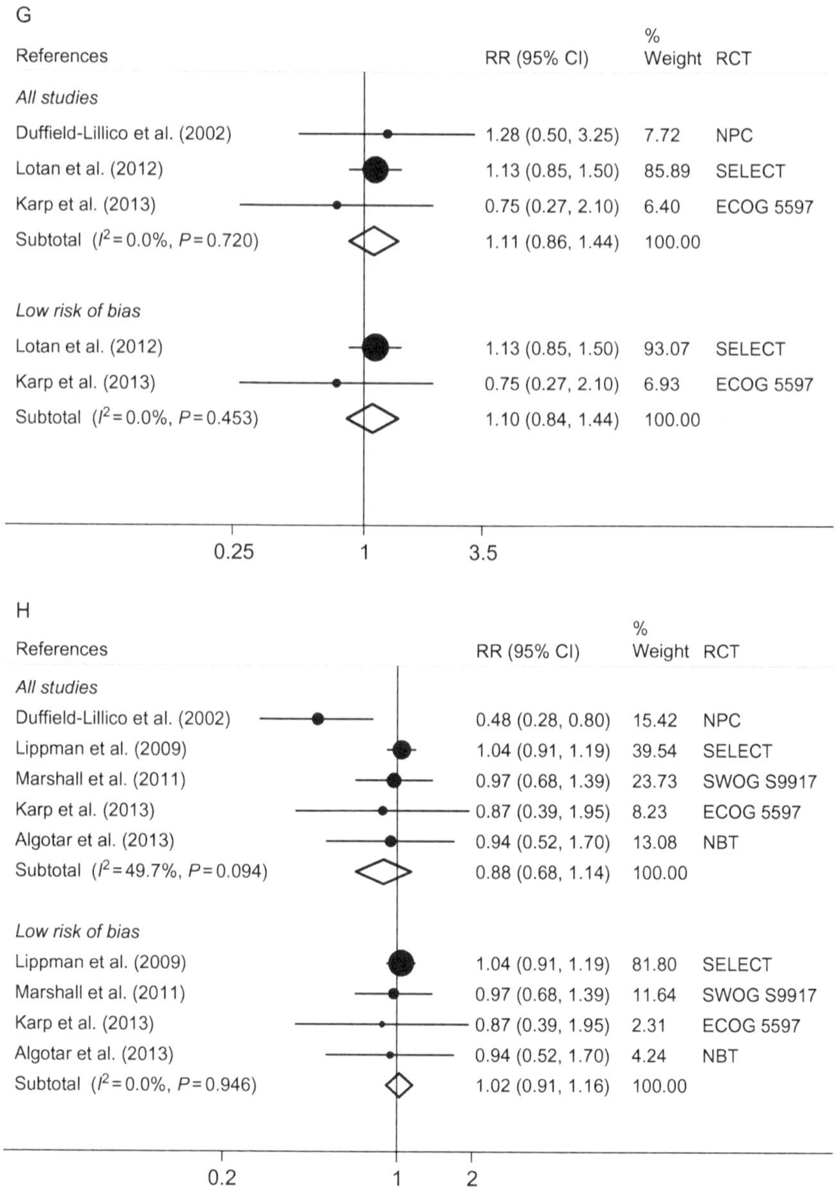
G
References
RR (95% CI)
% Weight
RCT
All studies
Duffield-Lillico et al. (2002)
1.28 (0.50, 3.25)
7.72
NPC
Lotan et al. (2012)
1.13 (0.85, 1.50)
85.89
SELECT
Karp et al. (2013)
0.75 (0.27, 2.10)
6.40
ECOG 5597
Subtotal (I^2=0.0%, P=0.720)
1.11 (0.86, 1.44)
100.00
Low risk of bias
Lotan et al. (2012)
1.13 (0.85, 1.50)
93.07
SELECT
Karp et al. (2013)
0.75 (0.27, 2.10)
6.93
ECOG 5597
Subtotal (I^2=0.0%, P=0.453)
1.10 (0.84, 1.44)
100.00
0.25
1
3.5
H
References
RR (95% CI)
% Weight
RCT
All studies
Duffield-Lillico et al. (2002)
0.48 (0.28, 0.80)
15.42
NPC
Lippman et al. (2009)
1.04 (0.91, 1.19)
39.54
SELECT
Marshall et al. (2011)
0.97 (0.68, 1.39)
23.73
SWOG S9917
Karp et al. (2013)
0.87 (0.39, 1.95)
8.23
ECOG 5597
Algotar et al. (2013)
0.94 (0.52, 1.70)
13.08
NBT
Subtotal (I^2=49.7%, P=0.094)
0.88 (0.68, 1.14)
100.00
Low risk of bias
Lippman et al. (2009)
1.04 (0.91, 1.19)
81.80
SELECT
Marshall et al. (2011)
0.97 (0.68, 1.39)
11.64
SWOG S9917
Karp et al. (2013)
0.87 (0.39, 1.95)
2.31
ECOG 5597
Algotar et al. (2013)
0.94 (0.52, 1.70)
4.24
NBT
Subtotal (I^2=0.0%, P=0.946)
1.02 (0.91, 1.16)
100.00
0.2
1
2

Fig. 3—Cont'd

cancer (Algotar et al., 2013; Marshall et al., 2011; Stratton et al., 2003), subjects who received organ graft (Dreno et al., 2007) or subjects with resected non-small-cell lung cancer (Karp et al., 2013). These trials also covered a wide range of background selenium intakes. Interestingly, when we compare cancer risk in these trials to that reported in NPC for comparable baseline selenium status, we find several conflicting results. Subjects in the lowest baseline selenium status tertile in the NBT trial (Algotar et al., 2013) did not show a prostate cancer reduction despite being comparable (Table IV of that paper and Fig. 2) with the two NPC trial lowest and middle baseline selenium tertiles, which showed strong reductions in all cancer and prostate cancer incidence (Duffield-Lillico, Dalkin, et al., 2003; Duffield-Lillico et al., 2002). Moreover, the results of the SWOG S9917 trial did not indicate a modifying effect of baseline selenium levels on the effect of selenium supplementation on prostate cancer risk (Marshall et al., 2011). Although the lowest baseline plasma selenium quartile in the SWOG S9917 trial was <106 μg/L, a cutpoint almost identical to that of the lowest tertile of NPC trial for which a strongly decreased prostate cancer HR was seen (0.14, 95% CI 0.03–0.61 (Duffield-Lillico, Dalkin, et al., 2003)), little evidence of decreased risk was seen in the SWOG trial (RR 0.82, 95% CI 0.40–1.69). In addition, in the second quartile of the SWOG trial, population risk tended toward an increase (RR 1.38, 95% CI 0.68–2.78) despite the boundaries of this exposure category (106–132 μg/L of plasma selenium) overlapped the middle and part of the highest tertiles in the NPC trial, two categories for which an overall decreased incidence was seen (20 cases among selenium-treated participants vs 27 in the placebo ones (Duffield-Lillico, Dalkin, et al., 2003)). Finally, in a case-cohort study carried out within the SELECT cohort assessing the effect of selenium supplementation on prostate cancer risk according to baseline toenail selenium levels in 1739 cases and 3117 referent men, no decreased risk in subjects in the lowest baseline exposure categories emerged (Kristal et al., 2014). These results do not support the hypothesis (Lu et al., 2016; Rayman et al., 2009) that different baseline selenium status may explain the conflicting results between NPC and SELECT.

Overall, despite the heterogeneity of the study populations, the results of RCTs are consistent in ruling out a beneficial effect of selenium supplementation, further confirming the SELECT results. This is particularly true when limiting the analysis to the high-quality, low-bias studies (Fig. 3). Therefore, it is now clear that selenium intakes of around 250 μg/day or above have no effect on cancer risk compared with lower amounts, and they might also have adverse effects on the risk of cancer or other chronic diseases, contrary to previous claims about the possible benefits of supranutritional selenium intakes (Rayman, 2002).

However, it is still not entirely clear whether intakes of selenium below these levels may affect cancer risk, since these amounts have not been tested by randomized trials, and observational studies suffer from lack of reliability. However, the low amounts of selenium exposure shown to pose a health risk by the randomized trials coupled with the need to apply an adequate safety factor to the lowest observed adverse effect level identified by these RCTs, and the awareness of the powerful toxicity of selenium particularly in its inorganic forms or as selenomethionine, suggest caution in substantially increasing selenium intake above the 34/26 μg/day values recommended for males and females, respectively, by the World Health Organization (Joint FAO/WHO Expert Consultation on Human Vitamin and Mineral Requirements, 2004) and particularly above the 50–70 μg/day suggested by other agencies (EFSA NDA Panel, 2014; Institute of Medicine Food and Nutrition Board, 2000; Nordic Council of Ministers, 2014; Società Italiana di Nutrizione Umana, 2014). Organizations that have set the recommended dietary intakes of selenium in the 26–70 μg/day range have ruled out any relevance of changes in cancer risk associated with these amounts of selenium exposure when setting these limits (EFSA NDA Panel, 2014; Institute of Medicine Food and Nutrition Board, 2000; Joint FAO/WHO Expert Consultation on Human Vitamin and Mineral Requirements, 2004).

It should also be noted that most of the above-mentioned selenium RCTs were carried out exclusively, or in two cases predominantly (Duffield-Lillico et al., 2002; Karp et al., 2013), in men with one exception (Lubinski et al., 2011). This latter RCT was carried out in Poland in 1135 women with a highly specific genetic background, a mutation of the BRCA1 gene, who were therefore at high risk of breast cancer (Lubinski et al., 2011). After administering 250 μg/day of selenium in the inorganic tetravalent form (selenite), the investigators found an excess breast cancer risk in this population after a median follow-up of 35 months and based on 105 overall cases (HR 1.4, 95% CI 0.9–2.0). In that study, there was also a statistically imprecise excess risk of ovarian cancer (HR 1.3, 95% CI 0.6–2.7).

In the 2 other trials including female subjects and follow-up for breast cancer, ECOG 5597 and NPC, 8 breast cancer cases occurred in the first study among the 531 women randomized to selenium and 2 cases among the 271 women receiving placebo (RR 2.04, 95% CI 0.44–9.55 as computed by inputting the published data into the *csi* routine of STATA

14.2). In NPC, the HR of breast cancer associated with selenium administration was 1.89 (95% CI 0.69–5.14), based on 17 overall cases among the 318 women randomized in the trial. Therefore, despite the specificities of the different study populations (carrying the BRCA1 mutation in the Polish trial, with a history of resected stage I non-small-cell lung cancer in ECOG 5597 and nonmelanoma skin cancer in NPC), the consistency of these results appears to be concerning and calls for caution when considering administration of selenium in women.

8. A NATURAL EXPERIMENT ON SELENIUM AND CANCER

Though RCTs are the best design to identify a causal relation between a treatment and a health outcome, there is one peculiar observational study design which may share the strength of the experimental ones, the so-called natural experiment. As stated by Rothman and Greenland, "If the paradigm of scientific observation is the experiment, then the paradigm of nonexperimental epidemiologic research is the 'natural experiment,' in which nature emulates an experiment" (Rothman & Greenland, 1998). Such rare situations, by no means confined to "medical" epidemiology (Fine et al., 2013), mimic random allocation of subjects to conditions and can involve toxic exposures, which would be impossible to study with randomized trials.

There is an occurrence of a natural experiment for the relation between selenium and cancer that occurred in the city of Reggio Emilia in Northern Italy. In that community, for 16 years, a neighborhood was provided with municipal tapwater characterized by unusually high levels of inorganic hexavalent selenium, selenate, and no other distinctive chemical features compared to drinking water distributed in the remaining municipal territory (Vinceti et al., 1998, 1995). In addition, residents in that neighborhood had very similar sociodemographic characteristics compared to the remaining "unexposed" municipal population and were unaware of the selenium content in their drinking water, thus mirroring a blinded randomized study.

Both short- and long-term follow-ups have been performed in this community to assess the association between selenium exposure and cancer incidence (Vinceti, Ballotari, et al., 2016; Vinceti, Nacci, et al., 2000; Vinceti et al., 1995). So far there is no evidence of any beneficial effects for prostate cancer or other cancer types. In fact, a possible increased risk has emerged for some cancers such as neoplasms of the buccal cavity and pharynx, urinary tract, melanoma, and lymphoid cancers, and less clearly for other neoplasms (Vinceti, Ballotari, et al., 2016; Vinceti, Nacci, et al., 2000; Vinceti et al.,

1998, 1995). In addition, there was a regression toward the unity of the RRs over time, with the excess mortality from various cancers waning starting 10 years after the end of the selenium exposure through drinking water (when the well was closed), further supporting a causal relation between this naturally occurring selenium exposure and cancer risk.

So far, this study remains the only report which analyzed the effect of inorganic hexavalent selenium on cancer risk (Vinceti, Ballotari, et al., 2016). It must be noted that inorganic selenium compounds, such as selenate and selenite, are much more toxic than the organic forms (Ammar & Couri, 1981; Vinceti, Solovyev, et al., 2013). Notably, selenium sulfide, another inorganic form of selenium, has been identified as "reasonably anticipated to be a human carcinogen" by the United States National Toxicology Program (National Toxicology Program, 2011; Vinceti, Crespi, et al., 2013). These observations further indicate that the relation between selenium and cancer must always take into account the specific selenium compounds under investigation.

9. SELENIUM IN CANCER THERAPY

The use of selenium supplements as adjuvant therapy for cancer patients is an issue of growing interest (Fernandes & Gandin, 2015; Li et al., 2016; Misra et al., 2015, 2016; Plano et al., 2016; Wallenberg et al., 2014). The rationale for such studies is the antioxidant effects induced by selenium supplementation and in particular the ability of selenium species to inhibit cancer cells and tissue growth shown in laboratory studies (Fernandes & Gandin, 2015).

The mechanisms underlying the anticancer and antiproliferative properties of selenium compounds are not well established and are still under active investigation. They encompass a spectrum of effects including prooxidant and proteomic activities, which ultimately induce apoptosis, necrosis, and paraptosis (Bao et al., 2015; Fernandes & Gandin, 2015; Zhao et al., 2016). Therefore, the potential use of intermediate to high doses of selenium species in cancer therapy and more generally in human disease appears to be driven by its toxicological properties (Forceville, 2013; Jablonska & Vinceti, 2015; Nogueira & Rocha, 2011), and selenium compounds when used in this context must be considered drugs and tested as such in appropriate animal and human studies.

A few nonexperimental studies in cancer patients are available. These studies lack randomization and therefore lack adequate control of confounders. In addition, relatively low doses of selenium have been used. These studies have suggested that selenium supplementation after diagnosis of nonmetastatic prostate cancer may increase risk of prostate cancer mortality (Kenfield et al., 2015), has no effect on breast cancer or all cancer mortality in women with breast cancer (Greenlee et al., 2012), and has no effect on the quality of life in colorectal cancer survivors (Tsinovoi, Xun, & He, 2017). Lower levels of selenium-binding protein 1 have been recently shown to be associated with a poorer prognosis in patients affected by nasopharyngeal carcinoma (Chen et al., 2016), but an opposite trend has been found for thioredoxin reductase in patients with hepatocellular carcinoma (Li et al., 2015), and studies on the prognostic value of selenium-binding proteins and selenoproteins in cancer patients are lacking.

As far as experimental studies are concerned, phase II and III RCTs with either low or high doses of selenium have been performed to investigate the potential benefit of selenium in reducing the side effects of chemotherapy in cancer patients, such as reduction in rates of infection and oral mucositis (Asfour et al., 2006; Jahangard-Rafsanjani et al., 2013; Vieira et al., 2015).

As a potential anticancer agent, selenium compounds have been assessed in four randomized placebo-controlled trials (Table 3). The Watching Waiting Trial was carried out in men with localized prostate cancer in order to study disease progression after selenium supplementation for up to 5 years; results did not indicate a protective effect (Stratton et al., 2010). Women given selenium during radiation for gynecological cancer demonstrated a higher 5-year overall survival rate (91.9%) compared to placebo group (83.1%) (Muecke et al., 2014). Similarly, higher rates of regression of cervical intraepithelial neoplasia grade 1 were found in women after long-term selenium supplementation compared with placebo (Karamali et al., 2015).

Finally, a recently published large phase III trial was carried out in 14 Belgian hospitals, the SELEnium and BLAdder cancer Trial (SELEBLAT). This trial assessed the efficacy of 200 μg/selenium/day as selenized yeast in preventing the recurrence of noninvasive urothelial carcinoma in subjects with history of this cancer. In the selenium-supplemented arm, adjusted HR of invasive urothelial carcinoma was 0.88 (95% CI 0.58–1.35) in the intent-to-treat analysis and 1.01 (95% CI 0.65–1.57) in the per-protocol analysis, thus suggesting the lack of a beneficial effect of selenium in the therapy of this specific cancer type (Goossens et al., 2016).

We were able to identify two ongoing trials concerning the possible anticancer effects of selenium through the online database https://clinicaltrials.gov, using "selenium" AND "cancer" as search terms. The first trial

Table 3 Summary of Phase II and III Randomized Trials of Selenium (Se) Carried Out in Cancer Patients

References	Final Population Total (Se Group/Pl Group)	Se Supplement	Outcome	Side Effects[a]	Results
WWT: Watchful Waiting Trial (Stratton et al., 2010)	140 (47 + 47/46) men with localized prostate cancer aged <85 (years)	47 with 200 μg/day of selenized yeast Duration up to 5 years	Progression of prostate cancer measured using PSA velocity	Brittle hair and nails: 6/10 RR 0.59 (95% CI 0.23–1.48) Cataract: 6/7 RR 0.84 (95% CI 0.30–2.31) Diabetes: 1/2 RR 0.47 (95% CI 0.04–4.93)	Se supplement did not show a protective effect on PSA velocity. Se supplement increased PSA velocity in men with high baseline plasma Se
		47 with 800 μg/day of selenized yeast Duration up to 5 years		Brittle Hair and nails: 8/10 RR 0.78 (95% CI 0.34–1.81) Cataract: 9/7 RR 1.26 (95% CI 0.51–3.10) Diabetes: 3/2 RR 1.36 (95% CI 0.24–7.75)	
Muecke et al. (2014)	81 (39/42) women with carcinomas of the cervix or corpus uteri aged 31–80	Sodium selenite: 500 μg/day (days of radiation therapy) and 300 μg/day (days without treatment until the last day of radiation therapy)	Effect of Se supplement during adjuvant radiation therapy on patients' long-term survival	No Se-related side effects observed Diarrhea (grades 1–3) at week 4:	5 years disease-free survival rate of 80.1% and 83.2% in Se group and placebo group, respectively (median follow-up 49 months) 10 years disease-free survival rate of 80.1% and 83.2% in treated and placebo group, respectively (median follow-up 67 months)

Continued

Table 3 Summary of Phase II and III Randomized Trials of Selenium (Se) Carried Out in Cancer Patients—cont'd

References	Final Population Total (Se Group/Pl Group)	Se Supplement	Outcome	Side Effects	Results
		Average duration of intervention 38 days (cumulative selenite dose of 17 mg)		RR 0.52 (95% CI 0.30–0.88)	5 years overall survival rate of 91.9% and 83.1% in Se group and placebo group, respectively (median follow-up 51 months) 10 years overall survival rate of 55.3% and 42.7% in Se group and placebo group, respectively (median follow-up 70 months)
Karamali, Nourgostar, Zamani, Vahedpoor, and Asemi (2015)	56 (28/28) women diagnosed with CIN1 with mean aged 38	200 μg/day of selenized yeast. Duration 6 months	Primary endpoint: regression of CIN1 lesions Secondary endpoints: modification of metabolic status	—	Se group showed higher regression rate for CIN1 (88% vs 56%) Se group resulted in the decreased level of fasting plasma glucose (−0.37 (SD 0.32) vs +0.07 (SD 0.63) mmol/L); insulin (−28.8 (SD 31.2) vs +13.2 (SD 40.2) pmol/L); triglycerides (−0.14 (SD 0.55) vs +0.15 (SD 0.38) mmol/L) and an increased level of HDL-cholesterol (+0.13 (SD 0.21) vs −0.01 (SD 0.15) mmol/L)
SELEBLAT: Selenium and Bladder Cancer Trial (Goossens et al., 2016)	292 (151/141) subjects with history of invasive urothelial carcinoma aged 46–91	200 μg/day selenized yeast. Duration up to 3 years	Recurrence of noninvasive urothelial carcinoma	Fatigue: 1/2 RR 0.47 (95% CI 0.04–5.09) Sum of gastrointestinal effects: 4/6 RR 0.62 (95% CI 0.18–2.16)	HR 0.88 (95% CI 0.58–1.35) intent-to-treat analysis HR 1.01 (95% CI 0.65–1.57) per-protocol analysis

[a]Number of events in Se and placebo groups. Relative risk (RR) computed using the *csi* routine of STATA 14.2 (Stata Corp., College Station, TX).

addresses the safety and efficacy of a combination of axitinib and selenomethionine in adult patients with advanced metastatic clear-cell renal cell carcinoma (trial code NCT02535533). The second trial aims to evaluate whether selenomethionine and vitamin E are effective in preventing nonmelanoma skin cancers in subjects having manifest arsenical skin lesions (NCT00392561).

10. CONCLUSION

The recent conclusion of high-quality randomized trials had represented a tremendous advance in the understanding of the relation between selenium and human cancer, compared with what had been suggested by observational studies and laboratory investigations. Results of these trials indicate that raising exposure of organic selenium to supranutritional levels of about 300 μg/day is not effective in reducing cancer risk and may even increase it, and it also appears to increase the incidence of type 2 diabetes. As far as lower amounts of selenium intake (falling in the nutritional range) and their relation with cancer risk are concerned, no evidence from high-quality trials is available and no conclusion can be drawn from observational studies due to their methodological limitations, such as unmeasured confounding and exposure misclassification.

Evidence from a natural experiment indicates that even low-level overexposure to inorganic hexavalent selenium may increase cancer risk at some sites, without having a beneficial effect at the other cancer sites. Therefore, no cancer-preventive effect may be currently attributed to selenium and caution should be used when considering any increase of its exposure above the typical levels experienced by Western populations, or to the inorganic forms of the element.

In contrast, the potential for using some selenium compounds as antineoplastic or adjuvant agents in cancer therapy represents an interesting area of future research. Thus far very few trials have been carried out on this issue, and further experimental evidence must be collected to reliably assess the effectiveness and adverse effects of selenium in cancer therapy. When assessing this issue and more generally the relation of selenium with human health and disease, attention must be given to the species of selenium involved, due to the major differences between the various chemical forms of this element and their biological effects.

ACKNOWLEDGMENTS

Financial support to this study was provided by Lega Italiana per la Lotta contro i Tumori (LILT), Reggio Emilia Section, and Associazione Sostegno Oncologia Ematologia Pediatrica (ASEOP) of Modena (to Dr. Vinceti); the National Institute of Health Grant No. NIH CA016042 (to Dr. Crespi).

REFERENCES

Albanes, D., Till, C., Klein, E. A., Goodman, P. J., Mondul, A. M., Weinstein, S. J., et al. (2014). Plasma tocopherols and risk of prostate cancer in the Selenium and Vitamin E Cancer Prevention Trial (SELECT). *Cancer Prevention Research*, 7(9), 886–895.

Algotar, A. M., Stratton, M. S., Ahmann, F. R., Ranger-Moore, J., Nagle, R. B., Thompson, P. A., et al. (2013). Phase 3 clinical trial investigating the effect of selenium supplementation in men at high-risk for prostate cancer. *Prostate*, *73*(3), 328–335.

Ammar, E. M., & Couri, D. (1981). Acute toxicity of sodium selenite and selenomethionine in mice after ICV or IV administration. *Neurotoxicology*, *2*(2), 383–386.

Arner, E. S. J. (2016). Foreward. In D. Hatfield, U. Schweizer, & V. N. Gladyshev (Eds.), *Selenium: Its molecular biology and role in human health* (4th ed., pp. IX–XI). New York: Springer Science + Business Media.

Asfour, I. A., El Shazly, S., Fayek, M. H., Hegab, H. M., Raouf, S., & Moussa, M. A. (2006). Effect of high-dose sodium selenite therapy on polymorphonuclear leukocyte apoptosis in non-Hodgkin's lymphoma patients. *Biological Trace Element Research*, *110*(1), 19–32.

Ashton, K., Hooper, L., Harvey, L. J., Hurst, R., Casgrain, A., & Fairweather-Tait, S. J. (2009). Methods of assessment of selenium status in humans: A systematic review. *The American Journal of Clinical Nutrition*, *89*(6), 2025S–2039S.

Bao, P., Chen, Z., Tai, R. Z., Shen, H. M., Martin, F. L., & Zhu, Y. G. (2015). Selenite-induced toxicity in cancer cells is mediated by metabolic generation of endogenous selenium nanoparticles. *Journal of Proteome Research*, *14*(2), 1127–1136.

Beane Freeman, L. E., Karagas, M. R., Baris, D., Schwenn, M., Johnson, A. T., Colt, J. S., et al. (2015). Is the inverse association between selenium and bladder cancer due to confounding by smoking? *American Journal of Epidemiology*, *181*(7), 488–495.

Birt, D. F., Julius, A. D., Runice, C. E., White, L. T., Lawson, T., & Pour, P. M. (1988). Enhancement of BOP-induced pancreatic carcinogenesis in selenium-fed Syrian golden hamsters under specific dietary conditions. *Nutrition and Cancer*, *11*(1), 21–33.

Borella, P., Bargellini, A., & Medici, C. I. (1996). Chemical form of selenium greatly affects metal uptake and responses by cultured human lymphocytes. *Biological Trace Element Research*, *51*(1), 43–54.

Brigelius-Flohe, R., & Flohe, L. (2017). Selenium and redox signaling. *Archives of Biochemistry and Biophysics*, *617*, 48–59.

Brigelius-Flohe, R., & Kipp, A. P. (2016). Glutathione peroxidase 2, a selenoprotein exhibiting a dual personality in preventing and promoting cancer. In D. Hatfield, U. Schweizer, & V. N. Gladyshev (Eds.), *Selenium: Its molecular biology and role in human health* (4th ed., pp. 451–462). New York: Springer Science + Business Media.

Bruhn, R. L., Stamer, W. D., Herrygers, L. A., Levine, J. M., & Noecker, R. J. (2009). Relationship between glaucoma and selenium levels in plasma and aqueous humour. *The British Journal of Ophthalmology*, *93*(9), 1155–1158.

Casey, C. E. (1988). Selenophilia. *The Proceedings of the Nutrition Society*, *47*(4), 55–62.

Chan, J. M., Oh, W. K., Xie, W., Regan, M. M., Stampfer, M. J., King, I. B., et al. (2009). Plasma selenium, manganese superoxide dismutase, and intermediate- or high-risk prostate cancer. *Journal of Clinical Oncology*, *27*(22), 3577–3583.

Chen, F., Chen, C., Qu, Y., Xiang, H., Ai, Q., Yang, F., et al. (2016). Selenium-binding protein 1 in head and neck cancer is low-expression and associates with the prognosis of nasopharyngeal carcinoma. *Medicine (Baltimore), 95*(35), e4592.

Chen, J., Geissler, C., Parpia, B., Li, J., & Campbell, T. C. (1992). Antioxidant status and cancer mortality in China. *International Journal of Epidemiology, 21*(4), 625–635.

Chen, X., Mikhail, S. S., Ding, Y. W., Yang, G. y., Bondoc, F., & Yang, C. S. (2000). Effects of vitamin E and selenium supplementation on esophageal adenocarcinogenesis in a surgical model with rats. *Carcinogenesis, 21*(8), 1531–1536.

Clark, L. C., Cantor, K. P., & Allaway, W. H. (1991). Selenium in forage crops and cancer mortality in U.S. counties. *Archives of Environmental Health, 46*(1), 37–42.

Clark, L. C., Combs, G. F. J., Turnbull, B. W., Slate, E. H., Chalker, D. K., Chow, J., et al. (1996). Effects of selenium supplementation for cancer prevention in patients with carcinoma of the skin. A randomized controlled trial. Nutritional Prevention of Cancer Study Group. *JAMA, 276*(24), 1957–1963.

Clayton, C. C., & Baumann, C. A. (1949). Diet and azo dye tumors: Effect of diet during a period when the dye is not fed. *Cancer Research, 9*(10), 575–582.

Coates, R. J., Weiss, N. S., Daling, J. R., Morris, J. S., & Labbe, R. F. (1988). Serum levels of selenium and retinol and the subsequent risk of cancer. *American Journal of Epidemiology, 128*(3), 515–523.

Dalkin, B. L., Lillico, A. J., Reid, M. E., Jacobs, E. T., Combs, G. F., Slate, E., et al. (2001). In *Selenium and chemoprevention against prostate cancer: An update of the Clark results 92nd meeting of the American association for cancer research.* New Orleans, LA: American Association for Cancer Research.

Dreno, B., Euvrard, S., Frances, C., Moyse, D., & Nandeuil, A. (2007). Effect of selenium intake on the prevention of cutaneous epithelial lesions in organ transplant recipients. *European Journal of Dermatology, 17*(2), 140–145.

Duffield-Lillico, A. J., Dalkin, B. L., Reid, M. E., Turnbull, B. W., Slate, E. H., Jacobs, E. T., et al. (2003). Selenium supplementation, baseline plasma selenium status and incidence of prostate cancer: An analysis of the complete treatment period of the Nutritional Prevention of Cancer Trial. *BJU International, 91*(7), 608–612.

Duffield-Lillico, A. J., Reid, M. E., Turnbull, B. W., Combs, G. F. J., Slate, E. H., Fischbach, L. A., et al. (2002). Baseline characteristics and the effect of selenium supplementation on cancer incidence in a randomized clinical trial: A summary report of the Nutritional Prevention of Cancer Trial. *Cancer Epidemiology, Biomarkers & Prevention, 11*(7), 630–639.

Duffield-Lillico, A. J., Slate, E. H., Reid, M. E., Turnbull, B. W., Wilkins, P. A., Combs, G. F., Jr., et al. (2003). Selenium supplementation and secondary prevention of nonmelanoma skin cancer in a randomized trial. *Journal of the National Cancer Institute, 95*(19), 1477–1481.

EFSA NDA Panel. (2014). Scientific opinion on dietary reference values for selenium. *EFSA Journal, 12*(10), 3846.

Epplein, M., Burk, R. F., Cai, Q., Hargreaves, M. K., & Blot, W. J. (2014). A prospective study of plasma selenoprotein P and lung cancer risk among low-income adults. *Cancer Epidemiology, Biomarkers & Prevention, 23*(7), 1238–1244.

Fairweather-Tait, S. J., Bao, Y., Broadley, M. R., Collings, R., Ford, D., Hesketh, J. E., et al. (2011). Selenium in human health and disease. *Antioxidants & Redox Signaling, 14*(7), 1337–1383.

Fan, A., & Vinceti, M. (2015). Selenium and its compounds. In R. D. Harbison, M. M. Bourgeois, & G. T. Johnson (Eds.), *Hamilton & Hardy's industrial toxicology* (6th ed.). Hoboken, NJ: John Wiley & Sons, Inc.

Fernandes, A. P., & Gandin, V. (2015). Selenium compounds as therapeutic agents in cancer. *Biochimica et Biophysica Acta, 1850*(8), 1642–1660.

Filippini, T., Ferrari, A., Michalke, B., Grill, P., Vescovi, L., Salvia, C., et al. (2017). Toenail selenium as an indicator of environmental exposure: A cross-sectional study. *Molecular Medicine Reports*, *15*(5), 3405–3412.

Fine, P., Victora, C. G., Rothman, K. J., Moore, P. S., Chang, Y., Curtis, V., et al. (2013). John Snow's legacy: Epidemiology without borders. *Lancet*, *381*(9874), 1302–1311.

Flohe, L., Gunzler, W. A., & Schock, H. H. (1973). Glutathione peroxidase: A selenoenzyme. *FEBS Letters*, *32*(1), 132–134.

Forceville, X. (2013). The effect of selenium therapy on mortality in patients with sepsis syndrome: Simple selenium supplementation or real (5 H2O)·Na2SeO3 pharmacological effect? *Critical Care Medicine*, *41*(6), 1591–1592.

Frost, D. V. (1972). The two faces of selenium—Can selenophobia be cured? *CRC Critical Reviews in Toxicology*, *1*(4), 467–514.

Gammelgaard, B., Rasmussen, L. H., Gabel-Jensen, C., & Steffansen, B. (2012). Estimating intestinal absorption of inorganic and organic selenium compounds by in vitro flux and biotransformation studies in Caco-2 cells and ICP-MS detection. *Biological Trace Element Research*, *145*(2), 248–256.

Garland, M., Morris, J. S., Stampfer, M. J., Colditz, G. A., Spate, V. L., Baskett, C. K., et al. (1995). Prospective study of toenail selenium levels and cancer among women. *Journal of the National Cancer Institute*, *87*(7), 497–505.

Geybels, M. S., van den Brandt, P. A., Schouten, L. J., van Schooten, F. J., van Breda, S. G., Rayman, M. P., et al. (2014). Selenoprotein gene variants, toenail selenium levels, and risk for advanced prostate cancer. *Journal of the National Cancer Institute*, *106*(3), dju003.

Goodman, P. J., Hartline, J. A., Tangen, C. M., Crowley, J. J., Minasian, L. M., Klein, E. A., et al. (2013). Moving a randomized clinical trial into an observational cohort. *Clinical Trials*, *10*(1), 131–142.

Goodman, G. E., Schaffer, S., Bankson, D. D., Hughes, M. P., & Omenn, G. S. (2001). Predictors of serum selenium in cigarette smokers and the lack of association with lung and prostate cancer risk. *Cancer Epidemiology, Biomarkers & Prevention*, *10*(10), 1069–1076.

Goossens, M. E., Zeegers, M. P., van Poppel, H., Joniau, S., Ackaert, K., Ameye, F., et al. (2016). Phase III randomised chemoprevention study with selenium on the recurrence of non-invasive urothelial carcinoma. The SELEnium and BLAdder cancer trial. *European Journal of Cancer*, *69*, 9–18.

Graff, R. E., Judson, G., Ahearn, T. U., Fiorentino, M., Loda, M., Giovannucci, E. L., et al. (2017). Circulating antioxidant levels and risk of prostate cancer by TMPRSS2:ERG. *Prostate*, 77(6), 647–653.

Greenlee, H., Kwan, M. L., Kushi, L. H., Song, J., Castillo, A., Weltzien, E., et al. (2012). Antioxidant supplement use after breast cancer diagnosis and mortality in the Life After Cancer Epidemiology (LACE) cohort. *Cancer*, *118*(8), 2048–2058.

Guo, W., Chow, W., Zheng, W., Li, J., & Blot, W. J. (1994). Diet, serum markers and breast cancer mortality in China. *Japanese Journal of Cancer Research*, *85*(6), 572–577.

Guo, C. H., Hsia, S., Hsiung, D. Y., & Chen, P. C. (2015). Supplementation with selenium yeast on the prooxidant-antioxidant activities and anti-tumor effects in breast tumor xenograft-bearing mice. *The Journal of Nutritional Biochemistry*, *26*(12), 1568–1579.

Guo, W., Hsing, A. W., Li, J., Chen, J., Chow, W., & Blot, W. J. (1994). Correlation of cervical cancer mortality with reproductive and dietary factors, and serum markers in China. *International Journal of Epidemiology*, *23*(6), 1127–1132.

Hafeman, D. G., Sunde, R. A., & Hoekstra, W. G. (1974). Effect of dietary selenium on erythrocyte and liver glutathione peroxidase in the rat. *Journal of Nutrition*, *104*(5), 580–587.

Haldimann, M., Venner, T. Y., & Zimmerli, B. (1996). Determination of selenium in the serum of healthy Swiss adults and correlation to dietary intake. *Journal of Trace Elements in Medicine and Biology*, *10*(1), 31–45.

Hatfield, D. L., Tsuji, P. A., Carlson, B. A., & Gladyshev, V. N. (2014). Selenium and selenocysteine: Roles in cancer, health, and development. *Trends in Biochemical Sciences*, *39*(3), 112–120.

Hatfield, D. L., Yoo, M. H., Carlson, B. A., & Gladyshev, V. N. (2009). Selenoproteins that function in cancer prevention and promotion. *Biochimica et Biophysica Acta*, *1790*(11), 1541–1545.

Higgins, J. P. T., & Green, S. (Eds.), (2011). Cochrane handbook for systematic reviews of interventions version 5.1.0 [updated March 2011]. The Cochrane Collaboration. Available from www.handbook.cochrane.org.

Higgins, J. P. T., Sterne, J. A. C., Savovic, J., Page, M. J., Hrobjartsson, A., Boutron, I., et al. (2016). A revised tool for assessing risk of bias in randomized trials. *Cochrane Database of Systematic Reviews*, *10*(Suppl. 1), 29.

Higgins, J. P., Thompson, S. G., Deeks, J. J., & Altman, D. G. (2003). Measuring inconsistency in meta-analyses. *BMJ*, *327*(7414), 557–560.

Hughes, D. J., Duarte-Salles, T., Hybsier, S., Trichopoulou, A., Stepien, M., Aleksandrova, K., et al. (2016). Prediagnostic selenium status and hepatobiliary cancer risk in the European Prospective Investigation into Cancer and Nutrition cohort. *The American Journal of Clinical Nutrition*, *104*(2), 406–414.

Hughes, D. J., Fedirko, V., Jenab, M., Schomburg, L., Meplan, C., Freisling, H., et al. (2015). Selenium status is associated with colorectal cancer risk in the European Prospective Investigation of Cancer and Nutrition cohort. *International Journal of Cancer*, *136*(5), 1149–1161.

Institute of Medicine Food and Nutrition Board. (2000). *Dietary references intakes for vitamin C, vitamin E, selenium, and carotenoids*. Washington, DC: National Academy Press.

IPCS INCHEM. (1987). *Environmental health criteria, 58. Selenium*. Geneva, Switzerland: World Health Organization.

Jablonska, E., Gromadzinska, J., Sobala, W., Reszka, E., & Wasowicz, W. (2008). Lung cancer risk associated with selenium status is modified in smoking individuals by Sep15 polymorphism. *European Journal of Nutrition*, *47*(1), 47–54.

Jablonska, E., & Vinceti, M. (2015). Selenium and human health: Witnessing a Copernican revolution? *Journal of Environmental Science and Health. Part C, Environmental Carcinogenesis & Ecotoxicology Reviews*, *33*(3), 328–368.

Jahangard-Rafsanjani, Z., Gholami, K., Hadjibabaie, M., Shamshiri, A. R., Alimoghadam, K., Sarayani, A., et al. (2013). The efficacy of selenium in prevention of oral mucositis in patients undergoing hematopoietic SCT: A randomized clinical trial. *Bone Marrow Transplantation*, *48*(6), 832–836.

Joint FAO/WHO Expert Consultation on Human Vitamin and Mineral Requirements. (2004). *Vitamin and mineral requirements in human nutrition* (2nd ed.). Geneva, Switzerland: World Health Organization and Food and Agriculture Organization of the United Nations.

Jossa, F., Trevisan, M., Krogh, V., Farinaro, E., Giumetti, D., Fusco, G., et al. (1991). Serum selenium and coronary heart disease risk factors in southern Italian men. *Atherosclerosis*, *87*(2–3), 129–134.

Kabuto, M., Imai, H., Yonezawa, C., Neriishi, K., Akiba, S., Kato, H., et al. (1994). Prediagnostic serum selenium and zinc levels and subsequent risk of lung and stomach cancer in Japan. *Cancer Epidemiology, Biomarkers & Prevention*, *3*(6), 465–469.

Karamali, M., Nourgostar, S., Zamani, A., Vahedpoor, Z., & Asemi, Z. (2015). The favourable effects of long-term selenium supplementation on regression of cervical tissues and metabolic profiles of patients with cervical intraepithelial neoplasia: A randomised, double-blind, placebo-controlled trial. *The British Journal of Nutrition, 114*(12), 2039–2045.

Karp, D. D., Lee, S. J., Keller, S. M., Wright, G. S., Aisner, S., Belinsky, S. A., et al. (2013). Randomized, double-blind, placebo-controlled, phase III chemoprevention trial of selenium supplementation in patients with resected stage I non-small-cell lung cancer: ECOG 5597. *Journal of Clinical Oncology, 31*(33), 4179–4187.

Kasaikina, M. V., Turanov, A. A., Avanesov, A., Schweizer, U., Seeher, S., Bronson, R. T., et al. (2013). Contrasting roles of dietary selenium and selenoproteins in chemically induced hepatocarcinogenesis. *Carcinogenesis, 34*(5), 1089–1095.

Kenfield, S. A., Van Blarigan, E. L., DuPre, N., Stampfer, M. J., L Giovannucci, E. L., & Chan, J. M. (2015). Selenium supplementation and prostate cancer mortality. *Journal of the National Cancer Institute, 107*(1), 360.

Keshavarzi, B., Moore, F., Najmeddin, A., & Rahmani, F. (2012). The role of selenium and selected trace elements in the etiology of esophageal cancer in high risk Golestan province of Iran. *Science of the Total Environment, 433*, 89–97.

Klein, E. A., Thompson, I. M., Jr., Tangen, C. M., Crowley, J. J., Lucia, M. S., Goodman, P. J., et al. (2011). Vitamin E and the risk of prostate cancer: The Selenium and Vitamin E Cancer Prevention Trial (SELECT). *JAMA, 306*(14), 1549–1556.

Knekt, P., Aromaa, A., Maatela, J., Alfthan, G., Aaran, R. K., Hakama, M., et al. (1990). Serum selenium and subsequent risk of cancer among Finnish men and women. *Journal of the National Cancer Institute, 82*(10), 864–868.

Knekt, P., Marniemi, J., Teppo, L., Heliovaara, M., & Aromaa, A. (1998). Is low selenium status a risk factor for lung cancer? *American Journal of Epidemiology, 148*(10), 975–982.

Kneller, R. W., Guo, W., Hsing, A. W., Chen, J., Blot, W. J., Li, J., et al. (1992). Risk factors for stomach cancer in sixty-five Chinese countries. *Cancer Epidemiology, Biomarkers & Prevention, 1*(2), 113–118.

Kristal, A. R., Darke, A. K., Morris, J. S., Tangen, C. M., Goodman, P. J., Thompson, I. M., et al. (2014). Baseline selenium status and effects of selenium and vitamin e supplementation on prostate cancer risk. *Journal of the National Cancer Institute, 106*(3), djt456.

Labunskyy, V. M., Hatfield, D. L., & Gladyshev, V. N. (2014). Selenoproteins: Molecular pathways and physiological roles. *Physiological Reviews, 94*(3), 739–777.

Lance, P., Alberts, D. S., Thompson, P. A., Fales, L., Wang, F., San Jose, J., et al. (2017). Colorectal adenomas in participants of the SELECT randomized trial of selenium and vitamin E for prostate cancer prevention. *Cancer Prevention Research, 10*(1), 45–54.

Lawson, K. A., Wright, M. E., Subar, A., Mouw, T., Hollenbeck, A., Schatzkin, A., et al. (2007). Multivitamin use and risk of prostate cancer in the National Institutes of Health-AARP Diet and Health Study. *Journal of the National Cancer Institute, 99*(10), 754–764.

Lazard, M., Dauplais, M., Blanquet, S., & Plateau, P. (2015). Trans-sulfuration pathway seleno-amino acids are mediators of selenomethionine toxicity in *Saccharomyces cerevisiae*. *The Journal of Biological Chemistry, 290*(17), 10741–10750.

Letavayova, L., Vlasakova, D., Spallholz, J. E., Brozmanova, J., & Chovanec, M. (2008). Toxicity and mutagenicity of selenium compounds in *Saccharomyces cerevisiae*. *Mutation Research, 638*(1–2), 1–10.

Li, W. G. (1992). Preliminary observations on effect of selenium yeast on high risk populations with primary liver cancer. *Zhonghua Yu Fang Yi Xue Za Zhi, 26*(5), 268–271.

Li, T., Li, F., Xiang, W., Yi, Y., Chen, Y., Cheng, L., et al. (2016). Selenium-containing amphiphiles reduced and stabilized gold nanoparticles: Kill cancer cells via reactive oxygen species. *ACS Applied Materials & Interfaces, 8*(34), 22106–22112.

Li, C., Peng, Y., Mao, B., & Qian, K. (2015). Thioredoxin reductase: A novel, independent prognostic marker in patients with hepatocellular carcinoma. *Oncotarget, 6*(19), 17792–17804.

Li, W., Zhu, Y., Yan, X., Zhang, Q., Li, X., Ni, Z., et al. (2000). The prevention of primary liver cancer by selenium in high risk populations. *Zhonghua Yu Fang Yi Xue Za Zhi, 34*(6), 336–338.

Lippman, S. M., Goodman, P. J., Klein, E. A., Parnes, H. L., Thompson, I. M., Jr., Kristal, A. R., et al. (2005). Designing the Selenium and Vitamin E Cancer Prevention Trial (SELECT). *Journal of the National Cancer Institute, 97*(2), 94–102.

Lippman, S. M., Klein, E. A., Goodman, P. J., Lucia, M. S., Thompson, I. M., Ford, L. G., et al. (2009). Effect of selenium and vitamin E on risk of prostate cancer and other cancers: The selenium and vitamin E cancer prevention trial (SELECT). *JAMA, 301*(1), 39–51.

Lockitch, G. (1989). Selenium: Clinical significance and analytical concepts. *Critical Reviews in Clinical Laboratory Sciences, 27*(6), 483–541.

Lotan, Y., Goodman, P. J., Youssef, R. F., Svatek, R. S., Shariat, S. F., Tangen, C. M., et al. (2012). Evaluation of vitamin E and selenium supplementation for the prevention of bladder cancer in SWOG coordinated SELECT. *The Journal of Urology, 187*(6), 2005–2010.

Lu, J., Zhang, J., Jiang, C., Deng, Y., Ozten, N., & Bosland, M. C. (2016). Cancer chemoprevention research with selenium in the post-SELECT era: Promises and challenges. *Nutrition and Cancer, 68*(1), 1–17.

Lubinski, J., Jaworska, K., Durda, K., Jakubowska, A., Huzarski, T., Byrski, T., et al. (2011). Selenium and the risk of cancer in BRCA1 carriers. *Hereditary Cancer in Clinical Practice, 9*(Suppl. 2), A5.

Ma, X., Yang, Y., Li, H. L., Zheng, W., Gao, J., Zhang, W., et al. (2017). Dietary trace element intake and liver cancer risk: Results from two population-based cohorts in China. *International Journal of Cancer, 140*(5), 1050–1059.

Mandrioli, J., Michalke, B., Solovyev, N., Grill, P., Violi, F., Lunetta, C., et al. (2017). Elevated levels of selenium species in cerebrospinal fluid of amyotrophic lateral sclerosis patients with disease-associated gene mutations. *Neurodegenerative Diseases, 17*(4–5), 171–180.

Marschall, T. A., Bornhorst, J., Kuehnelt, D., & Schwerdtle, T. (2016). Differing cytotoxicity and bioavailability of selenite, methylselenocysteine, selenomethionine, selenosugar 1 and trimethylselenonium ion and their underlying metabolic transformations in human cells. *Molecular Nutrition & Food Research, 60*(12), 2622–2632.

Marshall, J. R., Reid, M. E., & Lillico, A. (2002). *Nutritional prevention of cancer: Review of glaucoma data.* Tucson, AZ: University of Arizona.

Marshall, J. R., Tangen, C. M., Sakr, W. A., Wood, D. P., Jr., Berry, D. L., Klein, E. A., et al. (2011). Phase III trial of selenium to prevent prostate cancer in men with high-grade prostatic intraepithelial neoplasia: SWOG S9917. *Cancer Prevention Research, 4*(11), 1761–1769.

Martinez, E. E., Darke, A. K., Tangen, C. M., Goodman, P. J., Fowke, J. H., Klein, E. A., et al. (2014). A functional variant in NKX3.1 associated with prostate cancer risk in the Selenium and Vitamin E Cancer Prevention Trial (SELECT). *Cancer Prevention Research, 7*(9), 950–957.

Menkes, M. S., Comstock, G. W., Vuilleumier, J. P., Helsing, K. J., Rider, A. A., & Brookmeyer, R. (1986). Serum beta-carotene, vitamins A and E, selenium, and the risk of lung cancer. *The New England Journal of Medicine, 315*(20), 1250–1254.

Misra, S., Boylan, M., Selvam, A., Spallholz, J. E., & Bjornstedt, M. (2015). Redox-active selenium compounds—From toxicity and cell death to cancer treatment. *Nutrients*, 7(5), 3536–3556.

Misra, S., Selvam, A. K., Wallenberg, M., Ambati, A., Matolcsy, A., Magalhaes, I., et al. (2016). Selenite promotes all-trans retinoic acid-induced maturation of acute promyelocytic leukemia cells. *Oncotarget, 15*(7), 74686–74700.

Muecke, R., Micke, O., Schomburg, L., Glatzel, M., Reichl, B., Kisters, K., et al. (2014). Multicenter, phase III trial comparing selenium supplementation with observation in

gynecologic radiation oncology: Follow-up analysis of the survival data 6 years after cessation of randomization. *Integrative Cancer Therapies, 13*(6), 463–467.
Nakahara, S. B., Sanches, D. S., Caniceiro, B. D., Wysochi, H. L., Jr., da Silva, G. B., & Latorre, A. O. (2013). Effects of selenium on Pteridium aquilinum and urethane-induced lung carcinogenesis. *Immunopharmacology and Immunotoxicology, 35*(5), 605–614.
National Toxicology Program. (2011). Selenium sulfide. *Report on Carcinogens, 12*, 376–377.
Navia, B., Ortega, R. M., Perea, J. M., Aparicio, A., Lopez-Sobaler, A. M., Rodriguez-Rodriguez, E., et al. (2014). Selenium status in a group of schoolchildren from the region of Madrid, Spain. *Journal of Human Nutrition and Dietetics, 27*(Suppl. 2), 239–246.
Nelson, A. A., Fitzhugh, O. G., & Calvery, H. O. (1943). Liver tumors following cirrhosis caused by selenium in rats. *Cancer Research, 3*(4), 230–236.
Nogueira, C. W., & Rocha, J. B. (2011). Toxicology and pharmacology of selenium: Emphasis on synthetic organoselenium compounds. *Archives of Toxicology, 85*(11), 1313–1359.
Nordic Council of Ministers. (2014). *Nordic nutrition recommendations 2012* (5th ed.). Copenhagen: Nordic Co-operation.
Novoselov, S. V., Calvisi, D. F., Labunskyy, V. M., Factor, V. M., Carlson, B. A., Fomenko, D. E., et al. (2005). Selenoprotein deficiency and high levels of selenium compounds can effectively inhibit hepatocarcinogenesis in transgenic mice. *Oncogene, 24*(54), 8003–8011.
Outzen, M., Tjonneland, A., Larsen, E. H., Andersen, K. K., Christensen, J., Overvad, K., et al. (2015). The effect on selenium concentrations of a randomized intervention with fish and mussels in a population with relatively low habitual dietary selenium intake. *Nutrients*, 7(1), 608–624.
Panter, K. E., Hartley, W. J., James, L. F., Mayland, H. F., Stegelmeier, B. L., & Kechele, P. O. (1996). Comparative toxicity of selenium from seleno-DL-methionine, sodium selenate, and *Astragalus bisulcatus* in pigs. *Fundamental and Applied Toxicology, 32*(2), 217–223.
Park, S. Y., Haiman, C. A., Cheng, I., Park, S. L., Wilkens, L. R., Kolonel, L. N., et al. (2015). Racial/ethnic differences in lifestyle-related factors and prostate cancer risk: The multiethnic cohort study. *Cancer Causes & Control, 26*(10), 1507–1515.
Penney, K. L., Li, H., Mucci, L. A., Loda, M., Sesso, H. D., Stampfer, M. J., et al. (2013). Selenoprotein P genetic variants and mrna expression, circulating selenium, and prostate cancer risk and survival. *Prostate, 73*(7), 700–705.
Penney, K. L., Schumacher, F. R., Li, H., Kraft, P., Morris, J. S., Kurth, T., et al. (2010). A large prospective study of SEP15 genetic variation, interaction with plasma selenium levels, and prostate cancer risk and survival. *Cancer Prevention Research, 3*(5), 604–610.
Plano, D., Karelia, D. N., Pandey, M. K., Spallholz, J. E., Amin, S., & Sharma, A. K. (2016). Design, synthesis, and biological evaluation of novel selenium (Se-NSAID) molecules as anticancer agents. *Journal of Medicinal Chemistry, 59*(5), 1946–1959.
Porto, B. A. A., Moreira, L. M. C., Haddad-Ribeiro, F., Belo, L. C. M., & Neves, M. J. (2016). Inorganic selenium acts as a sensitizer to ionizing radiation in vivo *Saccharomyces cerevisiae* cells. *Journal of Radioanalytical and Nuclear Chemistry, 307*(1), 419–426.
Pounis, G., Costanzo, S., Persichillo, M., de Curtis, A., Sieri, S., Vinceti, M., et al. (2014). Mushroom and dietary selenium intakes in relation to fasting glucose levels in a free-living Italian adult population: The Moli-sani Project. *Diabetes & Metabolism, 40*(1), 34–42.
Ratnasinghe, D., Tangrea, J. A., Forman, M. R., Hartman, T., Gunter, E. W., Qiao, Y. L., et al. (2000). Serum tocopherols, selenium and lung cancer risk among tin miners in China. *Cancer Causes & Control, 11*(2), 129–135.
Rayman, M. P. (2012). Selenium and human health. *Lancet, 379*(9822), 1256–1268.

Rayman, M. P., Combs, G. F., Jr., & Waters, D. J. (2009). Selenium and vitamin E supplementation for cancer prevention. *JAMA*, *301*(18), 1876 [author reply 1877].
Rayman, M. P. (2002). The argument for increasing selenium intake. *The Proceedings of the Nutrition Society*, *61*(2), 203–215.
Reid, M. E., Duffield-Lillico, A. J., Garland, L., Turnbull, B. W., Clark, L. C., & Marshall, J. R. (2002). Selenium supplementation and lung cancer incidence: An update of the nutritional prevention of cancer trial. *Cancer Epidemiology, Biomarkers & Prevention*, *11*(11), 1285–1291.
Reid, M. E., Duffield-Lillico, A. J., Slate, E., Natarajan, N., Turnbull, B., Jacobs, E., et al. (2008). The nutritional prevention of cancer: 400 mcg per day selenium treatment. *Nutrition and Cancer*, *60*(2), 155–163.
Reid, M. E., Stratton, M. S., Lillico, A. J., Fakih, M., Natarajan, R., Clark, L. C., et al. (2004). A report of high-dose selenium supplementation: Response and toxicities. *Journal of Trace Elements in Medicine and Biology*, *18*(1), 69–74.
Rothman, K. J. (2014). Six persistent research misconceptions. *Journal of General Internal Medicine*, *29*(7), 1060–1064.
Rothman, K. J., & Greenland, S. (1998). *Modern epidemiology* (2nd ed.). Philadelphia: Lippincott-Raven.
Rotruck, J. T., Pope, A. L., Ganther, H. E., Swanson, A. B., Hafeman, D. G., & Hoekstra, W. G. (1973). Selenium: Biochemical role as a component of glutathione peroxidase. *Science*, *179*(73), 588–590.
Salbe, A. D., & Levander, O. A. (1990). Comparative toxicity and tissue retention of selenium in methionine-deficient rats fed sodium selenate or L-selenomethionine. *Journal of Nutrition*, *120*(2), 207–212.
Schrauzer, G. N., White, D. A., & Schneider, C. J. (1977). Cancer mortality correlation studies—III: Statistical associations with dietary selenium intakes. *Bioinorganic Chemistry*, 7(1), 23–31.
Schroeder, H. A., & Mitchener, M. (1972). Selenium and tellurium in mice. Effects on growth, survival, and tumors. *Archives of Environmental Health*, *24*(1), 66–71.
Seifter, J., Ehrich, W. E., Hudyma, G., & Mueller, G. (1946). Thyroid adenomas in rats receiving selenium. *Science*, *103*(2687), 762.
Shamberger, R. J., & Frost, D. V. (1969). Possible protective effect of selenium against human cancer. *Canadian Medical Association Journal*, *100*(14), 682.
Shamberger, R. J., & Rudolph, G. (1966). Protection against cocarcinogenesis by antioxidants. *Experientia*, *22*(2), 116.
Società Italiana di Nutrizione Umana. (2014). *LARN—Livelli di assunzione di riferimento di nutrienti e energia per la popolazione italiana—IV revisione*. Milano: SICS.
Sorensen, H. T., Lash, T. L., & Rothman, K. J. (2006). Beyond randomized controlled trials: A critical comparison of trials with nonrandomized studies. *Hepatology*, *44*(5), 1075–1082.
Spallholz, J. E. (1997). Free radical generation by selenium compounds and their prooxidant toxicity. *Biomedical and Environmental Sciences*, *10*(2–3), 260–270.
Steinbrenner, H., Speckmann, B., & Sies, H. (2013). Toward understanding success and failures in the use of selenium for cancer prevention. *Antioxidants & Redox Signaling*, *19*(2), 181–191.
Stranges, S., Marshall, J. R., Natarajan, R., Donahue, R. P., Trevisan, M., Combs, G. F., et al. (2007). Effects of long-term selenium supplementation on the incidence of type 2 diabetes: A randomized trial. *Annals of Internal Medicine*, *147*(4), 217–223.
Stranges, S., Sieri, S., Vinceti, M., Grioni, S., Guallar, E., Laclaustra, M., et al. (2010). A prospective study of dietary selenium intake and risk of type 2 diabetes. *BMC Public Health*, *10*, 564.
Stratton, M. S., Algotar, A. M., Ranger-Moore, J., Stratton, S. P., Slate, E. H., Hsu, C. H., et al. (2010). Oral selenium supplementation has no effect on prostate-specific antigen

velocity in men undergoing active surveillance for localized prostate cancer. *Cancer Prevention Research, 3*(8), 1035–1043.
Stratton, M. S., Reid, M. E., Schwartzberg, G., Minter, F. E., Monroe, B. K., Alberts, D. S., et al. (2003). Selenium and prevention of prostate cancer in high-risk men: The Negative Biopsy Study. *Anti-Cancer Drugs, 14*(8), 589–594.
Su, Y. P., Tang, J. M., Tang, Y., & Gao, H. Y. (2005). Histological and ultrastructural changes induced by selenium in early experimental gastric carcinogenesis. *World Journal of Gastroenterology, 11*(29), 4457–4460.
Suadicani, P., Hein, H. O., & Gyntelberg, F. (2012). Serum selenium level and risk of lung cancer mortality: A 16-year follow-up of the Copenhagen Male Study. *The European Respiratory Journal, 39*(6), 1443–1448.
Takata, Y., Kristal, A. R., Santella, R. M., King, I. B., Duggan, D. J., Lampe, J. W., et al. (2012). Selenium, selenoenzymes, oxidative stress and risk of neoplastic progression from Barrett's esophagus: Results from biomarkers and genetic variants. *PLoS One,* 7(6), e38612.
Tangen, C. M., Goodman, P. J., Till, C., Schenk, J. M., Lucia, M. S., & Thompson, I. M., Jr. (2016). Biases in recommendations for and acceptance of prostate biopsy significantly affect assessment of prostate cancer risk factors: Results from two large randomized clinical trials. *Journal of Clinical Oncology, 34*(36), 4338–4344.
The Alpha-Tocopherol Beta Carotene Cancer Prevention Study Group. (1994). The effect of vitamin E and beta carotene on the incidence of lung cancer and other cancers in male smokers. *The New England Journal of Medicine, 330*(15), 1029–1035.
Thompson, P. A., Ashbeck, E. L., Roe, D. J., Fales, L., Buckmeier, J., Wang, F., et al. (2016). Selenium supplementation for prevention of colorectal adenomas and risk of associated type 2 diabetes. *Journal of the National Cancer Institute, 108*(12), djw152.
Touat-Hamici, Z., Legrain, Y., Bulteau, A.-L., & Chavatte, L. (2014). Selective up-regulation of human selenoproteins in response to oxidative stress. *The Journal of Biological Chemistry, 289*(12), 14750–14761.
Tsinovoi, C. L., Xun, P., & He, K. (2017). Antioxidant supplementation is not associated with long-term quality of life in stage-II colorectal cancer survivors: A follow-up of the study of colorectal cancer survivors cohort. *Nutrition and Cancer, 69*(1), 159–166.
van den Brandt, P. A., Goldbohm, R. A., van 't Veer, P., Bode, P., Dorant, E., Hermus, R. J., et al. (1993). A prospective cohort study on selenium status and the risk of lung cancer. *Cancer Research, 53*(20), 4860–4865.
van Noord, P. A., Maas, M. J., van der Tweel, I., & Collette, C. (1993). Selenium and the risk of postmenopausal breast cancer in the DOM cohort. *Breast Cancer Research and Treatment, 25*(1), 11–19.
Varlamova, E. G., & Cheremushkina, I. V. (2017). Contribution of mammalian selenocysteine-containing proteins to carcinogenesis. *Journal of Trace Elements in Medicine and Biology, 39,* 76–85.
Vieira, M. L., Fonseca, F. L., Costa, L. G., Beltrame, R. L., Chaves, C. M., Cartum, J., et al. (2015). Supplementation with selenium can influence nausea, fatigue, physical, renal, and liver function of children and adolescents with cancer. *Journal of Medicinal Food, 18*(1), 109–117.
Vinceti, M., Ballotari, P., Steinmaus, C., Malagoli, C., Luberto, F., Malavolti, M., et al. (2016). Long-term mortality patterns in a residential cohort exposed to inorganic selenium in drinking water. *Environmental Research, 150,* 348–356.
Vinceti, M., Burlingame, B., Fillippini, T., Naska, A., Bargellini, A., & Borella, P. (2016). The epidemiology of selenium and human health. In D. Hatfield, U. Schweizer, & V. N. Gladyshev (Eds.), *Selenium: Its molecular biology and role in human health.* (4th ed., pp. 365–376). New York: Springer Science + Business Media.

Vinceti, M., Crespi, C. M., Malagoli, C., Bottecchi, I., Ferrari, A., Sieri, S., et al. (2012). A case-control study of the risk of cutaneous melanoma associated with three selenium exposure indicators. *Tumori*, *98*(3), 287–295.

Vinceti, M., Crespi, C. M., Malagoli, C., Del Giovane, C., & Krogh, V. (2013). Friend or foe? The current epidemiologic evidence on selenium and human cancer risk. *Journal of Environmental Science and Health. Part C, Environmental Carcinogenesis & Ecotoxicology Reviews*, *31*(4), 305–341.

Vinceti, M., Dennert, G., Crespi, C. M., Zwahlen, M., Brinkman, M., Zeegers, M. P., et al. (2014). Selenium for preventing cancer. *Cochrane Database of Systematic Reviews*, *3*, CD005195.

Vinceti, M., Filippini, T., Cilloni, S., Bargellini, A., Vergoni, A. V., Tsatsakis, A., et al. (2017). Health risk assessment of environmental selenium: Emerging evidence and challenges (Review). *Molecular Medicine Reports*, *15*(5), 3323–3335.

Vinceti, M., Filippini, T., Crippa, A., de Sesmaisons, A., Wise, L. A., & Orsini, N. (2016). Meta-analysis of potassium intake and the risk of stroke. *Journal of the American Heart Association*, *5*(10), e004210.

Vinceti, M., Grill, P., Malagoli, C., Filippini, T., Storani, S., Malavolti, M., et al. (2015). Selenium speciation in human serum and its implications for epidemiologic research: A cross-sectional study. *Journal of Trace Elements in Medicine and Biology*, *31*, 1–10.

Vinceti, M., Grioni, S., Alber, D., Consonni, D., Malagoli, C., Agnoli, C., et al. (2015). Toenail selenium and risk of type 2 diabetes: The ORDET cohort study. *Journal of Trace Elements in Medicine and Biology*, *29*, 145–150.

Vinceti, M., Mandrioli, J., Borella, P., Michalke, B., Tsatsakis, A., & Finkelstein, Y. (2014). Selenium neurotoxicity in humans: Bridging laboratory and epidemiologic studies. *Toxicology Letters*, *230*(2), 295–303.

Vinceti, M., Maraldi, T., Bergomi, M., & Malagoli, C. (2009). Risk of chronic low-dose selenium overexposure in humans: Insights from epidemiology and biochemistry. *Reviews on Environmental Health*, *24*(3), 231–248.

Vinceti, M., Nacci, G., Rocchi, E., Cassinadri, T., Vivoli, R., Marchesi, C., et al. (2000). Mortality in a population with long-term exposure to inorganic selenium via drinking water. *Journal of Clinical Epidemiology*, *53*(10), 1062–1068.

Vinceti, M., & Rothman, K. J. (2016). More results but no clear conclusion on selenium and cancer. *The American Journal of Clinical Nutrition*, *104*(2), 245–246.

Vinceti, M., Rothman, K. J., Bergomi, M., Borciani, N., Serra, L., & Vivoli, G. (1998). Excess melanoma incidence in a cohort exposed to high levels of environmental selenium. *Cancer Epidemiology, Biomarkers & Prevention*, 7(10), 853–856.

Vinceti, M., Rovesti, S., Bergomi, M., & Vivoli, G. (2000). The epidemiology of selenium and human cancer. *Tumori*, *86*(2), 105–118.

Vinceti, M., Rovesti, S., Gabrielli, C., Marchesi, C., Bergomi, M., Martini, M., et al. (1995). Cancer mortality in a residential cohort exposed to environmental selenium through drinking water. *Journal of Clinical Epidemiology*, *48*(9), 1091–1097.

Vinceti, M., Solovyev, N., Mandrioli, J., Crespi, C. M., Bonvicini, F., Arcolin, E., et al. (2013). Cerebrospinal fluid of newly diagnosed amyotrophic lateral sclerosis patients exhibits abnormal levels of selenium species including elevated selenite. *Neurotoxicology*, *38*, 25–32.

Vinceti, M., Wei, E. T., Malagoli, C., Bergomi, M., & Vivoli, G. (2001). Adverse health effects of selenium in humans. *Reviews on Environmental Health*, *16*(4), 233–251.

Wallenberg, M., Misra, S., & Bjornstedt, M. (2014). Selenium cytotoxicity in cancer. *Basic and Clinical Pharmacology and Toxicology*, *114*(5), 377–386.

Wang, X., Sun, K., Tan, Y., Wu, S., & Zhang, J. (2014). Efficacy and safety of selenium nanoparticles administered intraperitoneally for the prevention of growth of cancer cells in the peritoneal cavity. *Free Radical Biology & Medicine*, *72*, 1–10.

Waters, D. J., Shen, S., Kengeri, S. S., Chiang, E. C., Combs, G. F., Jr., Morris, J. S., et al. (2012). Prostatic response to supranutritional selenium supplementation: Comparison of the target tissue potency of selenomethionine vs. selenium-yeast on markers of prostatic homeostasis. *Nutrients*, *4*(11), 1650–1663.

Weekley, C. M., & Harris, H. H. (2013). Which form is that? The importance of selenium speciation and metabolism in the prevention and treatment of disease. *Chemical Society Reviews*, *42*(23), 8870–8894.

Woutersen, R. A., Appel, M. J., & Van Garderen-Hoetmer, A. (1999). Modulation of pancreatic carcinogenesis by antioxidants. *Food and Chemical Toxicology*, *37*(9–10), 981–984.

Wrobel, J. K., Wolff, G., Xiao, R., Power, R. F., & Toborek, M. (2016). Dietary selenium supplementation modulates growth of brain metastatic tumors and changes the expression of adhesion molecules in brain microvessels. *Biological Trace Element Research*, *172*(2), 395–407.

Yang, H., Jia, X., Chen, X., Yang, C. S., & Li, N. (2012). Time-selective chemoprevention of vitamin E and selenium on esophageal carcinogenesis in rats: The possible role of nuclear factor kappaB signaling pathway. *International Journal of Cancer*, *131*(7), 1517–1527.

Yang, G. Q., Wang, S. Z., Zhou, R. H., & Sun, S. Z. (1983). Endemic selenium intoxication of humans in China. *The American Journal of Clinical Nutrition*, *37*(5), 872–881.

Yu, S. Y., Zhu, Y. J., & Li, W. G. (1997). Protective role of selenium against hepatitis B virus and primary liver cancer in Qidong. *Biological Trace Element Research*, *56*(1), 117–124.

Yu, S. Y., Zhu, Y. J., Li, W. G., Huang, Q. S., Huang, C. Z., Zhang, Q. N., et al. (1991). A preliminary report on the intervention trials of primary liver cancer in high-risk populations with nutritional supplementation of selenium in China. *Biological Trace Element Research*, *29*(3), 289–294.

Zhang, Y., Coogan, P., Palmer, J. R., Strom, B. L., & Rosenberg, L. (2009). Vitamin and mineral use and risk of prostate cancer: The case-control surveillance study. *Cancer Causes & Control*, *20*(5), 691–698.

Zhao, J., Zhou, R., Hui, K., Yang, Y., Zhang, Q., Ci, Y., et al. (2016). Selenite inhibits glutamine metabolism and induces apoptosis by regulating GLS1 protein degradation via APC/C-CDH1 pathway in colorectal cancer cells. *Oncotarget*, *8*(12), 18832–18847.

CHAPTER TWO

Selenoproteins in Tumorigenesis and Cancer Progression

Sarah P. Short*, Christopher S. Williams*,†,‡,§,¶,1
*Vanderbilt University Medical Center, Nashville, TN, United States
†Vanderbilt University School of Medicine, Vanderbilt University, Nashville, TN, United States
‡Vanderbilt University, Nashville, TN, United States
§Vanderbilt Ingram Cancer Center, Vanderbilt University Medical Center, Nashville, TN, United States
¶Veterans Affairs Tennessee Valley HealthCare System, Nashville, TN, United States
[1]Corresponding author: e-mail address: christopher.williams@vanderbilt.edu

Contents

Abstract

Selenium is a micronutrient essential to human health and has long been associated with cancer prevention. Functionally, these effects are thought to be mediated by a class of selenium-containing proteins known as selenoproteins. Indeed, many selenoproteins have antioxidant activity which can attenuate cancer development by minimizing oxidative insult and resultant DNA damage. However, oxidative stress is increasingly being recognized for its "double-edged sword" effect in tumorigenesis, whereby it can mediate both negative and positive effects on tumor growth depending on the cellular context. In addition to their roles in redox homeostasis, recent work has also implicated selenoproteins in key oncogenic and tumor-suppressive pathways. Together, these data suggest that the overall contribution of selenoproteins to tumorigenesis is complicated and may be affected by a variety of factors. In this review, we discuss what is currently known about selenoproteins in tumorigenesis with a focus on their contextual roles in cancer development, growth, and progression.

Advances in Cancer Research, Volume 136
ISSN 0065-230X
http://dx.doi.org/10.1016/bs.acr.2017.08.002

1. INTRODUCTION

Selenium (Se) was discovered by J.J. Berzelius in 1817 and, while initially identified as a toxin when ingested in large amounts, it was determined to be an essential micronutrient in the 1950s (Boyd, 2011). Since then, Se has been determined to influence human health, and inadequate levels of Se can lead to deleterious effects. Indeed, Se deficiency has been identified as the underlying etiological precipitant of the cardiomyopathy known as Keshan disease and is also thought to be a key contributor to development of the deforming osteochondropathy, Kashin–Beck disease (Chen, 2012; Yao, Pei, & Kang, 2011). Low serum Se levels have also been associated with epilepsy, age-associated neurological disorders, and decreased survival following HIV infection, although a causal role for Se in these diseases has not been proven (Ashrafi et al., 2007; Baum et al., 1997; Zhang, Rocourt, & Cheng, 2010).

Once ingested, Se can be present as low molecular weight selenocompounds or by incorporation into proteins as modified amino acids. One such amino acid modification is selenomethionine (SeMet), which is the primary chemical form of dietary selenium (Burk & Hill, 2015). SeMet then enters the body's methionine pool where it is incorporated randomly into proteins in place of methionine in proportion to its concentration. However, SeMet can also undergo metabolism in the liver (and other tissues) to produce the selenium-modified cysteine, selenocysteine (Sec), which is degraded by selenocysteine lyase, yielding selenide. The selenide is then metabolized by selenophosphate synthetase to produce selenophosphate, the selenium of which is inserted into serine by selenocysteine synthase to yield Sec attached to the specialized tRNA. The selenocystyl-tRNA then serves to insert the selenocysteine into the nascent protein at the UGA. Collectively, selenocysteine-containing proteins are referred to as selenoproteins (SePs) and are believed to be the primary functional effectors of Se's effect on human homeostasis and disease. To-date, 25 SePs have been identified in the human genome and 24 in the mouse, including those in the glutathione peroxidase (GPx), iodothyronine deiodinase (DIO), and thioredoxin reductase (TrxR) protein families (Gladyshev et al., 2016).

SePs are well known for their antioxidant roles, as many catalyze oxidation–reduction reactions using the Sec as an active site. The Sec makes them particularly effective enzymes; the Sec selenol group is more fully ionized than the thiol of cysteine (Cys) at physiological pH and has a lower pK_a (~5.2) and reduction potential than Cys (Stadtman, 2000). Altogether, these render Sec more reactive than the Cys residue which comprises the active site of many nonselenoprotein enzymes (Stadtman, 1996). Thus, it is perhaps not surprising that the contributions of Se and SePs to human health are most often attributed to their antioxidant functions.

Altered redox homeostasis is believed to be a contributing factor to a number of diseases, including neuronal diseases such as Alzheimer's and Parkinson's, diabetes, fibrosis, inflammatory bowel diseases, and even cardiovascular disorders (Perez, Talens-Visconti, Rius-Perez, Finamor, & Sastre, 2017; Sarsour, Kumar, Chaudhuri, Kalen, & Goswami, 2009). However, it is particularly implicated in the development of cancer, as oxidative insult can lead to genomic instability, DNA mutation, tumor initiation, and progression (Panieri & Santoro, 2016). As Se may reduce oxidative stress by the antioxidant activities of its associated SePs, it was hypothesized that Se supplementation would confer protection against tumor initiation and progression. Indeed, animal tumor models have demonstrated that Se supplementation can reduce the incidence and severity of liver, esophageal, pancreatic, prostatic, colon, and mammary carcinogenesis (Baines et al., 2000; Daoud & Griffin, 1980; Ip, Thompson, & Ganther, 2000; van Rensburg, Hall, & Gathercole, 1986; Wang, Bonorden, et al., 2009; Woutersen, Appel, & Van Garderen-Hoetmer, 1999). Unfortunately, large clinical trials in human subjects have yielded mixed results; some suggest that Se supplementation and/or higher Se status may reduce cancer risk (Clark et al., 1998; Duffield-Lillico et al., 2003; Jacobs et al., 2004), while others fail to correlate serum Se levels with cancer development (Klein et al., 2011; Lippman et al., 2009; Papaioannou et al., 2011; Wallace et al., 2003). Of note, a large international case–control study testing the efficacy of Se and vitamin E supplementation in cancer (SELECT) did not demonstrate a protective effect of Se supplementation on risk of prostate cancer (Klein et al., 2011; Lippman et al., 2009), and trial meta-analysis determined that oral administration of Se was not effective in preventing colorectal neoplasia (Lance et al., 2017; Papaioannou et al., 2011). Furthermore, while not statistically significant, Se supplementation was associated with increased incidence of type-2 diabetes. Thus, the impact of Se supplementation on cancer is now recognized to be a complex issue and careful consideration should be given to untoward "off-target" effects. More detail on the epidemiology of Se in cancer as well as the association with type-2 diabetes can be found in chapter "The epidemiology of selenium and human cancer" by Vinceti et al.

Interestingly, one potential explanation for the inconsistent effects of Se supplementation in the above studies is the action of the SePs themselves. Indeed, while SePs have long been thought to protect against cancer development and progression by antioxidant and antiinflammatory roles, other signaling roles for SePs are becoming increasingly recognized. These other activities, such as modulation of angiogenesis, growth factor signaling, and apoptosis inhibition, may contribute to tumorigenesis in both positive and negative fashions. In this review, we will discuss what is known about the role of a subset of individual SePs in tumorigenesis, with a focus on how these proteins contribute to both pro- and antitumorigenic pathways in order to better understand the full effect of Se supplementation on cancer development and growth.

2. SELENOPROTEIN PRODUCTION

Expression of SePs is tightly controlled by the SeP translational process, which is highly specialized and with dedicated machinery. Sites of Sec incorporation are identified by the opal codon (UGA) and require the presence of a *cis*-acting stem-loop sequence in the selenoprotein mRNA 3′ UTR known as a selenocysteine insertion sequence, or SECIS (Walczak, Westhof, Carbon, & Krol, 1996). This SECIS site is responsible for recruiting a complex of specialized *trans*-acting factors dedicated to Sec incorporation, including the Sec-specific tRNA (Sec tRNA) and its required elongation factors (Driscoll & Copeland, 2003; Lee et al., 1990; Low, Grundner-Culemann, Harney, & Berry, 2000). It is important to note that the SeP translation process is highly dependent on the presence of Se, as the Sec-charged tRNA is only produced upon Se availability (Hatfield, Lee, Hampton, & Diamond, 1991). In times of Se deficiency, limited amounts of Sec tRNA result in the UGA codon being read as a premature stop codon, which recruits nonsense-mediated decay factors and induces mRNA turnover (Howard, Carlson, Anderson, & Hatfield, 2013; Seyedali & Berry, 2014). More recently, it has been shown that the Sec tRNA can bind cysteine when Se is deficient, resulting in the incorporation of cysteine at UGAs intended to designate Sec (Xu et al., 2010). Whether this process is significant is unknown.

However, in times of selenium deficiency, SeP production is not universally decreased, but rather a "hierarchy" of selenoprotein expression is maintained (Barnes, Evenson, Raines, & Sunde, 2009; Schriever, Barnes, Evenson, Raines, & Sunde, 2009). This is due in part to differential SeP mRNA stability based on resistance to nonsense-mediated decay, and SePs responsible for housekeeping functions, such as glutathione peroxidase 4 (GPx4), generally having more stable mRNA than SePs induced in response to stress and injury, such as glutathione peroxidase 1 (GPx1) (Hill, Lyons, & Burk, 1992; Seyedali & Berry, 2014). Selenoprotein synthesis is also modulated by differential expression of two Sec tRNA isoforms, which are distinguished by the presence or absence of 2′-*O*-methylribose at position 34. Mice maintained on a normal Se diet have been observed to have increased percentages of mcm^5Um-containing Sec tRNA (Um34) as compared to those on Se-deficient chow, suggesting that the Um34 modification is dependent on Se availability (Diamond et al., 1993). Functionally, the two Sec tRNA isoforms are differentially associated with production of specific classes of SePs. The housekeeping SePs essential for survival, such as GPx4, are not dependent on the Um34 modification which allows higher expression in the context of decreased Se availability. Conversely, the stress-induced SePs such as GPx1 and glutathione peroxidase 3 (GPx3) are more responsive to Um34 Sec tRNA and thereby more responsive to Se levels (Hatfield, Carlson, Xu, Mix, & Gladyshev, 2006). Indeed, a mutation of the Sec tRNA gene (*TRSP*) which interferes with the Um34 modification has recently been described in a human subject, where authors note decreased expression of stress-related SePs, while expression of housekeeping SePs is largely preserved (Schoenmakers et al., 2016).

3. INVESTIGATING SELENOPROTEINS THROUGH SEC tRNA MUTATIONS

To determine whether Se functionally contributes to disease and tissue homeostasis through these Sec-containing proteins, early studies sought to specifically investigate the role of the SeP protein family by modifying expression of Sec tRNA (*Trsp*), thereby globally interfering with selenoprotein biosynthesis. The first such Sec tRNA mouse models [(i^6A^-) $tRNA^{[Ser]Sec}$] relied on transgenic expression of mutated Sec tRNA which specifically inhibits Um34 Sec tRNA through a dominant-negative effect (Moustafa et al., 2001). While physiologically normal, mice with global transgene expression were determined to have decreased levels of stress-related SePs, such as GPx1 (Moustafa et al., 2001). Interestingly, studies using this mouse model were the first to show that SeP expression could directly modify the development of tumorigenesis (Table 1). In the gut, Sec tRNA transgenic mice were observed to have increased numbers of aberrant

Table 1 Effect of *Trsp* Alteration in Mouse Models of Carcinogenesis

Trsp Alteration	Genotype	Tissue	Oncogenic Driver	Phenotype	Citation
Transgenic	(i^6A^-) $tRNA^{[Ser]Sec}$	Prostate	SV40 [C3(1)/Tag]	Accelerated tumor development and increased aggressiveness	Diwadkar-Navsariwala et al. (2006)
	(i^6A^-) $tRNA^{[Ser]Sec}$	Colon	Azoxymethane	Increased preneoplastic lesions (aberrant crypt foci)	Irons, Carlson, Hatfield, and Davis (2006)
	(i^6A^-) $tRNA^{[Ser]Sec}$	Liver	TGF-α	No effect on liver tumorigenesis, and increased pyogranuloma formation	Moustafa et al. (2013)
Knockout	$Trsp^{fl}$; *MMTV-Cre*	Mammary gland	DMBA	Higher tumor incidence and shortened survival	Hudson et al. (2012)
	$Trsp^{fl}$; *PB-Cre*	Prostate	n/a	Prostatic intraepithelial neoplasias by 6 weeks, progressing to adenocarcinomas by 30 weeks	Luchman, Villemaire, Bismar, Carlson, and Jirik (2014)
	$Trsp^{fl}$; *LysM-Cre*	Myeloid cells	3LL Lewis lung carcinoma cells	Increased metastasis of ectopically injected lung cancer cells	Hiramoto et al. (2014)

Studies of tumor growth using either transgenic or knockout Trsp mouse lines.

crypt foci (ACF), a type of preneoplastic colonic lesion, after exposure to the mutagenic agent azoxymethane (AOM) (Irons et al., 2006). Interbreeding of Sec tRNA transgenic mice with a model of prostate cancer (*C3(1)/Tag*) resulted in the accelerated development of prostatic intraepithelial neoplasias (PIN), as well as more aggressive high-grade lesions (Diwadkar-Navsariwala et al., 2006). However, these results were not universal, as Sec tRNA transgenic mice had no change in hepatocellular tumor number or size as compared to WT mice when crossed onto a tumor-prone TGF-α transgenic background (Moustafa et al., 2013). Thus, the role of stress-related SePs in tumorigenesis appears to be dependent on organ and cancer type. Furthermore, these results may also point to differential SeP function in the context of specific oncogenic drivers or undetermined "environmental" cofactors, such as inflammation or microbiota.

The Sec tRNA transgenic model only interfered with translation of a subset of SePs and did not globally inhibit SeP production. However, global knockout (KO) of the mammalian Sec tRNA gene (*Trsp*) is embryonically lethal; thus a conditional approach has been used to analyze the effects of complete SeP loss (Bosl, Takaku, Oshima, Nishimura, & Taketo, 1997). Interestingly, even restricting *Trsp* KO to certain tissues often yielded severe effects, and SeP expression in many tissues, such as the endothelium, cardiac muscle, liver, and skin, is apparently required for survival (Carlson et al., 2004; Sengupta et al., 2010; Shrimali et al., 2007). Nevertheless, the conditional *Trsp* KO model has allowed assessment of SeP contribution to tissue homeostasis as well as a number of malignancies (Table 1). In the mammary gland, homozygous *Trsp* loss was associated with increased tumor incidence and shorter survival time after treatment with the carcinogen, 7,12-dimethylbenzyl[*a*]antracene (DMBA) (Hudson et al., 2012). Interestingly, while *Trsp* loss was not sufficient to induce tumors in non-DMBA-treated mice, it altered expression of BRCA and p53 in the mammary epithelium and suggests that SePs can affect oncogenic pathways in premalignant tissue (Kumaraswamy et al., 2003). In the prostate, *Trsp* loss alone was sufficient to induce tumorigenesis even in the absence of additional oncogenic alteration (Luchman et al., 2014). These mice presented with widespread PIN lesions by 6 weeks of age, which soon progressed to microinvasive carcinomas characterized by increased proliferation and activation of the mitogen-activated protein kinase (MAPK) pathway. While mice do not live long enough to assess susceptibility to cancer development, *Trsp* KO in the skin also led to development of widespread proliferation and hyperplasia, which are considered early events in tumor development (Sengupta et al., 2010). Finally, the *Trsp* KO model has also been utilized to gain insight into the function of SePs in immune cell population which, in turn, may also influence tumor development. *Trsp* knockout in myeloid lineages through a *LysM-Cre* driver led to increased oxidative stress, upregulated transcription of antioxidant enzymes, accumulation of reactive oxygen species (ROS), altered expression of extracellular matrix-related genes, diminished migration through matrix, and increased severity in a model of experimental colitis (Carlson et al., 2009; Kaushal et al., 2014; Suzuki et al., 2008). Furthermore, myeloid-specific *Trsp* knockout increased lung metastasis of ectopically injected Lewis lung carcinoma cells (Hiramoto et al., 2014). While the effect of myeloid-specific *Trsp* loss on tumor development has not been directly analyzed, collectively these results point to a widespread tumor-suppressive role for SePs and suggest that they inhibit key oncogenic molecular pathways.

4. ROLE OF SPECIFIC SELENOPROTEINS IN TUMORIGENESIS

4.1 Selenoprotein P

While studies of Se supplementation and global SeP modification suggest that SePs broadly suppress tumorigenesis, they do not address the role of specific selenium-containing proteins. Therefore, researchers have generated numerous mouse models (Table 2) and cell culture systems to investigate the contributions of individual SePs to tumorigenesis, such as selenoprotein P (known also as SELENOP, SEPP1, and several other designations (Gladyshev et al., 2016)). Unlike the majority of SePs which are best characterized by their enzymatic activity, SELENOP is better known for its roles in Se transport. In addition to GPx3, SELENOP is one of the few secreted SePs and is primarily produced in the liver at the main site of Se metabolism (Burk & Hill, 2005). In comparison to most SePs which only have 1 Sec residue, SELENOP is able to incorporate Se in up to 10 Sec residues within its primary structure, with the majority harbored within a Se-rich C-terminal domain. After Sec incorporation, SELENOP is secreted into the plasma and travels to distant tissues where it is taken up via endocytosis and lysosomally degraded to free Se for synthesis of other SePs (Kurokawa, Hill, McDonald, & Burk, 2012; Saito et al., 2004). Indeed, liver-specific *Selenop* KO revealed a striking 95% reduction in plasma SELENOP levels, which greatly reduced Se levels in distant tissues (Hill et al., 2012). Further highlighting the role for this protein in Se transport, many of the severe

Table 2 Carcinogenesis Studies in Selenoprotein Mouse Models

Gene	Alteration	Tissue	Oncogenic Driver	Phenotype	Citation
Selenop	Knockout	Intestine	AOM/DSS	Increased tumorigenesis in heterozygous mice associated with altered macrophage polarization, and decreased tumor burden in *Selenop*-null animals	Barrett et al. (2015)
	Transport domain mutant	Intestine	AOM/DSS	Increased tumorigenesis in homozygous animals	Barrett et al. (2015)
	Antioxidant domain mutant	Intestine	AOM/DSS	Increased tumorigenesis in homozygous animals	Barrett et al. (2015)
Gpx1	Transgenic	Skin	DMBA/TPA	Enhanced tumorigenesis and delayed tumor regression	Lu et al. (1997)
Gpx2	Knockout	Intestine	AOM/DSS	More severe inflammation and increased tumorigenesis	Krehl et al. (2012)
	Knockout	Intestine	AOM	Decreased tumor development on normal chow	Muller et al. (2013)
Gpx1/ *Gpx2*	Double knockout	Intestine	n/a	Spontaneous tumor development dependent on intestinal microflora	Chu et al. (2004)
Gpx3	Knockout	Prostate	TRAMP	Increased tumor incidence and higher WNT pathway activation	Chang, Lee, Oh, and Park (2016)
	Knockout	Intestine	AOM/DSS	Increased tumor number and degree of dysplasia	Barrett et al. (2013)
Selenof	Knockout	Intestine	AOM	Decreased numbers of preneoplastic lesions (aberrant crypt foci)	Tsuji et al. (2012)

Mouse lines with altered selenoprotein expression analyzed in models of tumorigenesis.

phenotypic consequences of *Selenop* loss can be circumvented via a supranutritional Se diet (Hill et al., 2012, 2007, 2003). However, it is important to note that SELENOP still retains antioxidant activity through a single N-terminal Sec, which exists within a UXXC motif that catalyzes the oxidation of glutathione (GSH) (Kurokawa, Eriksson, et al., 2014; Saito et al., 1999). Thus, both N- and C-terminal domains contribute to the overall function of SELENOP, making it vital for the production of other selenoproteins within target organs while still serving an antioxidant function.

Because such a large proportion of Se is incorporated into SELENOP, it is frequently measured as a reliable marker for overall Se levels. Thus, a number of epidemiological studies have analyzed SELENOP serum levels and tissue expression to determine their correlation with cancer incidence, aggressiveness, and survival. On the whole, these studies have determined SELENOP to be decreased in various tumor types, including hepatocellular carcinomas, gallbladder and biliary tract cancers, gastric adenocarcinomas, colorectal cancer, and prostate cancer (Al-Taie et al., 2004; Calvo et al., 2002; Gonzalez-Moreno et al., 2011; Hughes et al., 2016; Murawaki et al., 2008; Wang, Gong, et al., 2009). Indeed, *SELENOP* transcript levels are reduced as early as the premalignant adenoma stage in the colon (Barrett et al., 2015; Smith et al., 2010). SELENOP expression was also inversely correlated with disease stage in renal, prostate, and colon cancer patients (Meyer et al., 2012; Murawaki et al., 2008; Outzen et al., 2016). Interestingly, several single nucleotide polymorphisms (SNPs) have also been identified in *SELENOP* that may attenuate its expression or function, and these SNPs have been associated with cancer risk in the colon, breast, and prostate (Al-Taie et al., 2002; Meplan et al., 2013, 2010; Pellatt et al., 2013; Peters et al., 2008; Steinbrecher et al., 2010). However, SELENOP expression is not universally decreased in all tumor types. It has been reported to be increased in metastatic melanoma samples as compared to normal melanocytes, and its upregulation has also been associated with poorly differentiated prostate cancer which is more likely to metastasize (Hassona et al., 2013; Riker et al., 2008).

Despite a wealth of expression and epidemiological data, very few studies have directly analyzed SELENOP's contribution to tumor formation and progression. Using an siRNA approach, Gonzalez-Moreno et al. determined that reductions in SELENOP increased ROS in prostate cancer cells (Gonzalez-Moreno et al., 2011). In line with the expression studies, these data support a role in tumor suppression by decreasing ROS-mediated mutation and tumor initiation. However, this increase in ROS also rendered cells more susceptible to growth inhibition upon stimulation with hydrogen peroxide (H_2O_2). Thus, this illustrates a facet of the "double-edged sword" effect of oxidative stress which is highly relevant to studies of SePs in cancer progression; increased oxidative stress can enhance tumorigenesis through genomic instability and mutation, but excess can be toxic and induce tumor cell death (Fig. 1A; Perez et al., 2017). This dichotomy is also highlighted in the studies by Barrett et al., in which *Selenop* WT, heterozygous, and KO mice maintained on a supranutritional Se diet were placed on an azoxymethane/dextran sodium sulfate (AOM/DSS) protocol to induce inflammatory carcinogenesis in the colon (Barrett et al., 2015). Here, heterozygous *Selenop* loss augmented tumor number, size, and proliferation, while *Selenop*-null mice were relatively protected from tumorigenesis and displayed smaller tumors with elevated DNA damage and increased apoptosis.

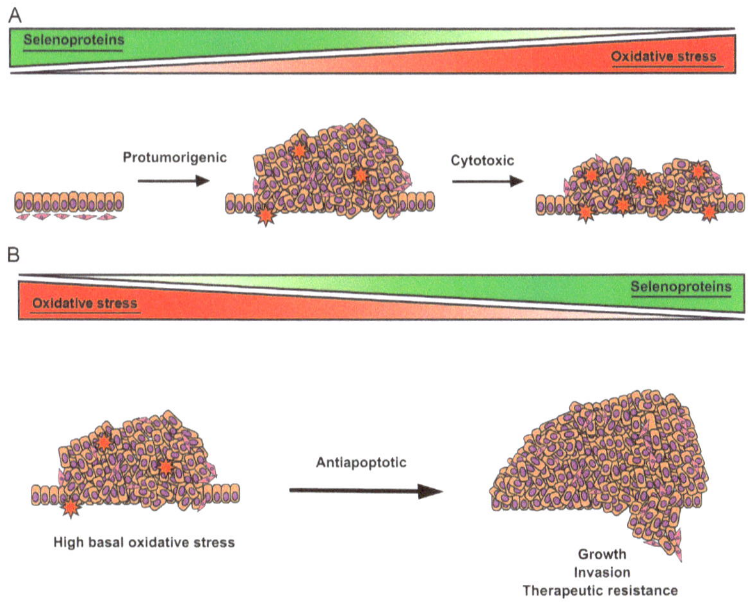

Fig. 1 Context-dependent effects of selenoprotein expression on oxidative stress and tumorigenesis. (A) In normal epithelium, selenoprotein loss induces a moderate increase in oxidative stress which can increase DNA damage and favor tumor initiation, although high levels of oxidative stress can promote cytotoxicity resulting in tumor cell clearance. (B) In tumor cells with higher baseline levels of oxidative stress, increased expression of selenoproteins can protect from cytotoxic effects and can lead to increased tumor growth, progression, and therapeutic resistance.

Due to its roles in Se transport, SELENOP is crucial for formation of other SePs, which may further affect net antioxidant function. Therefore, it can be difficult to determine whether the protumorigenic effects observed with SELENOP reduction are truly specific to SELENOP, or instead driven by loss of other selenium-containing proteins. This question was addressed by utilizing two additional mutant mouse models of SELENOP which express modified genes deficient in either the transport or antioxidant functions (Hill et al., 2007; Kurokawa, Eriksson, et al., 2014). These studies observed increased tumor numbers with loss of either functional domain, similar to the results with the heterozygous mice. Interestingly, not only do these results suggest that both antioxidant and transport capabilities functionally contribute to tumorigenesis, but suggest that it is their combined loss which increases oxidative stress to the point of cytotoxicity. Thus, it seems that moderate, but not complete, loss of SELENOP and antioxidant capability is the most tumorigenic, which further highlights a potential bimodal effect of SePs and oxidative stress.

While much of SELENOP's effect on tumor development has been attributed to increased oxidative DNA damage, it can also mediate microenvironmental changes which can further influence tumor cell growth. For example, SELENOP has been found to modify cytokine production, and *SELENOP* knockdown in the HepG2 hepatoma cell line induced expression of the proinflammatory cytokine TNF-α (Yi, Park, Byeon, Kwon, & Jung, 2003). Overexpression of *SELENOP* attenuates TGF-β-mediated transition of fibroblasts to myofibroblasts and reduces their secretion of HGF, VEGF, and IL-6, which can induce tumor cell invasion (Cat et al., 2006). Interestingly, local SELENOP expression is regulated by inflammatory cytokines, as TGF-β, IL-1β, TNF-α, and IFN-γ inhibit *SELENOP* transcription, while IL-10 induces expression (Bosschaerts et al., 2008; Mostert, Dreher, Kohrle, & Abel, 1999; Speckmann et al., 2010). SELENOP can also contribute to a protumorigenic microenvironment through altering macrophage polarization and function. Gene expression analysis has determined that *SELENOP* is one of the most upregulated genes in the M2 macrophage phenotype, which are widely believed to be more protumorigenic than their classically activated M1 counterparts (Bosschaerts et al., 2008; Solinas et al., 2010). Indeed, macrophages from *Selenop* heterozygous mice show altered macrophage polarization both in vivo and in vitro and show enhanced macrophage migration in tumors from AOM/DSS-treated mice (Barrett et al., 2015).

In order to take up extracellular SELENOP, cells rely on two known receptors which are differentially expressed based on the tissue type. In tissues such as the brain and testes, SELENOP is taken up by apoER2-mediated endocytosis, although in other tissues such as the kidney, the primary SELENOP receptor is megalin. Both receptors are members of the lipoprotein receptor (LRP) family (Burk et al., 2007; Hill et al., 2007; Olson, Winfrey, Nagdas, Hill, & Burk, 2007). However, it is important to note that these receptors are not specific to SELENOP, and each has additional functions that may impact tumor growth. ApoER2, for example, is known to regulate cell motility through interactions with Reelin and Cdc42 (Leemhuis & Bock, 2011), and its knockdown inhibits cell migration in pancreatic cancer cells (Sato, Fukushima, Chang, Matsubayashi, & Goggins, 2006). Likewise, megalin is increased in some cancer types such as cerebral lymphomas and melanomas, and in these cases, its loss decreases tumor cell proliferation and survival (Andersen et al., 2015; Pedersen, Hansen, Nielsen, & Penkowa, 2010). On the other hand, megalin mediates uptake of Vitamin D (which has been associated with a reduced cancer risk), chemotherapeutic drugs, and drug-sensitizing agents in addition to SELENOP (Christensen & Birn, 2002; Christensen & Willnow, 1999; Deeb, Trump, & Johnson, 2007). Thus, it is possible that SELENOP may further modulate tumorigenic signaling pathways downstream of these receptors by competitive binding or modifying receptor internalization. Another compelling question is whether SELENOP can bind to other members of the LRP family, as the β-propeller region bound by SELENOP is found on all LRP family members (Kurokawa, Bellinger, Hill, Burk, & Berry, 2014). This is especially intriguing in the case of LRP5/6, which functions as the primary receptor for the morphogen WNT3a which elicits numerous protumorigenic growth responses. Interestingly, Se deficiency induces WNT pathway activation, and reduced *Selenop* enhanced expression of WNT targets and stem cell phenotypes in intestinal tumors and intestinal organoid models, respectively (Barrett et al., 2015; Kipp et al., 2009). However, whether these effects are mediated by SELENOP interaction with LRP family receptors has yet to be addressed.

4.2 Glutathione Peroxidases

The glutathione peroxidase family includes eight members, GPx1–8, and constitutes some of the SePs most thoroughly studied for their contributions to tumorigenesis to date. These proteins catalyze hydroperoxide reduction by using GSH as a reductant, and their antioxidant capabilities are well established. However, not all GPx family members contain the selenocysteine residue required for classification as a SeP. In humans, the selenium-containing GPx family members consist of GPx1, GPx2, GPx3, GPx4, and GPx6, while GPx5, GPx7, and GPx8 contain a Cys in place of Sec (Brigelius-Flohe & Maiorino, 2013). While a SeP in humans, murine GPx6 does not contain a Sec, and its restriction to the adult olfactory system renders its contribution to tumorigenesis unlikely (Dear, Campbell, & Rabbitts, 1991; Kryukov et al., 2003). Therefore, for this review we will focus on GPx1-4, although the role of GPx family proteins in cancer risk and progression is covered in more depth in the accompanying chapter "Selenium-dependent glutathione peroxidases during tumor development" by Kipp.

Of the GPx family members, GPx1 is the most abundant and is widely expressed in the cytosol and mitochondria of various tissue types (Lubos, Loscalzo, & Handy, 2011). While all GPxs catalyze peroxide reduction, GPx1 serves as the family prototype as it contains the full complement of residues for optimal GSH binding (Flohe, Toppo, Cozza, & Ursini, 2011). As one of the lowest SePs in the "hierarchy of selenoproteins," GPx1 is highly responsive to Se levels and its expression decreases dramatically in the case of Se deficiency or Sec tRNA modification (Barnes et al., 2009; Moustafa et al., 2001). Thus, GPx1 may be one of the SePs most easily modified by Se supplementation. Indeed, GPx1 expression is decreased in many tumor types, and multiple studies have suggested a protective role for GPx1 (Lubos et al., 2011). For example, overexpression of GPx1 has been found to reduce growth of cancer cells in vitro and in vivo, inhibit prosurvival signaling through the AKT pathway, and attenuate expression of COX2-derived, proinflammatory mediators (Capdevila et al., 1995; Handy et al., 2009; Liu et al., 2004; Nasr, Fedele, Esser, & Diamond, 2004). On the other hand, studies using mouse models of skin cancer have found that GPx1-overexpressing mice had higher tumor numbers and tumor growth rate, and GPx1 expression has been linked to resistance to chemotherapy and radiation (Gan et al., 2014; Lu et al., 1997; Yang, Shen, Wei, & Liu, 2015). Interestingly, it is possible that this disparity can be attributed to a further complication of oxidative stress in tumorigenesis (Fig. 1B). Indeed, established tumors often have higher levels of basal oxidative stress, and in this case, SeP expression can enhance protection from ROS-based apoptosis. Thus, the cumulative effect of GPx1 expression in cancer is not yet clear.

In contrast to GPx1, glutathione peroxidase 2 (GPx2) shows a more localized expression pattern and is most highly expressed in the proliferative crypt compartment of the intestinal epithelium (Florian et al., 2010). Like

GPx1, it too appears to have an antiinflammatory role and can reduce expression of proinflammatory cytokines (Banning, Florian, et al., 2008). Furthermore, GPx1 and GPx2 appear to synergize, as mice that lack both GPx1 and GPx2 develop ileocolitis despite little phenotype being observed with loss of either GPx individually (Esworthy et al., 2001). Interestingly, these mice also develop spontaneous tumorigenesis, which was dependent on inflammation driven by interactions with pathogenic microflora, as germ-free mice remained tumor free (Chu et al., 2004). Placing GPx2-null mice on an inflammatory carcinogenesis protocol also determined that GPx2-null mice had more severe inflammation as well as higher tumor incidence in the colon (Krehl et al., 2012), overall indicating a protective role for GPx2 in colitis and colitis-associated tumorigenesis. However, the effect of GPx2 in noninflammatory-driven cancers appears to be the opposite. GPx2 is often found to be overexpressed in sporadic (i.e. nonhereditary and noninflammatory) colon cancers as well as cancers from the esophagus and liver, and it appears to promote carcinogenesis in this context (Florian et al., 2001; Lei, Tian, Zhang, Zhao, & Su, 2016; Suzuki et al., 2013). GPx2-null mice had decreased ACF formation after AOM treatment, and its knockdown has been shown to decrease growth of HT-29 colon cancer xenografts, patient-derived colon tumor spheroids, and castration-resistant prostate cancer cells (Banning, Kipp, et al., 2008; Emmink et al., 2014; Muller et al., 2013; Naiki et al., 2014).

While the roles of GPx1 and GPx2 in tumorigenesis appear to be tissue or context specific, perhaps the GPx which most clearly functions as a tumor suppressor is the extracellular family member, GPx3. This GPx is most highly expressed in the kidneys and secreted extracellularly into the plasma, although expression can also be detected in other tissue types (Avissar et al., 1994; Whitin, Bhamre, Tham, & Cohen, 2002). In tumor cells, GPx3 is a frequent target of hypermethylation, and its downregulation and/or methylation is associated with poorer prognosis and chemotherapeutic resistance in a wide range of cancer types, a topic discussed in depth in chapter "Selenium and epigenetics in cancer: Focus on DNA methylation" by Ewa and Edyta (Falck et al., 2010; Pelosof et al., 2017; Yu et al., 2007). Functionally, GPx3 overexpression has been reported to decrease anchorage-independent growth, xenograft size, and metastasis of prostate cancer cells (Yu et al., 2007). Interestingly, these authors also report downregulation of c-Met, a tyrosine kinase receptor which mediates signaling downstream of hepatic growth factor signaling. Other studies have also determined that GPx3's antitumorigenic function may be through direct interaction with p53-induced gene 3 (PIG3), a p53 target gene which regulates cell death following DNA damage (Wang et al., 2012). In vivo, GPx3 maintained these tumor-suppressive functions in the TRAMP model of prostate cancer and the AOM/DSS inflammatory carcinogenesis model of the colon, and in both cases, *Gpx3*-null mice displayed greater tumor number (Barrett et al., 2013; Chang et al., 2016). It is worth noting, however, that knockdown of *GPX3* in an established colon cancer line, CaCo-2, both induced apoptosis (in the context of H_2O_2) and decreased growth in soft agar, supporting a protumorigenic role for GPx3 in these cells (Barrett et al., 2013). Interestingly, CaCo-2 cells may be an outlier as they express a mutated form of p53, which may interfere with signaling through PIG3.

GPx4 is unique among SePs due to its localization to cellular membranes where it inhibits lipid peroxidation (Maiorino, Thomas, Girotti, & Ursini, 1991; Ursini, Maiorino, Valente, Ferri, & Gregolin, 1982). Because of this function, it is required for cellular homeostasis and its loss cannot be compensated by other members of the GPx family. Indeed, loss of GPx4 in vivo is not tolerated during development or in the adult mouse (Yant et al., 2003; Yoo et al., 2012). While conditional GPx4 knockout mice have been generated, the effects of tissue-specific GPx4 knockout on cancer development have not yet been reported (Sengupta et al., 2013). However, studies in cell lines have determined GPx4 may function in a tumor-suppressive role, and overexpression reduces tumor growth in fibrosarcoma and pancreatic cancer cells (Heirman et al., 2006; Liu et al., 2006). On the other hand, recent years have also seen exciting developments on the role of GPx4 in a novel non-apoptotic, iron-based cell death pathway known as ferroptosis. In this context, loss of GPx4 has been shown to render tumor cells more sensitive to ferroptosis-inducing compounds and thus may be a target of therapeutic intervention (Yang et al., 2014; Zhu et al., 2017). Collectively, these results suggest that loss of GPx4 expression functionally mediates tumor cell growth, although further research is needed to determine the full scope of GPx4's contributions to tumorigenesis.

4.3 Thioredoxin Reductases

In mammals, the thioredoxin reductase (TrxR) family consists of three family members: thioredoxin reductase 1 (TrxR1), thioredoxin reductase 2 (TrxR2), and thioredoxin reductase 3 (TrxR3). While structurally similar, they are broadly distinguished by their differential expression; TrxR1 is predominantly localizes to the cytosol, TrxR2 to the mitochondria, and TrxR3 to the testis, although TrxR1 may also be secreted and TrxR3 can be lowly expressed in other cell types (Soderberg, Sahaf, & Rosen, 2000; Su et al., 2005; Sun et al., 2005). Each TrxR family enzyme contains two redox domains, with the C-terminal position harboring a Sec required for activity (Gasdaska, Gasdaska, Cochran, & Powis, 1995; Gladyshev, Jeang, & Stadtman, 1996; Lee et al., 2000). Collectively, this group of selenoproteins is responsible for reduction of thioredoxins (Trx), which are ubiquitous small oxidoreductase enzymes capable of reducing disulfide bonds. The Trx enzymes are present in all known organisms and critical for redox-mediated regulation of a wide range of cellular substrates, including ribonuclease reductase, peroxiredoxins, glucocorticoid receptors, and transcription factors such as AP-1 (Powis, Mustacich, & Coon, 2000). As TrxRs are the only known reducer of thioredoxins, they are required for Trx activity and thus indispensable for redox homeostasis. In light of this housekeeping function, it is perhaps not surprising that the expression of TrxR1 and TrxR2 was not as greatly reduced in the Sec tRNA transgenic mouse model as SePs such as GPx1 or GPx3, although their activity is reduced by Se deficiency (Hill, McCollum, Boeglin, & Burk, 1997; Moustafa et al., 2001). TrxRs are also responsive Se supplementation, which increases TrxR protein levels and activity and augments antioxidant defense mechanisms (Berggren, Mangin, Gasdaka, & Powis, 1999; Miller et al., 2001). While TrxR is the only reducer of Trx, it is also important to note that TrxR reduces substrates other than Trx. Interestingly, this includes selenite, selenodiglutathione, and methylseleninate (Bjornstedt, Kumar, & Holmgren, 1992; Gromer & Gross, 2002; Kumar, Bjornstedt, & Holmgren, 1992). Therefore, in addition to being a selenoprotein, TrxR has a critical function in Se metabolism, supports SeP production, and may globally modify SeP function on tumorigenesis.

As TrxR expression is critical to maintaining redox balance, it was first hypothesized to protect against malignant transformation. Indeed, TrxR has been shown to modify many tumor-suppressive pathways, either alone or as a component of the Trx system. For example, TrxR1 promotes maturation of p53 and is able to reactivate oxidized, inactive forms of protein kinase C and PTEN, both of which typically function as tumor suppressors (Kahlos, Zhang, Block, & Patel, 2003; Lee et al., 2002; Moos, Edes, Cassidy, Massuda, & Fitzpatrick, 2003). Furthermore, TrxR family members can activate protein tyrosine phosphatases and thereby inhibit receptor tyrosine kinase-induced growth signals, and in this manner, TrxR1 has been shown to inhibit Src tyrosine kinase after activation by H_2O_2 (Shinozaki, Koizumi, Ohno, Nagao, & Inoue, 2006). TrxR may also inhibit microtubule assembly and thus abrogate tumor cell migration and invasion (Khan &

Luduena, 1991). Finally, the increase in vascular endothelial growth factor (VEGF) observed after Se depletion is also associated with TrxR inhibition (Streicher, Sylte, Johnson, & Sordillo, 2004). Therefore, it appears that TrxR can suppress multiple stages of tumor progression, from initiation to growth, invasion, and metastasis.

Surprisingly, in spite of these many tumor-suppressive functions, expression of TrxR family members is instead increased in a variety of cancers and often correlated with more severe disease (Lincoln, Ali Emadi, Tonissen, & Clarke, 2003). On one hand, this upregulation is likely due to TrxR induction by the higher levels of oxidative stress observed in tumor cells, and so does not rule out an antitumorigenic role. Yet studies in cell culture systems widely report that this increase in TrxR indeed contributes to tumor cell survival. For example, in multiple myeloma cells, pharmacological inhibition of TrxR with the gold compound, Auranofin, increased ROS-induced apoptosis and decreased activation of NF-κB and sensitization to NF-κB-based therapies (Raninga, Di Trapani, Vuckovic, Bhatia, & Tonissen, 2015; Raninga, Di Trapani, Vuckovic, & Tonissen, 2016). This is in part due to TrxR's function as a potent inhibitor of oxidative stress-induced apoptosis via regulation of apoptosis signal-regulating kinase 1 and other redox-sensitive members of the MAPK pathway (Matsuzawa & Ichijo, 2008; Rundlof & Arner, 2004). However, even in studies which do not report widespread apoptosis, TrxR1 loss is often associated with growth arrest, alterations in morphology, and/or anchorage-independent growth (Gan, Yang, Liu, & Xu, 2005; Yoo, Xu, Carlson, Gladyshev, & Hatfield, 2006). In one such study, Yoo et al. report that TrxR1 was still required for tumorigenesis, cells with loss of *TrxR1* were much less efficient at forming tumor and metastasis in vivo, and tumors that did form still expressed TrxR1 due to loss of the targeting construct (Yoo et al., 2006). In breast cancer cells, TrxR1 appears to mediate the epithelial-to-mesenchymal transition, as loss of TrxR1 pushes cells toward an epithelial phenotype, while overexpression enhanced invasion in Boyden chamber assays (Bhatia et al., 2016; Dong et al., 2016). Together, these results offer strong evidence that TrxR's primarily function is that of an oncogene in malignant cells.

Taken together, the effect of TrxR inhibition on tumor cell survival and aggressiveness is so robust that it is currently considered a candidate for therapeutic targeting. Indeed, a TrxR inhibitor, motexafin gadolinium (proposed trade name Xcytrin), was recently evaluated in phase 2 clinical trials for treatment of brain metastases in lung cancer patients. While preclinical data found success in multiple cancer types, it was ultimately not granted

FDA approval as it did not offer a significant survival benefit (Brachman et al., 2015; Richards & Mehta, 2007). Gold compounds, such as Auranofin or Aurothioglucose, are other highly potent and well-characterized inhibitors of TrxR (Kean & Kean, 2008). These compounds are currently approved for treatment of rheumatoid arthritis, although studies have also found them to induce apoptosis in a wide variety of cancer cell lines even with fairly low concentrations (Cox, Brown, Arner, & Hampton, 2008). Interestingly, both motexafin gadolinium and various gold compounds show little to no adverse effects, which is surprising considering the Trx system required for maintaining homeostasis. Thus, the story of pharmacological inhibition of TrxR in cancer is likely not over and is discussed more thoroughly in the accompanying chapter "Targeting the selenoprotein thioredoxin reductase 1 for anticancer therapy" by Arnér. It is also worth noting that a number of approved cancer pharmaceuticals may derive activity in part by TrxR inhibition. For example, both platinum derivatives (e.g., cisplatin) and chemotherapeutic alkylating agents have been determined to inhibit TrxR, which likely accounts for some degree of their cytotoxicity (Arner et al., 2001; Wang, Zhang, & Xu, 2008). Taken together, these results suggest that through TrxR activation, Se supplementation may actually promote tumor cell survival in primary and metastatic cancers.

4.4 Selenoprotein F

While the above SePs function primarily as cytoplasmic antioxidants, a number of SePs localize to the endoplasmic reticulum (ER) and instead can function as regulators of protein folding. Such Se-containing proteins include selenoprotein F (SELENOF, also known as SEP15), type 2 iodothyronine deiodinase (DIO2), selenoprotein M (SELENOM), selenoprotein N (SELENON), selenoprotein K (SELENOK), selenoprotein T (SELENOT), and selenoprotein S (SELENOS) (Davis, Tsuji, & Milner, 2012). While the role of the majority of these ER-localized SePs in tumorigenesis is still unclear, one, SELENOF, has been reported to mediate cancer cell growth and development.

SELENOF is a 15-kDa thioredoxin-like oxidoreductase which interacts with the ER folding sensor UDP-glucose:glycoprotein glucosyltransferase in order to mediate quality control of protein disulfide bonds. While originally described in human T cells, it is also expressed in numerous epithelial tissues such as the liver and prostate (Gladyshev, Jeang, Wootton, & Hatfield, 1998). Interestingly, SELENOF is located on a locus of chromosome 1p31 which is often deleted in human cancers, and SELENOF is downregulated in gastric cancers, mesotheliomas, hepatocellular carcinomas, prostate cancers, and colorectal cancers (Apostolou et al., 2004; Irons et al., 2010; Kumaraswamy et al., 2000; Lan et al., 2016). *SELENOF* also harbors two polymorphic sites at positions 811 (C/T) and 1125 (G/A), with the latter localized to the SECIS insertion element (Gladyshev et al., 1998). Thus, the A^{1125} appears to have lower Sec incorporation efficiency and is less sensitive to Se levels (Apostolou et al., 2004). This particular allele is associated with increased risk of breast and head and neck cancers in African Americans, although among smokers, those harboring a GG or GA1125 variant were at higher risk for lung cancer than those with AA1125 (Hu et al., 2001; Jablonska, Gromadzinska, Sobala, Reszka, & Wasowicz, 2008). SELENOF is a functional homolog of SELENOM (Reeves & Hoffmann, 2009), which is downregulated in cholangiocarcinoma and low expression is correlated with worse prognosis (Dai et al., 2016). However, SELENOM's functional contribution to tumorigenesis remains largely unknown.

Despite being one of the most recently identified SePs, the effect of SELENOF expression on tumor growth has been investigated in a number of cancer models. Despite frequent reports of its reduction in tumor cells and tumor cell lines, these studies uniformly report that SELENOF loss attenuates protumorigenic phenotypes. Downregulation of SELENOF in cancer cell lines resulted in upregulation of the cell cycle inhibitors p21 and p27 and led to decreased proliferation and growth arrest in colon and liver cancer cells (Bang, Huh, et al., 2015; Irons et al., 2010; Tsuji et al., 2011). These studies also determined that SELENOF loss decreased growth in soft agar and induced aberrant activation of Rho signaling (Bang, Jang, et al., 2015; Irons et al., 2010; Tsuji et al., 2011). Colon cancer cells with SELENOF knockdown were also less likely to form tumors and had a striking decrease in pulmonary metastases (Irons et al., 2010). Likewise, Tsuji et al. observed that SELENOF loss prevented the formation of ACF following AOM injection and induced IFN-γ- and STAT1-regulated gene expression (Tsuji et al., 2012). Collectively, these results are somewhat counterintuitive, as sustained ER stress and abnormal protein folding are often associated with aggressive features and drug resistance, and SELENOF KO indeed display phenotypes attributed to abnormal protein folding (Cubillos-Ruiz, Bettigole, & Glimcher, 2017; Kasaikina et al., 2011). However, these effects on protein folding may be overshadowed by robust activation of the STAT1 pathway, which induces apoptosis and antitumoral immunity. It also remains possible that effects of SELENOF are tissue specific, as its knockdown in mesothelioma cells abrogated apoptosis and decreased cell growth induced by treatment with Se, while loss of SELENOF in lung cancer cells had no discernible effects (Apostolou et al., 2004; Irons et al., 2010). Thus, additional studies are necessary to determine the full contribution of SELENOF to multiple cancer types.

5. CONSIDERATIONS AND CONCLUSIONS

A wealth of data exist correlating levels of Se and/or expression of selenium-containing proteins with cancer risk and tumor aggressiveness. While these data suggest that the effects of Se on tumor biology can partially be attributed to SePs, mechanistic studies to determine exactly how these proteins modify tumor growth are still comparatively lacking. In this review, we have discussed what is known about just a fraction of the SeP family, which were chosen due to their more established connections to tumor growth. However, a number of SePs still remain relatively uncharacterized in the context of tumorigenesis, such as members of the iodothyronine deiodinase family (DIO1, DIO2, and DIO3), SELENOK, SELENOS, and SELENOT. Future studies may identify contributions of these proteins to oncogenic pathways, as well. It is also worth noting that Se can function through Se-binding proteins which do not incorporate a selenocysteine. One such protein, selenium-binding protein 1, has been reported to function as a tumor suppressor, and reduced levels are correlated with poor clinical outcome in various cancer types (Yang & Diamond, 2013). SECIS-binding protein 2 is a major component of the selenocysteine insertion machinery; thus its expression can broadly influence SeP production (Fletcher, Copeland, Driscoll, & Krol, 2001). While these proteins are not classified as SePs, their expression or activity may broadly impact tumorigenesis in a Se-dependent manner.

Overall, tissue-specific and context-specific analyses appear crucial to our understanding of SeP function in cancer progression. Many SePs, such as GPx1 and TrxR1, are widely expressed across cell and tissue types, while SELENOP and GPx3 are secreted into the plasma. Thus, these proteins can affect numerous cell types, although research suggests not all are influenced equally. For example, GPx1 overexpression in a mouse model of skin cancer found increased tumor growth, yet reduced tumor growth was observed in the context of pancreatic cancer cells. Knockdown of SELENOF induced apoptosis in mesothelioma cells yet had no effect on growth of lung cancer cells. As pathology, aggressiveness, and oncogenic drivers can vary across tissues, and even across molecular subtypes of the same cancer, the analysis of tissue-specific contributions will greatly aid our understanding of SePs' role in carcinogenesis. Further testing will also need to address disparities in the context of alternative molecular drivers and mutagenic alterations. For example, the broad tumor-suppressive abilities of GPx3 were lost in the context of p53 mutation (Barrett et al., 2013). However, the effect of p53 mutation on the proapoptotic function of other SePs remains unclear.

Collectively, these data suggest that SePs are potent modifiers of tumorigenesis, although their action can vary greatly. Numerous epidemiological, in vitro, and in vivo studies are in agreement that SePs can inhibit tumor development by abrogating oxidative insults and mutagenesis, particularly in the context of inflammatory-driven cancers. However, in a tumor cell with higher basal levels of oxidative injury, this SeP activity can render tumor cells resistant to apoptosis and even chemotherapeutic intervention. Furthermore, while linked by the presence of a Sec, SePs are quite varied in their structure and function, and can individually influence numerous oncogenic pathways, such as MAPK, AKT, VEGF, and c-Met (Fig. 2). Thus, Se

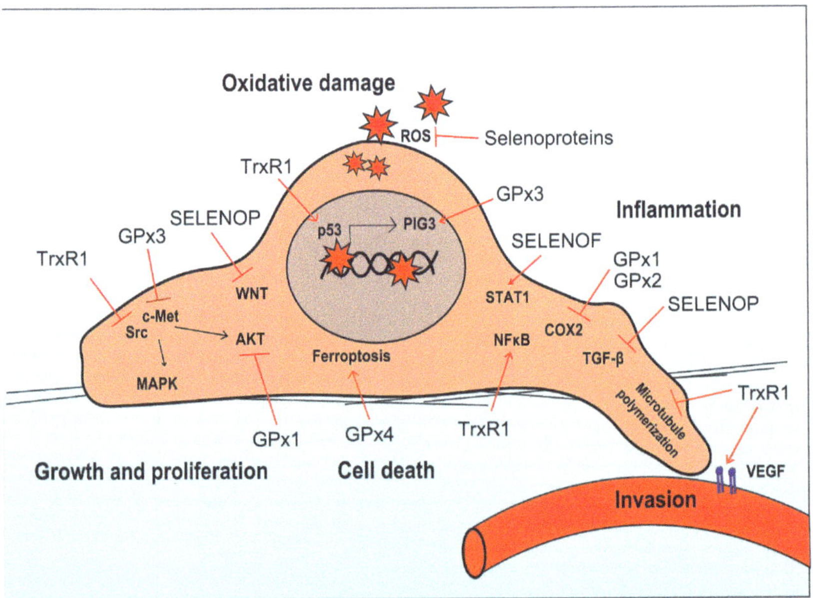

Fig. 2 Selenoproteins and their roles in tumor-associated signaling pathways. While numerous selenoproteins can inhibit reactive oxygen species (ROS), many have been implicated in additional tumor-associated signaling pathways in manners that are both pro- and antitumorigenic.

supplementation remains an area of considerable debate, as global modification of selenium-containing proteins may not always be beneficial, and indeed, it may be harmful in established tumors. Much work remains to be done to define the precise biologic mechanisms by which individual SePs influence oncogenic programs, taking into account molecular drivers, tissue types, and tumor stage. Once delineated, this knowledge could be overlaid with patient-specific characteristics, such as tissue Se status, SeP levels in target organs, and the presence of genetic polymorphisms in individual SePs (Fig. 3). Ultimately, this could determine the benefit of Se supplementation for individual disease prevention and treatment and, while complex, provide enormous benefit for human health.

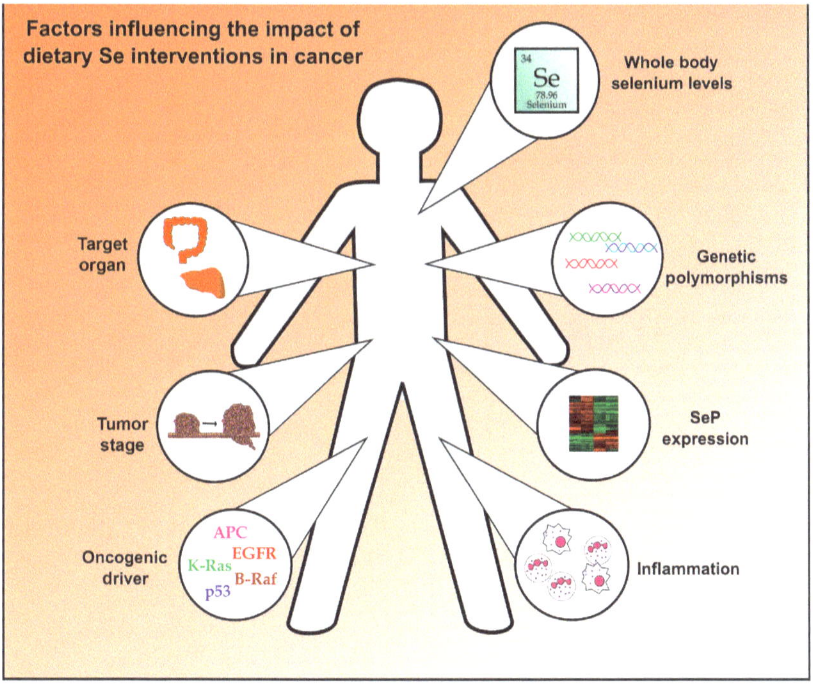

Fig. 3 Factors influencing the impact of dietary selenium in cancer. As selenium (Se) supplementation on tumor risk and growth has yielded mixed results in human trials, future studies will be necessary to elucidate the context-specific nature of selenium and selenoprotein function. Ultimately, this will allow the benefit of Se supplementation for individual disease prevention and treatment to be determined based on patient-specific parameters, such as overall Se levels, the presence of genetic polymorphisms, expression of individual selenoproteins, degree of inflammation, oncogenic driver, tumor stage, and tumor location.

ACKNOWLEDGMENTS

The authors gratefully acknowledge Dr. Raymond Burk for his expertise and insightful comments on this manuscript.

REFERENCES

Al-Taie, O. H., Seufert, J., Mork, H., Treis, H., Mentrup, B., Thalheimer, A., et al. (2002). A complex DNA-repeat structure within the selenoprotein P promoter contains a functionally relevant polymorphism and is genetically unstable under conditions of mismatch repair deficiency. *European Journal of Human Genetics*, *10*, 499–504.

Al-Taie, O. H., Uceyler, N., Eubner, U., Jakob, F., Mork, H., Scheurlen, M., et al. (2004). Expression profiling and genetic alterations of the selenoproteins GI-GPx and SePP in colorectal carcinogenesis. *Nutrition and Cancer*, *48*, 6–14.

Andersen, R. K., Hammer, K., Hager, H., Christensen, J. N., Ludvigsen, M., Honore, B., et al. (2015). Melanoma tumors frequently acquire LRP2/megalin expression, which modulates melanoma cell proliferation and survival rates. *Pigment Cell & Melanoma Research*, *28*, 267–280.

Apostolou, S., Klein, J. O., Mitsuuchi, Y., Shetler, J. N., Poulikakos, P. I., Jhanwar, S. C., et al. (2004). Growth inhibition and induction of apoptosis in mesothelioma cells by selenium and dependence on selenoprotein SEP15 genotype. *Oncogene*, *23*, 5032–5040.

Arner, E. S., Nakamura, H., Sasada, T., Yodoi, J., Holmgren, A., & Spyrou, G. (2001). Analysis of the inhibition of mammalian thioredoxin, thioredoxin reductase, and glutaredoxin by cis-diamminedichloroplatinum (II) and its major metabolite, the glutathione-platinum complex. *Free Radical Biology & Medicine*, *31*, 1170–1178.

Ashrafi, M. R., Shams, S., Nouri, M., Mohseni, M., Shabanian, R., Yekaninejad, M. S., et al. (2007). A probable causative factor for an old problem: Selenium and glutathione peroxidase appear to play important roles in epilepsy pathogenesis. *Epilepsia*, *48*, 1750–1755.

Avissar, N., Ornt, D. B., Yagil, Y., Horowitz, S., Watkins, R. H., Kerl, E. A., et al. (1994). Human kidney proximal tubules are the main source of plasma glutathione peroxidase. *The American Journal of Physiology*, *266*, C367–75.

Baines, A. T., Holubec, H., Basye, J. L., Thorne, P., Bhattacharyya, A. K., Spallholz, J., et al. (2000). The effects of dietary selenomethionine on polyamines and azoxymethane-induced aberrant crypts. *Cancer Letters*, *160*, 193–198.

Bang, J., Huh, J. H., Na, J. W., Lu, Q., Carlson, B. A., Tobe, R., et al. (2015). Cell proliferation and motility are inhibited by G1 phase arrest in 15-kDa selenoprotein-deficient chang liver cells. *Molecules and Cells*, *38*, 457–465.

Bang, J., Jang, M., Huh, J. H., Na, J. W., Shim, M., Carlson, B. A., et al. (2015). Deficiency of the 15-kDa selenoprotein led to cytoskeleton remodeling and non-apoptotic membrane blebbing through a RhoA/ROCK pathway. *Biochemical and Biophysical Research Communications*, *456*, 884–890.

Banning, A., Florian, S., Deubel, S., Thalmann, S., Muller-Schmehl, K., Jacobasch, G., et al. (2008). GPx2 counteracts PGE2 production by dampening COX-2 and mPGES-1 expression in human colon cancer cells. *Antioxidants & Redox Signaling*, *10*, 1491–1500.

Banning, A., Kipp, A., Schmitmeier, S., Lowinger, M., Florian, S., Krehl, S., et al. (2008). Glutathione peroxidase 2 inhibits cyclooxygenase-2-mediated migration and invasion of HT-29 adenocarcinoma cells but supports their growth as tumors in nude mice. *Cancer Research*, *68*, 9746–9753.

Barnes, K. M., Evenson, J. K., Raines, A. M., & Sunde, R. A. (2009). Transcript analysis of the selenoproteome indicates that dietary selenium requirements of rats based on selenium-regulated selenoprotein mRNA levels are uniformly less than those based on glutathione peroxidase activity. *The Journal of Nutrition*, *139*, 199–206.

Barrett, C. W., Ning, W., Chen, X., Smith, J. J., Washington, M. K., Hill, K. E., et al. (2013). Tumor suppressor function of the plasma glutathione peroxidase gpx3 in colitis-associated carcinoma. *Cancer Research, 73*, 1245–1255.

Barrett, C. W., Reddy, V. K., Short, S. P., Motley, A. K., Lintel, M. K., Bradley, A. M., et al. (2015). Selenoprotein P influences colitis-induced tumorigenesis by mediating stemness and oxidative damage. *The Journal of Clinical Investigation, 125*, 2646–2660.

Baum, M. K., Shor-Posner, G., Lai, S., Zhang, G., Lai, H., Fletcher, M. A., et al. (1997). High risk of HIV-related mortality is associated with selenium deficiency. *Journal of Acquired Immune Deficiency Syndromes and Human Retrovirology, 15*, 370–374.

Berggren, M. M., Mangin, J. F., Gasdaka, J. R., & Powis, G. (1999). Effect of selenium on rat thioredoxin reductase activity: Increase by supranutritional selenium and decrease by selenium deficiency. *Biochemical Pharmacology, 57*, 187–193.

Bhatia, M., Lovitt, C. J., Raninga, P. V., Avery, V. M., Di Trapani, G., & Tonissen, K. F. (2016). Expression of the thioredoxin system in an in vivo-like cancer cell environment upon auranofin treatment. *European Journal of Cell Biology, 95*, 378–388.

Bjornstedt, M., Kumar, S., & Holmgren, A. (1992). Selenodiglutathione is a highly efficient oxidant of reduced thioredoxin and a substrate for mammalian thioredoxin reductase. *The Journal of Biological Chemistry, 267*, 8030–8034.

Bosl, M. R., Takaku, K., Oshima, M., Nishimura, S., & Taketo, M. M. (1997). Early embryonic lethality caused by targeted disruption of the mouse selenocysteine tRNA gene (Trsp). *Proceedings of the National Academy of Sciences of the United States of America, 94*, 5531–5534.

Bosschaerts, T., Guilliams, M., Noel, W., Herin, M., Burk, R. F., Hill, K. E., et al. (2008). Alternatively activated myeloid cells limit pathogenicity associated with African trypanosomiasis through the IL-10 inducible gene selenoprotein P. *Journal of Immunology, 180*, 6168–6175.

Boyd, R. (2011). Selenium stories. *Nature Chemistry, 3*, 570.

Brachman, D. G., Pugh, S. L., Ashby, L. S., Thomas, T. A., Dunbar, E. M., Narayan, S., et al. (2015). Phase 1/2 trials of temozolomide, motexafin gadolinium, and 60-Gy fractionated radiation for newly diagnosed supratentorial glioblastoma multiforme: Final results of RTOG 0513. *International Journal of Radiation Oncology, Biology, Physics, 91*, 961–967.

Brigelius-Flohe, R., & Maiorino, M. (2013). Glutathione peroxidases. *Biochimica et Biophysica Acta, 1830*, 3289–3303.

Burk, R. F., & Hill, K. E. (2005). Selenoprotein P: An extracellular protein with unique physical characteristics and a role in selenium homeostasis. *Annual Review of Nutrition, 25*, 215–235.

Burk, R. F., & Hill, K. E. (2015). Regulation of selenium metabolism and transport. *Annual Review of Nutrition, 35*, 109–134.

Burk, R. F., Hill, K. E., Olson, G. E., Weeber, E. J., Motley, A. K., Winfrey, V. P., et al. (2007). Deletion of apolipoprotein E receptor-2 in mice lowers brain selenium and causes severe neurological dysfunction and death when a low-selenium diet is fed. *The Journal of Neuroscience, 27*, 6207–6211.

Calvo, A., Xiao, N., Kang, J., Best, C. J., Leiva, I., Emmert-Buck, M. R., et al. (2002). Alterations in gene expression profiles during prostate cancer progression: Functional correlations to tumorigenicity and down-regulation of selenoprotein-P in mouse and human tumors. *Cancer Research, 62*, 5325–5335.

Capdevila, J. H., Morrow, J. D., Belosludtsev, Y. Y., Beauchamp, D. R., DuBois, R. N., & Falck, J. R. (1995). The catalytic outcomes of the constitutive and the mitogen inducible isoforms of prostaglandin H2 synthase are markedly affected by glutathione and glutathione peroxidase(s). *Biochemistry, 34*, 3325–3337.

Carlson, B. A., Novoselov, S. V., Kumaraswamy, E., Lee, B. J., Anver, M. R., Gladyshev, V. N., et al. (2004). Specific excision of the selenocysteine tRNA[Ser]Sec (Trsp) gene in mouse liver demonstrates an essential role of selenoproteins in liver function. *The Journal of Biological Chemistry, 279,* 8011–8017.

Carlson, B. A., Yoo, M. H., Sano, Y., Sengupta, A., Kim, J. Y., Irons, R., et al. (2009). Selenoproteins regulate macrophage invasiveness and extracellular matrix-related gene expression. *BMC Immunology, 10,* 57.

Cat, B., Stuhlmann, D., Steinbrenner, H., Alili, L., Holtkotter, O., Sies, H., et al. (2006). Enhancement of tumor invasion depends on transdifferentiation of skin fibroblasts mediated by reactive oxygen species. *Journal of Cell Science, 119,* 2727–2738.

Chang, S. N., Lee, J. M., Oh, H., & Park, J. H. (2016). Glutathione peroxidase 3 inhibits prostate tumorigenesis in TRAMP mice. *Prostate, 76,* 1387–1398.

Chen, J. (2012). An original discovery: Selenium deficiency and Keshan disease (an endemic heart disease). *Asia Pacific Journal of Clinical Nutrition, 21,* 320–326.

Christensen, E. I., & Birn, H. (2002). Megalin and cubilin: Multifunctional endocytic receptors. *Nature Reviews. Molecular Cell Biology, 3,* 256–266.

Christensen, E. I., & Willnow, T. E. (1999). Essential role of megalin in renal proximal tubule for vitamin homeostasis. *Journal of the American Society of Nephrology: JASN, 10,* 2224–2236.

Chu, F. F., Esworthy, R. S., Chu, P. G., Longmate, J. A., Huycke, M. M., Wilczynski, S., et al. (2004). Bacteria-induced intestinal cancer in mice with disrupted Gpx1 and Gpx2 genes. *Cancer Research, 64,* 962–968.

Clark, L. C., Dalkin, B., Krongrad, A., Combs, G. F., Jr., Turnbull, B. W., Slate, E. H., et al. (1998). Decreased incidence of prostate cancer with selenium supplementation: Results of a double-blind cancer prevention trial. *British Journal of Urology, 81,* 730–734.

Cox, A. G., Brown, K. K., Arner, E. S., & Hampton, M. B. (2008). The thioredoxin reductase inhibitor auranofin triggers apoptosis through a Bax/Bak-dependent process that involves peroxiredoxin 3 oxidation. *Biochemical Pharmacology, 76,* 1097–1109.

Cubillos-Ruiz, J. R., Bettigole, S. E., & Glimcher, L. H. (2017). Tumorigenic and immunosuppressive effects of endoplasmic reticulum stress in cancer. *Cell, 168,* 692–706.

Dai, X., Thongchot, S., Dokduang, H., Loilome, W., Khuntikeo, N., Titapun, A., et al. (2016). Potential of selenium compounds as new anticancer agents for cholangiocarcinoma. *Anticancer Research, 36,* 5981–5988.

Daoud, A. H., & Griffin, A. C. (1980). Effect of retinoic acid, butylated hydroxytoluene, selenium and sorbic acid on azo-dye hepatocarcinogenesis. *Cancer Letters, 9,* 299–304.

Davis, C. D., Tsuji, P. A., & Milner, J. A. (2012). Selenoproteins and cancer prevention. *Annual Review of Nutrition, 32,* 73–95.

Dear, T. N., Campbell, K., & Rabbitts, T. H. (1991). Molecular cloning of putative odorant-binding and odorant-metabolizing proteins. *Biochemistry, 30,* 10376–10382.

Deeb, K. K., Trump, D. L., & Johnson, C. S. (2007). Vitamin D signalling pathways in cancer: Potential for anticancer therapeutics. *Nature Reviews. Cancer, 7,* 684–700.

Diamond, A. M., Choi, I. S., Crain, P. F., Hashizume, T., Pomerantz, S. C., Cruz, R., et al. (1993). Dietary selenium affects methylation of the wobble nucleoside in the anticodon of selenocysteine tRNA([Ser]Sec). *The Journal of Biological Chemistry, 268,* 14215–14223.

Diwadkar-Navsariwala, V., Prins, G. S., Swanson, S. M., Birch, L. A., Ray, V. H., Hedayat, S., et al. (2006). Selenoprotein deficiency accelerates prostate carcinogenesis in a transgenic model. *Proceedings of the National Academy of Sciences of the United States of America, 103,* 8179–8184.

Dong, C., Zhang, L., Sun, R., Liu, J., Yin, H., Li, X., et al. (2016). Role of thioredoxin reductase 1 in dysplastic transformation of human breast epithelial cells triggered by chronic oxidative stress. *Scientific Reports, 6,* 36860.

Driscoll, D. M., & Copeland, P. R. (2003). Mechanism and regulation of selenoprotein synthesis. *Annual Review of Nutrition, 23*, 17–40.

Duffield-Lillico, A. J., Dalkin, B. L., Reid, M. E., Turnbull, B. W., Slate, E. H., Jacobs, E. T., et al. (2003). Selenium supplementation, baseline plasma selenium status and incidence of prostate cancer: An analysis of the complete treatment period of the Nutritional Prevention of Cancer Trial. *BJU International, 91*, 608–612.

Emmink, B. L., Laoukili, J., Kipp, A. P., Koster, J., Govaert, K. M., Fatrai, S., et al. (2014). GPx2 suppression of H2O2 stress links the formation of differentiated tumor mass to metastatic capacity in colorectal cancer. *Cancer Research, 74*, 6717–6730.

Esworthy, R. S., Aranda, R., Martin, M. G., Doroshow, J. H., Binder, S. W., & Chu, F. F. (2001). Mice with combined disruption of Gpx1 and Gpx2 genes have colitis. *American Journal of Physiology. Gastrointestinal and Liver Physiology, 281*, G848–55.

Falck, E., Karlsson, S., Carlsson, J., Helenius, G., Karlsson, M., & Klinga-Levan, K. (2010). Loss of glutathione peroxidase 3 expression is correlated with epigenetic mechanisms in endometrial adenocarcinoma. *Cancer Cell International, 10*, 46.

Fletcher, J. E., Copeland, P. R., Driscoll, D. M., & Krol, A. (2001). The selenocysteine incorporation machinery: Interactions between the SECIS RNA and the SECIS-binding protein SBP2. *RNA*, 7, 1442–1453.

Flohe, L., Toppo, S., Cozza, G., & Ursini, F. (2011). A comparison of thiol peroxidase mechanisms. *Antioxidants & Redox Signaling, 15*, 763–780.

Florian, S., Krehl, S., Loewinger, M., Kipp, A., Banning, A., Esworthy, S., et al. (2010). Loss of GPx2 increases apoptosis, mitosis, and GPx1 expression in the intestine of mice. *Free Radical Biology & Medicine, 49*, 1694–1702.

Florian, S., Wingler, K., Schmehl, K., Jacobasch, G., Kreuzer, O. J., Meyerhof, W., et al. (2001). Cellular and subcellular localization of gastrointestinal glutathione peroxidase in normal and malignant human intestinal tissue. *Free Radical Research, 35*, 655–663.

Gan, X., Chen, B., Shen, Z., Liu, Y., Li, H., Xie, X., et al. (2014). High GPX1 expression promotes esophageal squamous cell carcinoma invasion, migration, proliferation and cisplatin-resistance but can be reduced by vitamin D. *International Journal of Clinical and Experimental Medicine*, 7, 2530–2540.

Gan, L., Yang, X. L., Liu, Q., & Xu, H. B. (2005). Inhibitory effects of thioredoxin reductase antisense RNA on the growth of human hepatocellular carcinoma cells. *Journal of Cellular Biochemistry, 96*, 653–664.

Gasdaska, P. Y., Gasdaska, J. R., Cochran, S., & Powis, G. (1995). Cloning and sequencing of a human thioredoxin reductase. *FEBS Letters, 373*, 5–9.

Gladyshev, V. N., Arner, E. S., Berry, M. J., Brigelius-Flohe, R., Bruford, E. A., Burk, R. F., et al. (2016). Selenoprotein gene nomenclature. *The Journal of Biological Chemistry, 291*, 24036–24040.

Gladyshev, V. N., Jeang, K. T., & Stadtman, T. C. (1996). Selenocysteine, identified as the penultimate C-terminal residue in human T-cell thioredoxin reductase, corresponds to TGA in the human placental gene. *Proceedings of the National Academy of Sciences of the United States of America, 93*, 6146–6151.

Gladyshev, V. N., Jeang, K. T., Wootton, J. C., & Hatfield, D. L. (1998). A new human selenium-containing protein. Purification, characterization, and cDNA sequence. *The Journal of Biological Chemistry, 273*, 8910–8915.

Gonzalez-Moreno, O., Boque, N., Redrado, M., Milagro, F., Campion, J., Endermann, T., et al. (2011). Selenoprotein-P is down-regulated in prostate cancer, which results in lack of protection against oxidative damage. *Prostate, 71*, 824–834.

Gromer, S., & Gross, J. H. (2002). Methylseleninate is a substrate rather than an inhibitor of mammalian thioredoxin reductase. Implications for the antitumor effects of selenium. *The Journal of Biological Chemistry, 277*, 9701–9706.

Handy, D. E., Lubos, E., Yang, Y., Galbraith, J. D., Kelly, N., Zhang, Y. Y., et al. (2009). Glutathione peroxidase-1 regulates mitochondrial function to modulate redox-dependent cellular responses. *The Journal of Biological Chemistry*, *284*, 11913–11921.

Hassona, Y., Cirillo, N., Lim, K. P., Herman, A., Mellone, M., Thomas, G. J., et al. (2013). Progression of genotype-specific oral cancer leads to senescence of cancer-associated fibroblasts and is mediated by oxidative stress and TGF-beta. *Carcinogenesis*, *34*, 1286–1295.

Hatfield, D. L., Carlson, B. A., Xu, X. M., Mix, H., & Gladyshev, V. N. (2006). Selenocysteine incorporation machinery and the role of selenoproteins in development and health. *Progress in Nucleic Acid Research and Molecular Biology*, *81*, 97–142.

Hatfield, D., Lee, B. J., Hampton, L., & Diamond, A. M. (1991). Selenium induces changes in the selenocysteine tRNA[Ser]Sec population in mammalian cells. *Nucleic Acids Research*, *19*, 939–943.

Heirman, I., Ginneberge, D., Brigelius-Flohe, R., Hendrickx, N., Agostinis, P., Brouckaert, P., et al. (2006). Blocking tumor cell eicosanoid synthesis by GP x 4 impedes tumor growth and malignancy. *Free Radical Biology & Medicine*, *40*, 285–294.

Hill, K. E., Lyons, P. R., & Burk, R. F. (1992). Differential regulation of rat liver selenoprotein mRNAs in selenium deficiency. *Biochemical and Biophysical Research Communications*, *185*, 260–263.

Hill, K. E., McCollum, G. W., Boeglin, M. E., & Burk, R. F. (1997). Thioredoxin reductase activity is decreased by selenium deficiency. *Biochemical and Biophysical Research Communications*, *234*, 293–295.

Hill, K. E., Wu, S., Motley, A. K., Stevenson, T. D., Winfrey, V. P., Capecchi, M. R., et al. (2012). Production of selenoprotein P (Sepp1) by hepatocytes is central to selenium homeostasis. *The Journal of Biological Chemistry*, *287*, 40414–40424.

Hill, K. E., Zhou, J., Austin, L. M., Motley, A. K., Ham, A. J., Olson, G. E., et al. (2007). The selenium-rich C-terminal domain of mouse selenoprotein P is necessary for the supply of selenium to brain and testis but not for the maintenance of whole body selenium. *The Journal of Biological Chemistry*, *282*, 10972–10980.

Hill, K. E., Zhou, J., McMahan, W. J., Motley, A. K., Atkins, J. F., Gesteland, R. F., et al. (2003). Deletion of selenoprotein P alters distribution of selenium in the mouse. *The Journal of Biological Chemistry*, *278*, 13640–13646.

Hiramoto, K., Satoh, H., Suzuki, T., Moriguchi, T., Pi, J., Shimosegawa, T., et al. (2014). Myeloid lineage-specific deletion of antioxidant system enhances tumor metastasis. *Cancer Prevention Research (Philadelphia, Pa.)*, *7*, 835–844.

Howard, M. T., Carlson, B. A., Anderson, C. B., & Hatfield, D. L. (2013). Translational redefinition of UGA codons is regulated by selenium availability. *The Journal of Biological Chemistry*, *288*, 19401–19413.

Hu, Y. J., Korotkov, K. V., Mehta, R., Hatfield, D. L., Rotimi, C. N., Luke, A., et al. (2001). Distribution and functional consequences of nucleotide polymorphisms in the 3'-untranslated region of the human Sep15 gene. *Cancer Research*, *61*, 2307–2310.

Hudson, T. S., Carlson, B. A., Hoeneroff, M. J., Young, H. A., Sordillo, L., Muller, W. J., et al. (2012). Selenoproteins reduce susceptibility to DMBA-induced mammary carcinogenesis. *Carcinogenesis*, *33*, 1225–1230.

Hughes, D. J., Duarte-Salles, T., Hybsier, S., Trichopoulou, A., Stepien, M., Aleksandrova, K., et al. (2016). Prediagnostic selenium status and hepatobiliary cancer risk in the European Prospective Investigation into Cancer and Nutrition cohort. *The American Journal of Clinical Nutrition*, *104*, 406–414.

Ip, C., Thompson, H. J., & Ganther, H. E. (2000). Selenium modulation of cell proliferation and cell cycle biomarkers in normal and premalignant cells of the rat mammary gland. *Cancer Epidemiology, Biomarkers & Prevention*, *9*, 49–54.

Irons, R., Carlson, B. A., Hatfield, D. L., & Davis, C. D. (2006). Both selenoproteins and low molecular weight selenocompounds reduce colon cancer risk in mice with genetically impaired selenoprotein expression. *The Journal of Nutrition, 136*, 1311–1317.

Irons, R., Tsuji, P. A., Carlson, B. A., Ouyang, P., Yoo, M. H., Xu, X. M., et al. (2010). Deficiency in the 15-kDa selenoprotein inhibits tumorigenicity and metastasis of colon cancer cells. *Cancer Prevention Research (Philadelphia, Pa.), 3*, 630–639.

Jablonska, E., Gromadzinska, J., Sobala, W., Reszka, E., & Wasowicz, W. (2008). Lung cancer risk associated with selenium status is modified in smoking individuals by Sep15 polymorphism. *European Journal of Nutrition, 47*, 47–54.

Jacobs, E. T., Jiang, R., Alberts, D. S., Greenberg, E. R., Gunter, E. W., Karagas, M. R., et al. (2004). Selenium and colorectal adenoma: Results of a pooled analysis. *Journal of the National Cancer Institute, 96*, 1669–1675.

Kahlos, K., Zhang, J., Block, E. R., & Patel, J. M. (2003). Thioredoxin restores nitric oxide-induced inhibition of protein kinase C activity in lung endothelial cells. *Molecular and Cellular Biochemistry, 254*, 47–54.

Kasaikina, M. V., Fomenko, D. E., Labunskyy, V. M., Lachke, S. A., Qiu, W., Moncaster, J. A., et al. (2011). Roles of the 15-kDa selenoprotein (Sep15) in redox homeostasis and cataract development revealed by the analysis of Sep 15 knockout mice. *The Journal of Biological Chemistry, 286*, 33203–33212.

Kaushal, N., Kudva, A. K., Patterson, A. D., Chiaro, C., Kennett, M. J., Desai, D., et al. (2014). Crucial role of macrophage selenoproteins in experimental colitis. *Journal of Immunology, 193*, 3683–3692.

Kean, W. F., & Kean, I. R. (2008). Clinical pharmacology of gold. *Inflammopharmacology, 16*, 112–125.

Khan, I. A., & Luduena, R. F. (1991). Possible regulation of the in vitro assembly of bovine brain tubulin by the bovine thioredoxin system. *Biochimica et Biophysica Acta, 1076*, 289–297.

Kipp, A., Banning, A., van Schothorst, E. M., Meplan, C., Schomburg, L., Evelo, C., et al. (2009). Four selenoproteins, protein biosynthesis, and Wnt signalling are particularly sensitive to limited selenium intake in mouse colon. *Molecular Nutrition & Food Research, 53*, 1561–1572.

Klein, E. A., Thompson, I. M., Jr., Tangen, C. M., Crowley, J. J., Lucia, M. S., Goodman, P. J., et al. (2011). Vitamin E and the risk of prostate cancer: The Selenium and Vitamin E Cancer Prevention Trial (SELECT). *JAMA, 306*, 1549–1556.

Krehl, S., Loewinger, M., Florian, S., Kipp, A. P., Banning, A., Wessjohann, L. A., et al. (2012). Glutathione peroxidase-2 and selenium decreased inflammation and tumors in a mouse model of inflammation-associated carcinogenesis whereas sulforaphane effects differed with selenium supply. *Carcinogenesis, 33*, 620–628.

Kryukov, G. V., Castellano, S., Novoselov, S. V., Lobanov, A. V., Zehtab, O., Guigo, R., et al. (2003). Characterization of mammalian selenoproteomes. *Science, 300*, 1439–1443.

Kumar, S., Bjornstedt, M., & Holmgren, A. (1992). Selenite is a substrate for calf thymus thioredoxin reductase and thioredoxin and elicits a large non-stoichiometric oxidation of NADPH in the presence of oxygen. *European Journal of Biochemistry, 207*, 435–439.

Kumaraswamy, E., Carlson, B. A., Morgan, F., Miyoshi, K., Robinson, G. W., Su, D., et al. (2003). Selective removal of the selenocysteine tRNA [Ser]Sec gene (Trsp) in mouse mammary epithelium. *Molecular and Cellular Biology, 23*, 1477–1488.

Kumaraswamy, E., Malykh, A., Korotkov, K. V., Kozyavkin, S., Hu, Y., Kwon, S. Y., et al. (2000). Structure-expression relationships of the 15-kDa selenoprotein gene. Possible role of the protein in cancer etiology. *The Journal of Biological Chemistry, 275*, 35540–35547.

Kurokawa, S., Bellinger, F. P., Hill, K. E., Burk, R. F., & Berry, M. J. (2014). Isoform-specific binding of selenoprotein P to the beta-propeller domain of apolipoprotein E receptor 2 mediates selenium supply. *The Journal of Biological Chemistry, 289*, 9195–9207.

Kurokawa, S., Eriksson, S., Rose, K. L., Wu, S., Motley, A. K., Hill, S., et al. (2014). Sepp1 (UF) forms are N-terminal selenoprotein P truncations that have peroxidase activity when coupled with thioredoxin reductase-1. *Free Radical Biology & Medicine, 69*, 67–76.

Kurokawa, S., Hill, K. E., McDonald, W. H., & Burk, R. F. (2012). Long isoform mouse selenoprotein P (Sepp1) supplies rat myoblast L8 cells with selenium via endocytosis mediated by heparin binding properties and apolipoprotein E receptor-2 (ApoER2). *The Journal of Biological Chemistry, 287*, 28717–28726.

Lan, X., Xing, J., Gao, H., Li, S., Quan, L., Jiang, Y., et al. (2016). Decreased Expression of Selenoproteins as a Poor Prognosticator of Gastric Cancer in Humans. *Biological Trace Element Research*.

Lance, P., Alberts, D. S., Thompson, P. A., Fales, L., Wang, F., San Jose, J., et al. (2017). Colorectal adenomas in participants of the SELECT randomized trial of selenium and vitamin E for prostate cancer prevention. *Cancer Prevention Research (Philadelphia, Pa.), 10*, 45–54.

Lee, S. R., Bar-Noy, S., Kwon, J., Levine, R. L., Stadtman, T. C., & Rhee, S. G. (2000). Mammalian thioredoxin reductase: Oxidation of the C-terminal cysteine/selenocysteine active site forms a thioselenide, and replacement of selenium with sulfur markedly reduces catalytic activity. *Proceedings of the National Academy of Sciences of the United States of America, 97*, 2521–2526.

Lee, B. J., Rajagopalan, M., Kim, Y. S., You, K. H., Jacobson, K. B., & Hatfield, D. (1990). Selenocysteine tRNA[Ser]Sec gene is ubiquitous within the animal kingdom. *Molecular and Cellular Biology, 10*, 1940–1949.

Lee, S. R., Yang, K. S., Kwon, J., Lee, C., Jeong, W., & Rhee, S. G. (2002). Reversible inactivation of the tumor suppressor PTEN by H2O2. *The Journal of Biological Chemistry, 277*, 20336–20342.

Leemhuis, J., & Bock, H. H. (2011). Reelin modulates cytoskeletal organization by regulating Rho GTPases. *Communicative & Integrative Biology, 4*, 254–257.

Lei, Z., Tian, D., Zhang, C., Zhao, S., & Su, M. (2016). Clinicopathological and prognostic significance of GPX2 protein expression in esophageal squamous cell carcinoma. *BMC Cancer, 16*, 410.

Lincoln, D. T., Ali Emadi, E. M., Tonissen, K. F., & Clarke, F. M. (2003). The thioredoxin-thioredoxin reductase system: Over-expression in human cancer. *Anticancer Research, 23*, 2425–2433.

Lippman, S. M., Klein, E. A., Goodman, P. J., Lucia, M. S., Thompson, I. M., Ford, L. G., et al. (2009). Effect of selenium and vitamin E on risk of prostate cancer and other cancers: The Selenium and Vitamin E Cancer Prevention Trial (SELECT). *JAMA, 301*, 39–51.

Liu, J., Du, J., Zhang, Y., Sun, W., Smith, B. J., Oberley, L. W., et al. (2006). Suppression of the malignant phenotype in pancreatic cancer by overexpression of phospholipid hydroperoxide glutathione peroxidase. *Human Gene Therapy, 17*, 105–116.

Liu, J., Hinkhouse, M. M., Sun, W., Weydert, C. J., Ritchie, J. M., Oberley, L. W., et al. (2004). Redox regulation of pancreatic cancer cell growth: Role of glutathione peroxidase in the suppression of the malignant phenotype. *Human Gene Therapy, 15*, 239–250.

Low, S. C., Grundner-Culemann, E., Harney, J. W., & Berry, M. J. (2000). SECIS-SBP2 interactions dictate selenocysteine incorporation efficiency and selenoprotein hierarchy. *The EMBO Journal, 19*, 6882–6890.

Lu, Y. P., Lou, Y. R., Yen, P., Newmark, H. L., Mirochnitchenko, O. I., Inouye, M., et al. (1997). Enhanced skin carcinogenesis in transgenic mice with high expression of glutathione peroxidase or both glutathione peroxidase and superoxide dismutase. *Cancer Research, 57*, 1468–1474.

Lubos, E., Loscalzo, J., & Handy, D. E. (2011). Glutathione peroxidase-1 in health and disease: From molecular mechanisms to therapeutic opportunities. *Antioxidants & Redox Signaling, 15*, 1957–1997.

Luchman, H. A., Villemaire, M. L., Bismar, T. A., Carlson, B. A., & Jirik, F. R. (2014). Prostate epithelium-specific deletion of the selenocysteine tRNA gene Trsp leads to early onset intraepithelial neoplasia. *The American Journal of Pathology, 184*, 871–877.

Maiorino, M., Thomas, J. P., Girotti, A. W., & Ursini, F. (1991). Reactivity of phospholipid hydroperoxide glutathione peroxidase with membrane and lipoprotein lipid hydroperoxides. *Free Radical Research Communications, 12-13*(Pt. 1), 131–135.

Matsuzawa, A., & Ichijo, H. (2008). Redox control of cell fate by MAP kinase: Physiological roles of ASK1-MAP kinase pathway in stress signaling. *Biochimica et Biophysica Acta, 1780*, 1325–1336.

Meplan, C., Dragsted, L. O., Ravn-Haren, G., Tjonneland, A., Vogel, U., & Hesketh, J. (2013). Association between polymorphisms in glutathione peroxidase and selenoprotein P genes, glutathione peroxidase activity, HRT use and breast cancer risk. *PLoS One, 8*, e73316.

Meplan, C., Hughes, D. J., Pardini, B., Naccarati, A., Soucek, P., Vodickova, L., et al. (2010). Genetic variants in selenoprotein genes increase risk of colorectal cancer. *Carcinogenesis, 31*, 1074–1079.

Meyer, H. A., Endermann, T., Stephan, C., Stoedter, M., Behrends, T., Wolff, I., et al. (2012). Selenoprotein P status correlates to cancer-specific mortality in renal cancer patients. *PLoS One, 7*, e46644.

Miller, S., Walker, S. W., Arthur, J. R., Nicol, F., Pickard, K., Lewin, M. H., et al. (2001). Selenite protects human endothelial cells from oxidative damage and induces thioredoxin reductase. *Clinical Science (London, England), 100*, 543–550.

Moos, P. J., Edes, K., Cassidy, P., Massuda, E., & Fitzpatrick, F. A. (2003). Electrophilic prostaglandins and lipid aldehydes repress redox-sensitive transcription factors p53 and hypoxia-inducible factor by impairing the selenoprotein thioredoxin reductase. *The Journal of Biological Chemistry, 278*, 745–750.

Mostert, V., Dreher, I., Kohrle, J., & Abel, J. (1999). Transforming growth factor-beta1 inhibits expression of selenoprotein P in cultured human liver cells. *FEBS Letters, 460*, 23–26.

Moustafa, M. E., Carlson, B. A., Anver, M. R., Bobe, G., Zhong, N., Ward, J. M., et al. (2013). Selenium and selenoprotein deficiencies induce widespread pyogranuloma formation in mice, while high levels of dietary selenium decrease liver tumor size driven by TGFalpha. *PLoS One, 8*, e57389.

Moustafa, M. E., Carlson, B. A., El-Saadani, M. A., Kryukov, G. V., Sun, Q. A., Harney, J. W., et al. (2001). Selective inhibition of selenocysteine tRNA maturation and selenoprotein synthesis in transgenic mice expressing isopentenyladenosine-deficient selenocysteine tRNA. *Molecular and Cellular Biology, 21*, 3840–3852.

Muller, M. F., Florian, S., Pommer, S., Osterhoff, M., Esworthy, R. S., Chu, F. F., et al. (2013). Deletion of glutathione peroxidase-2 inhibits azoxymethane-induced colon cancer development. *PLoS One, 8*, e72055.

Murawaki, Y., Tsuchiya, H., Kanbe, T., Harada, K., Yashima, K., Nozaka, K., et al. (2008). Aberrant expression of selenoproteins in the progression of colorectal cancer. *Cancer Letters, 259*, 218–230.

Naiki, T., Naiki-Ito, A., Asamoto, M., Kawai, N., Tozawa, K., Etani, T., et al. (2014). GPX2 overexpression is involved in cell proliferation and prognosis of castration-resistant prostate cancer. *Carcinogenesis, 35*, 1962–1967.

Nasr, M. A., Fedele, M. J., Esser, K., & Diamond, A. M. (2004). GPx-1 modulates Akt and P70S6K phosphorylation and Gadd45 levels in MCF-7 cells. *Free Radical Biology & Medicine, 37*, 187–195.

Olson, G. E., Winfrey, V. P., Nagdas, S. K., Hill, K. E., & Burk, R. F. (2007). Apolipoprotein E receptor-2 (ApoER2) mediates selenium uptake from selenoprotein P by the mouse testis. *The Journal of Biological Chemistry, 282*, 12290–12297.

Outzen, M., Tjonneland, A., Larsen, E. H., Friis, S., Larsen, S. B., Christensen, J., et al. (2016). Selenium status and risk of prostate cancer in a Danish population. *The British Journal of Nutrition*, *115*, 1669–1677.

Panieri, E., & Santoro, M. M. (2016). ROS homeostasis and metabolism: A dangerous liason in cancer cells. *Cell Death & Disease*, 7, e2253.

Papaioannou, D., Cooper, K. L., Carroll, C., Hind, D., Squires, H., Tappenden, P., et al. (2011). Antioxidants in the chemoprevention of colorectal cancer and colorectal adenomas in the general population: A systematic review and meta-analysis. *Colorectal Disease*, *13*, 1085–1099.

Pedersen, M. O., Hansen, P. B., Nielsen, S. L., & Penkowa, M. (2010). Metallothionein-I + II and receptor megalin are altered in relation to oxidative stress in cerebral lymphomas. *Leukemia & Lymphoma*, *51*, 314–328.

Pellatt, A. J., Wolff, R. K., John, E. M., Torres-Mejia, G., Hines, L. M., Baumgartner, K. B., et al. (2013). SEPP1 influences breast cancer risk among women with greater native American ancestry: The breast cancer health disparities study. *PLoS One*, *8*, e80554.

Pelosof, L., Yerram, S., Armstrong, T., Chu, N., Danilova, L., Yanagisawa, B., et al. (2017). GPX3 promoter methylation predicts platinum sensitivity in colorectal cancer. *Epigenetics*, *12*, 540–550.

Perez, S., Talens-Visconti, R., Rius-Perez, S., Finamor, I., & Sastre, J. (2017). Redox signaling in the gastrointestinal tract. *Free Radical Biology & Medicine*, *104*, 75–103.

Peters, U., Chatterjee, N., Hayes, R. B., Schoen, R. E., Wang, Y., Chanock, S. J., et al. (2008). Variation in the selenoenzyme genes and risk of advanced distal colorectal adenoma. *Cancer Epidemiology, Biomarkers & Prevention*, *17*, 1144–1154.

Powis, G., Mustacich, D., & Coon, A. (2000). The role of the redox protein thioredoxin in cell growth and cancer. *Free Radical Biology & Medicine*, *29*, 312–322.

Raninga, P. V., Di Trapani, G., Vuckovic, S., Bhatia, M., & Tonissen, K. F. (2015). Inhibition of thioredoxin 1 leads to apoptosis in drug-resistant multiple myeloma. *Oncotarget*, *6*, 15410–15424.

Raninga, P. V., Di Trapani, G., Vuckovic, S., & Tonissen, K. F. (2016). TrxR1 inhibition overcomes both hypoxia-induced and acquired bortezomib resistance in multiple myeloma through NF-small ka, Cyrillicbeta inhibition. *Cell Cycle*, *15*, 559–572.

Reeves, M. A., & Hoffmann, P. R. (2009). The human selenoproteome: Recent insights into functions and regulation. *Cellular and Molecular Life Sciences*, *66*, 2457–2478.

Richards, G. M., & Mehta, M. P. (2007). Motexafin gadolinium in the treatment of brain metastases. *Expert Opinion on Pharmacotherapy*, *8*, 351–359.

Riker, A. I., Enkemann, S. A., Fodstad, O., Liu, S., Ren, S., Morris, C., et al. (2008). The gene expression profiles of primary and metastatic melanoma yields a transition point of tumor progression and metastasis. *BMC Medical Genomics*, *1*, 13.

Rundlof, A. K., & Arner, E. S. (2004). Regulation of the mammalian selenoprotein thioredoxin reductase 1 in relation to cellular phenotype, growth, and signaling events. *Antioxidants & Redox Signaling*, *6*, 41–52.

Saito, Y., Hayashi, T., Tanaka, A., Watanabe, Y., Suzuki, M., Saito, E., et al. (1999). Selenoprotein P in human plasma as an extracellular phospholipid hydroperoxide glutathione peroxidase. Isolation and enzymatic characterization of human selenoprotein p. *The Journal of Biological Chemistry*, *274*, 2866–2871.

Saito, Y., Sato, N., Hirashima, M., Takebe, G., Nagasawa, S., & Takahashi, K. (2004). Domain structure of bi-functional selenoprotein P. *The Biochemical Journal*, *381*, 841–846.

Sarsour, E. H., Kumar, M. G., Chaudhuri, L., Kalen, A. L., & Goswami, P. C. (2009). Redox control of the cell cycle in health and disease. *Antioxidants & Redox Signaling*, *11*, 2985–3011.

Sato, N., Fukushima, N., Chang, R., Matsubayashi, H., & Goggins, M. (2006). Differential and epigenetic gene expression profiling identifies frequent disruption of the RELN pathway in pancreatic cancers. *Gastroenterology, 130*, 548–565.

Schoenmakers, E., Carlson, B., Agostini, M., Moran, C., Rajanayagam, O., Bochukova, E., et al. (2016). Mutation in human selenocysteine transfer RNA selectively disrupts selenoprotein synthesis. *The Journal of Clinical Investigation, 126*, 992–996.

Schriever, S. C., Barnes, K. M., Evenson, J. K., Raines, A. M., & Sunde, R. A. (2009). Selenium requirements are higher for glutathione peroxidase-1 mRNA than gpx1 activity in rat testis. *Experimental Biology and Medicine (Maywood, N.J.), 234*, 513–521.

Sengupta, A., Lichti, U. F., Carlson, B. A., Cataisson, C., Ryscavage, A. O., Mikulec, C., et al. (2013). Targeted disruption of glutathione peroxidase 4 in mouse skin epithelial cells impairs postnatal hair follicle morphogenesis that is partially rescued through inhibition of COX-2. *The Journal of Investigative Dermatology, 133*, 1731–1741.

Sengupta, A., Lichti, U. F., Carlson, B. A., Ryscavage, A. O., Gladyshev, V. N., Yuspa, S. H., et al. (2010). Selenoproteins are essential for proper keratinocyte function and skin development. *PLoS One, 5*, e12249.

Seyedali, A., & Berry, M. J. (2014). Nonsense-mediated decay factors are involved in the regulation of selenoprotein mRNA levels during selenium deficiency. *RNA, 20*, 1248–1256.

Shinozaki, Y., Koizumi, S., Ohno, Y., Nagao, T., & Inoue, K. (2006). Extracellular ATP counteracts the ERK1/2-mediated death-promoting signaling cascades in astrocytes. *Glia, 54*, 606–618.

Shrimali, R. K., Weaver, J. A., Miller, G. F., Starost, M. F., Carlson, B. A., Novoselov, S. V., et al. (2007). Selenoprotein expression is essential in endothelial cell development and cardiac muscle function. *Neuromuscular Disorders, 17*, 135–142.

Smith, J. J., Deane, N. G., Wu, F., Merchant, N. B., Zhang, B., Jiang, A., et al. (2010). Experimentally derived metastasis gene expression profile predicts recurrence and death in patients with colon cancer. *Gastroenterology, 138*, 958–968.

Soderberg, A., Sahaf, B., & Rosen, A. (2000). Thioredoxin reductase, a redox-active selenoprotein, is secreted by normal and neoplastic cells: Presence in human plasma. *Cancer Research, 60*, 2281–2289.

Solinas, G., Schiarea, S., Liguori, M., Fabbri, M., Pesce, S., Zammataro, L., et al. (2010). Tumor-conditioned macrophages secrete migration-stimulating factor: A new marker for M2-polarization, influencing tumor cell motility. *Journal of Immunology, 185*, 642–652.

Speckmann, B., Pinto, A., Winter, M., Forster, I., Sies, H., & Steinbrenner, H. (2010). Proinflammatory cytokines down-regulate intestinal selenoprotein P biosynthesis via NOS2 induction. *Free Radical Biology & Medicine, 49*, 777–785.

Stadtman, T. C. (1996). Selenocysteine. *Annual Review of Biochemistry, 65*, 83–100.

Stadtman, T. C. (2000). Selenium biochemistry. Mammalian selenoenzymes. *Annals of the New York Academy of Sciences, 899*, 399–402.

Steinbrecher, A., Meplan, C., Hesketh, J., Schomburg, L., Endermann, T., Jansen, E., et al. (2010). Effects of selenium status and polymorphisms in selenoprotein genes on prostate cancer risk in a prospective study of European men. *Cancer Epidemiology, Biomarkers & Prevention, 19*, 2958–2968.

Streicher, K. L., Sylte, M. J., Johnson, S. E., & Sordillo, L. M. (2004). Thioredoxin reductase regulates angiogenesis by increasing endothelial cell-derived vascular endothelial growth factor. *Nutrition and Cancer, 50*, 221–231.

Su, D., Novoselov, S. V., Sun, Q. A., Moustafa, M. E., Zhou, Y., Oko, R., et al. (2005). Mammalian selenoprotein thioredoxin-glutathione reductase. Roles in disulfide bond formation and sperm maturation. *The Journal of Biological Chemistry, 280*, 26491–26498.

Sun, Q. A., Su, D., Novoselov, S. V., Carlson, B. A., Hatfield, D. L., & Gladyshev, V. N. (2005). Reaction mechanism and regulation of mammalian thioredoxin/glutathione reductase. *Biochemistry*, *44*, 14528–14537.

Suzuki, T., Kelly, V. P., Motohashi, H., Nakajima, O., Takahashi, S., Nishimura, S., et al. (2008). Deletion of the selenocysteine tRNA gene in macrophages and liver results in compensatory gene induction of cytoprotective enzymes by Nrf2. *The Journal of Biological Chemistry*, *283*, 2021–2030.

Suzuki, S., Pitchakarn, P., Ogawa, K., Naiki-Ito, A., Chewonarin, T., Punfa, W., et al. (2013). Expression of glutathione peroxidase 2 is associated with not only early hepatocarcinogenesis but also late stage metastasis. *Toxicology*, *311*, 115–123.

Tsuji, P. A., Carlson, B. A., Naranjo-Suarez, S., Yoo, M. H., Xu, X. M., Fomenko, D. E., et al. (2012). Knockout of the 15 kDa selenoprotein protects against chemically-induced aberrant crypt formation in mice. *PLoS One*, *7*, e50574.

Tsuji, P. A., Naranjo-Suarez, S., Carlson, B. A., Tobe, R., Yoo, M. H., & Davis, C. D. (2011). Deficiency in the 15 kDa selenoprotein inhibits human colon cancer cell growth. *Nutrients*, *3*, 805–817.

Ursini, F., Maiorino, M., Valente, M., Ferri, L., & Gregolin, C. (1982). Purification from pig liver of a protein which protects liposomes and biomembranes from peroxidative degradation and exhibits glutathione peroxidase activity on phosphatidylcholine hydroperoxides. *Biochimica et Biophysica Acta*, *710*, 197–211.

van Rensburg, S. J., Hall, J. M., & Gathercole, P. S. (1986). Inhibition of esophageal carcinogenesis in corn-fed rats by riboflavin, nicotinic acid, selenium, molybdenum, zinc, and magnesium. *Nutrition and Cancer*, *8*, 163–170.

Walczak, R., Westhof, E., Carbon, P., & Krol, A. (1996). A novel RNA structural motif in the selenocysteine insertion element of eukaryotic selenoprotein mRNAs. *RNA*, *2*, 367–379.

Wallace, K., Byers, T., Morris, J. S., Cole, B. F., Greenberg, E. R., Baron, J. A., et al. (2003). Prediagnostic serum selenium concentration and the risk of recurrent colorectal adenoma: A nested case-control study. *Cancer Epidemiology, Biomarkers & Prevention*, *12*, 464–467.

Wang, L., Bonorden, M. J., Li, G. X., Lee, H. J., Hu, H., Zhang, Y., et al. (2009). Methylselenium compounds inhibit prostate carcinogenesis in the transgenic adenocarcinoma of mouse prostate model with survival benefit. *Cancer Prevention Research (Philadelphia, Pa.)*, *2*, 484–495.

Wang, Q., Gong, L., Dong, R., Qiao, Q., He, X. L., Chu, Y. K., et al. (2009). Tissue microarray assessment of selenoprotein P expression in gastric adenocarcinoma. *The Journal of International Medical Research*, *37*, 169–174.

Wang, H., Luo, K., Tan, L. Z., Ren, B. G., Gu, L. Q., Michalopoulos, G., et al. (2012). p53-induced gene 3 mediates cell death induced by glutathione peroxidase 3. *The Journal of Biological Chemistry*, *287*, 16890–16902.

Wang, X., Zhang, J., & Xu, T. (2008). Thioredoxin reductase inactivation as a pivotal mechanism of ifosfamide in cancer therapy. *European Journal of Pharmacology*, *579*, 66–73.

Whitin, J. C., Bhamre, S., Tham, D. M., & Cohen, H. J. (2002). Extracellular glutathione peroxidase is secreted basolaterally by human renal proximal tubule cells. *American Journal of Physiology. Renal Physiology*, *283*, F20–8.

Woutersen, R. A., Appel, M. J., & Van Garderen-Hoetmer, A. (1999). Modulation of pancreatic carcinogenesis by antioxidants. *Food and Chemical Toxicology*, *37*, 981–984.

Xu, X. M., Turanov, A. A., Carlson, B. A., Yoo, M. H., Everley, R. A., Nandakumar, R., et al. (2010). Targeted insertion of cysteine by decoding UGA codons with mammalian selenocysteine machinery. *Proceedings of the National Academy of Sciences of the United States of America*, *107*, 21430–21434.

Yang, W., & Diamond, A. M. (2013). Selenium-binding protein 1 as a tumor suppressor and a prognostic indicator of clinical outcome. *Biomarker Research, 1*, 15.
Yang, W., Shen, Y., Wei, J., & Liu, F. (2015). MicroRNA-153/Nrf-2/GPx1 pathway regulates radiosensitivity and stemness of glioma stem cells via reactive oxygen species. *Oncotarget, 6*, 22006–22027.
Yang, W. S., SriRamaratnam, R., Welsch, M. E., Shimada, K., Skouta, R., Viswanathan, V. S., et al. (2014). Regulation of ferroptotic cancer cell death by GPX4. *Cell, 156*, 317–331.
Yant, L. J., Ran, Q., Rao, L., Van Remmen, H., Shibatani, T., Belter, J. G., et al. (2003). The selenoprotein GPX4 is essential for mouse development and protects from radiation and oxidative damage insults. *Free Radical Biology & Medicine, 34*, 496–502.
Yao, Y., Pei, F., & Kang, P. (2011). Selenium, iodine, and the relation with Kashin-Beck disease. *Nutrition, 27*, 1095–1100.
Yi, Y. S., Park, S. G., Byeon, S. M., Kwon, Y. G., & Jung, G. (2003). Hepatitis B virus X protein induces TNF-alpha expression via down-regulation of selenoprotein P in human hepatoma cell line, HepG2. *Biochimica et Biophysica Acta, 1638*, 249–256.
Yoo, S. E., Chen, L., Na, R., Liu, Y., Rios, C., Van Remmen, H., et al. (2012). Gpx4 ablation in adult mice results in a lethal phenotype accompanied by neuronal loss in brain. *Free Radical Biology & Medicine, 52*, 1820–1827.
Yoo, M. H., Xu, X. M., Carlson, B. A., Gladyshev, V. N., & Hatfield, D. L. (2006). Thioredoxin reductase 1 deficiency reverses tumor phenotype and tumorigenicity of lung carcinoma cells. *The Journal of Biological Chemistry, 281*, 13005–13008.
Yu, Y. P., Yu, G., Tseng, G., Cieply, K., Nelson, J., Defrances, M., et al. (2007). Glutathione peroxidase 3, deleted or methylated in prostate cancer, suppresses prostate cancer growth and metastasis. *Cancer Research, 67*, 8043–8050.
Zhang, S., Rocourt, C., & Cheng, W. H. (2010). Selenoproteins and the aging brain. *Mechanisms of Ageing and Development, 131*, 253–260.
Zhu, S., Zhang, Q., Sun, X., Zeh, H. J., Lotze, M. T., Kang, R., et al. (2017). HSPA5 regulates ferroptotic cell death in cancer cells. *Cancer Research, 77*, 2064–2077.

CHAPTER THREE

Selenoproteins and Metastasis

Michael P. Marciel, Peter R. Hoffmann[1]
John A. Burns School of Medicine, University of Hawaii, Honolulu, HI, United States
[1]Corresponding author: e-mail address: peterrh@hawaii.edu

Contents

Abstract

Cancer survival is largely impacted by the dissemination of cancer cells from the original tumor site to secondary tissues or organs through metastasis. Targets for antimetastatic therapies have recently become a focus of research, but progress will require a better understanding of the molecular mechanisms driving metastasis. Selenoproteins play important roles in many of the cellular activities underlying metastasis including cell adhesion, matrix degradation and migration, invasion into the blood and extravasation into secondary tissues, and subsequent proliferation into metastatic tumors along with the angiogenesis required for growth. In this review the roles identified for different selenoproteins in these steps and how they may promote or inhibit metastatic cancers is discussed. These roles include selenoenzyme modulation of redox tone and detoxification of reactive oxygen species, calcium homeostasis and unfolded protein responses regulated by endoplasmic reticulum selenoproteins, and the multiple physiological responses influenced by other selenoproteins.

Advances in Cancer Research, Volume 136
ISSN 0065-230X
http://dx.doi.org/10.1016/bs.acr.2017.07.008

1. INTRODUCTION

Cancer treatment approaches have conventionally focused on reducing cancer cell proliferation or "shrinking tumors." However, cancer patients most often die not from primary tumors but from metastasis, which involves the spread of cancer from primary sites to other tissues and accounts for greater than 90% of overall patient mortality (Gupta & Massague, 2006). Shrinking tumors is, of course, clinically important and represents one way to limit metastasis and thereby increase survival, but developing drugs that target other steps of the metastatic process can provide additional weapons to reduce cancer deaths. While such efforts have produced limited results thus far in generating effective treatments (Weber, 2013), researchers should not be deterred from delving deeply into the molecular underpinnings of metastasis so that targets can be identified and therapies pursued. The selenoprotein family is comprised of a diverse group of member proteins that may represent valuable targets for antimetastatic drugs. Some selenoproteins are well-characterized enzymes, while others may play nonenzymatic roles (Reeves & Hoffmann, 2009). The cellular activities regulated by selenoproteins involving redox reactions, calcium homeostasis, and stress mitigation suggest that many are functionally integrated into the steps of metastasis and also suggest that these proteins may be targeted for effective new anticancer therapies. This review will describe what is known to date regarding the role of selenoproteins in metastasis.

2. METASTASIS

Metastasis is a term to describe a set of biological steps promoting the dissemination of transformed cells from their original tissue to other sites in the body and the subsequent outgrowth of secondary tumors. As shown in Fig. 1, the main steps can be partitioned into the following: (1) local invasion of primary tumor cells through adjacent extracellular matrix (ECM), (2) intravasation into blood vessels or entry into lymphatic vessels, (3) survival during transit, (4) arrest at basement membrane (BM) of distant organ sites, (5) extravasation into parenchyma or entry into draining lymph nodes, (6) survival in an unfamiliar microenvironment, and (7) reestablishment of programs associated with proliferation including angiogenesis.

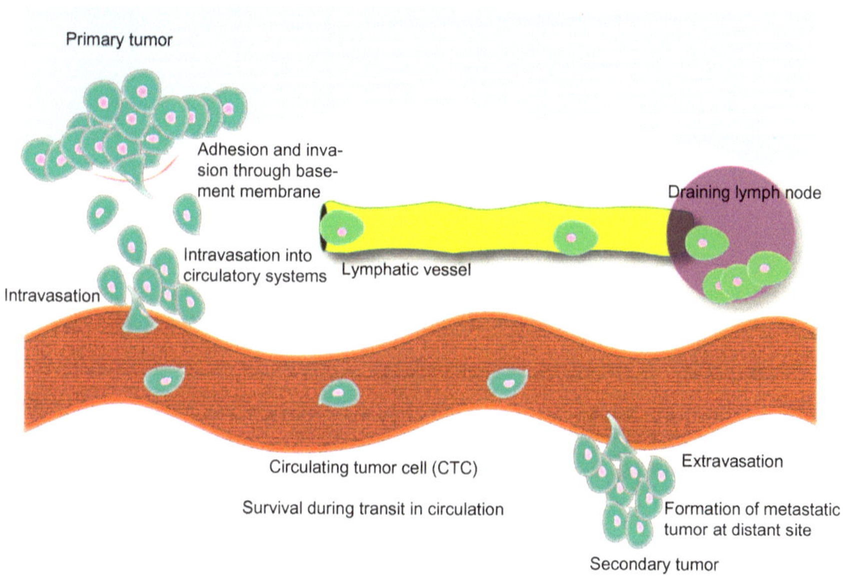

Fig. 1 The metastatic process. Metastases are formed via a multistep process occurring after an initial tumor has been established. During metastasis tumor cells escape their primary site by passage through extracellular matrix and intravasation into blood vessels or entry into lymphatic vessels. They then translocate via systemic circulatory systems protected by platelets (tumor cell embolus, adhesion to basement membrane, and extravasation) or they travel to draining lymph nodes. Once in distant tissue they proceed through metastatic deposit, angiogenesis, and growth.

2.1 Local Invasion

Local invasion involves the entry of cancer cells originating from the defined area of the primary tumor into adjacent normal tissue. An essential component characterizing local invasion is the ability of carcinoma cells to breach the BM, which is a specialized ECM component that organizes epithelial tissue. The organized structure of the BM acts as an inherent barrier to invasiveness that metastatic cancer cells must surmount. In fact, loss of expression of ECM proteins, such as type IV collagen, enhances cancer cell invasiveness in colorectal cancer (Ikeda et al., 2006). In general, active proteolysis performed by matrix metalloproteinases (MMPs) is a major contributor to the loss of the BM barrier, leading to invasion into adjacent tissue. MMPs are also implicated in other steps in the metastatic process, including intravasation and angiogenesis, and may be more relevant in the context of overall metastasis in those later steps (Stamenkovic, 2000). In addition to organizing epithelial tissue, the ECM contains a repository of growth factors that can be liberated via MMP actions, which can then be utilized by the carcinoma cells during invasion (Mott & Werb, 2004).

2.2 Intravasation

Intravasation is characterized by local invasion of the carcinoma cells into the lumina of lymph or blood vessels. At the location of intravasation, tumor cells make contact with endothelial cells via complex interactions involving proteins, lipids, and carbohydrates. For example, interactions between a complex of glycosphingolipids and α2β1 integrin receptors were shown to be responsible for metastatic behavior of C4-2B prostate cancer cells (Van Slambrouck et al., 2014). The mechanism of intravasation is also likely to be dependent on the formation of new blood vessels (neoangiogenesis) via the action of vascular endothelial growth factors (VEGF). Compared to normal vasculature, the new blood vessels originating from carcinoma cells are "leaky" and dynamic (Tong et al., 2004). Many mediators driving these early-stage vasculature dynamics originate from the cancer cells themselves. For example, cyclooxygenase-2 (COX-2) is expressed in 40% of human breast cancer and mediates production of prostaglandins, thereby promoting breast cancer metastasis to bone (Singh et al., 2007; Fig. 2).

2.3 Survival in Vasculature, Extravasation, and Formation of Micrometastases

After carcinoma cells have undergone intravasation into blood vessels, they have the ability to disseminate throughout the body and may now be considered circulating tumor cells (CTC). Alternatively, they may enter lymphatic vessels and travel to draining lymph nodes. In order for CTCs to survive in hematogenous circulation they must overcome a number of stresses including absence of anchorage, hemodynamic shear forces, and immune surveillance. In the absence of adhesion to the ECM, epithelial cells normally undergo a type of apoptosis referred to as anoikis (Guo & Giancotti, 2004). Mutations involving genes that regulate apoptosis and/or anoikis that promote survival would presumably enhance metastasis. For example, transfection of the antiapoptotic protein Bcl-xL into human breast cancer cell line MDA-MB-435 increased their metastatic ability in mice (Fernandez, Espana, Manas, Fabra, & Sierra, 2000).

Given that the average size of carcinoma cells is 20–30 μm and the luminal diameter of capillaries is approximately 8 μm, the cells must often squeeze through these spaces and CTCs eventually become trapped in various

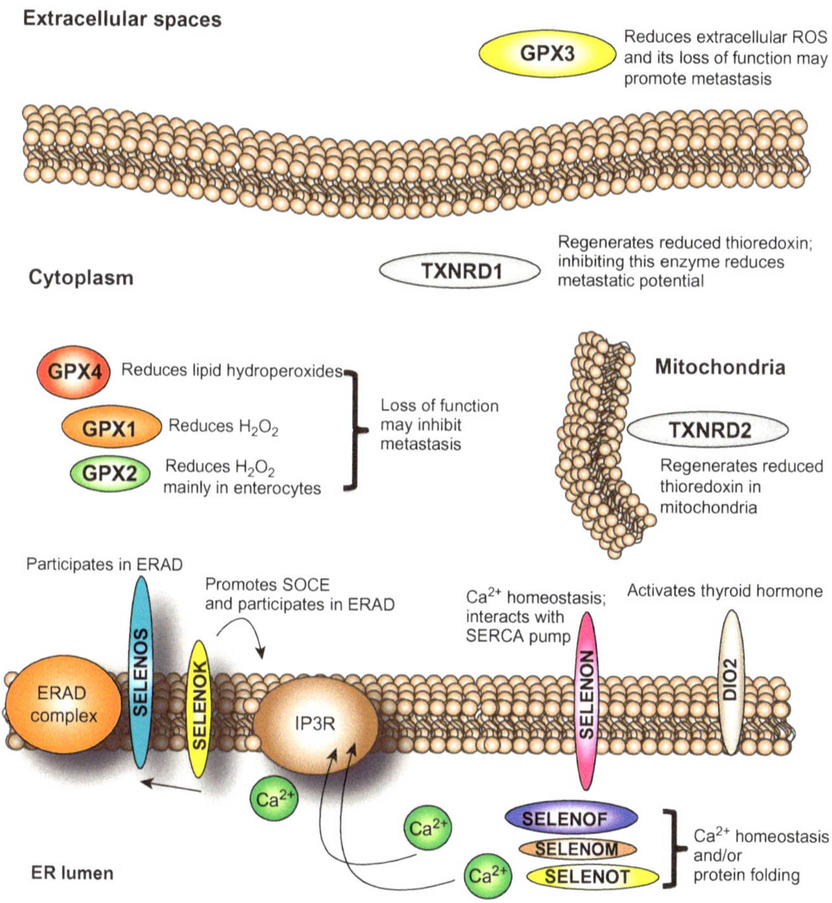

Fig. 2 Potential roles for selenoproteins in metastasis. Selenoproteins located in different sites within the cell or in extracellular spaces may promote or inhibit malignant cells as they progress through the steps or metastasis.

capillary beds. At this point CTCs can penetrate endothelial cells and migrate into new stromal microenvironment distal from the original tumor site (extravasation). Unlike the tumor-associated blood vessels available during intravasation, the vasculature at the site of extravasation is not "leaky," and therefore, additional mechanisms must be activated for the CTCs to induce permeability and extravasate. One example is neuropilin-2 (NRP-2), which is a receptor for VEGF and other growth factors that interacts with α5 integrin on endothelial cells to mediate extravasation in zebrafish and murine xenograft models of clear cell renal cell carcinoma and pancreatic adenocarcinoma (Cao et al., 2013). Furthermore, angiopoietin-like 4

(Angptl4), epidermal growth factor family member Epiregulin (EREG), COX-2, MMP-1, and MMP-2 disrupt endothelial cell–cell junctions in breast cancer cells in the lungs (Gupta et al., 2007). Similar to the primary tumors, the promotion of angiogenesis in secondary tumors influences the final stages of metastasis and affects patient outcome.

3. DIETARY SELENIUM AND SELENIUM SUPPLEMENTATION

Before discussing selenoproteins, one must first consider what is known about selenium metabolism and metastasis. This is because the biological effects of dietary selenium are largely exerted through the incorporation of this element into selenoproteins as selenocysteine, but when it comes to cancer biology, the intermediates of selenium metabolism can play significant roles (Jackson & Combs, 2012; Rayman, 2005). The notion that selenium metabolites affect carcinogenesis and tumor progression is supported by the observation that effective doses for anticancer effects are substantially greater than those required for maximal selenoprotein expression. In fact, selenium metabolites may in some cases assert effects on signaling pathways in cancer cells that are opposed to the actions of certain redox regulating selenoproteins (Gopalakrishna & Gundimeda, 2001). Thus, it is important to consider the effects of various selenium metabolites separately from the actions of selenoproteins when analyzing how dietary selenium influences cancer biology and metastasis. Extensive reviews have been published describing how different forms of selenium may impact carcinogenesis, tumor proliferation, angiogenesis, and metastasis (Bera, De Rosa, Rachidi, & Diamond, 2013; Brigelius-Flohe, 2008; Chen, Prabhu, & Mastro, 2013; Jackson & Combs, 2012), but there are some key issues worthy of discussion herein. First, it is important to note that data pertaining to selenium and metastasis in patients is scarce, and most of what is known has been obtained through mouse models and cell culture studies. Most of these data suggest that the form of selenium ingested and the type of metabolites they generate influence metastasis differently. One example is the finding that dietary supplementation with methylseleninic acid reduced spontaneous metastasis of Lewis lung carcinoma (LLC) in mice, while supplementation with selenomethionine was ineffective (Yan & DeMars, 2012). Selenomethionine is more readily metabolized to hydrogen selenide on its way to supporting selenoprotein synthesis, while methylseleninic acid generates the bioactive methylselenol. There are also differences in how these selenocompounds affect redox status and reactive oxygen species (ROS), which both can affect the proliferation and metastatic properties of cancer cells. Using the same LLC model, the antimetastatic effects of methylseleninic acid were abrogated by feeding the mice a high-fat diet (Yan & Combs, 2014), suggesting adipose-derived inflammatory mediators can impinge upon the effector mechanisms of selenium metabolites regarding cancer cell movement in vivo. These results suggest that obesity and chronic inflammatory conditions must be taken into consideration when investigating the effects of dietary selenium and metastasis.

In a metastasis model utilizing B16BL6 murine melanoma cells injected i.v. in C57BL/6 mice followed by analyses of tumors formed in the lung, both selenite and selenomethionine supplemented at high levels (2.5 and 5.0 ppm Se) were found to inhibit tumor size in the lungs, with selenite exhibiting a more effective reduction of lung tumor formation (Yan, Yee, Li, McGuire, & Graef, 1999). However, the high levels of selenium supplementation and model used in these studies may not reflect physiological effects of selenium on actual tumor metastasis. This same group found other forms of supplementation such as selenium-containing soy protein was effective in reducing pulmonary metastasis of melanoma in mice (Li, Graef, Yee, & Yan, 2004; Yan, Yee, McGuire, & Graef, 1997). A different model involving mice fed either selenium deficient or adequate diets (0.01 and 0.08 ppm Se, respectively), or diets supplemented with selenite (0.4 ppm Se), selenomethionine (3 ppm Se), or methylseleninic acid (3 ppm Se) were examined for metastasis using mammary gland inoculation of 4T1.2 murine breast cancer cells into mammary tissue followed by tumor size analyses in secondary organs (Chen, Prabhu, Das, & Mastro, 2013). While no differences were found between selenium deficient and adequate dietary groups, reduction of metastases to some organs was found with selenomethionine or methylseleninic acid, but selenite appeared to increase tumors in secondary tissues. Since the molar levels of selenium were not equivalent and bioavailability of selenium from each is different, it is difficult to compare these supplemented forms. However, the lack of differences between selenium deficient and adequate diets suggests that subtle, physiologically relevant differences in selenium intake may not have strong effects. Also, the variability in secondary tumor formation in supplemented groups also does not provide conclusive results regarding how selenium metabolites affect metastasis. In another study involving the 4T1.2 breast cancer model, selenium-enriched *Lactobacillus brevis* was found to reduce secondary tumor formation in the liver (Yazdi, Mahdavi, Setayesh, Esfandyar, & Shahverdi, 2013). The bioavailable levels from different bioactive forms of selenium leading to these results were not determined, but these results are consistent with the general notion that selenium supplementation can affect metastasis.

A number of studies have focused on how selenium compounds affect different steps of the metastatic process such as detachment of cells from the extracellular matrix or tumor tissue, degradation of extracellular matrix and migration, invasion into the blood and extravasation into secondary tissues, and subsequent proliferation into metastatic tumors. For example, in vitro studies using a human fibrosarcoma cell line, HT1080, showed that supplementation with selenite, but not selenite, decreased the invasiveness due to inhibition of cell–matrix interactions and alteration of protease expression involved in matrix degradation (Yoon, Kim, & Chung, 2001). Methylseleninic acid and methylselenol also were found to affect the expression profile of MMPs along with their inhibitors in HT1080 cell studies that resulted in decreased invasiveness (Park, Kim, Oh, & Chung, 2007; Zeng, Briske-Anderson, Idso, & Hunt, 2006). Another research group examined the effects of methylselenol generated from selenomethionine on cell proliferation, adhesion, and expression of integrins in murine melanoma B16F10 cells, which are metastatic in the lungs of syngeneic C57BL/6J mice (Kim, Oh, Park, & Chung, 2007). Treatment with both selenomethionine and the enzyme that converts it to methylselenol, methioninase, led to reduced integrin expression and inhibited melanoma–ECM adhesion. This detachment led to increased apoptosis and the approach used suggests that the anticancer effects of selenium supplementation can be enhanced by therapeutic manipulation of factors that regulate certain metabolic steps. Migration of melanoma cells was also decreased by selenite treatment by decreasing IL-18 gene and protein expression, which was utilized by the cells to promote migration (Song et al., 2011).

As these studies demonstrate, the relationship between dietary selenium and metastasis is quite complex. The physiology underlying how selenium status affects the stages of tumor development, metastasis, and clinical outcome depends on the type of cancer, the form of selenium ingested, and its interaction with several patient factors such as inflammatory conditions and others. The use of pharmacological levels of selenocompounds for therapy appears to show promise based on mouse models and cell studies. However, results from clinical trials and the wide variety of tissues affected by using high doses of selenocompounds to reduce tumor progression and metastasis do not provide a clear picture of how this therapeutic approach may be optimally utilized. This has led many researchers to turn their attention to better understanding roles that individual selenoproteins may play in cancer metastasis. Selenoproteins may serve as more selective targets for controlling certain steps of cancer progression and metastasis. In addition, underlying genetic factors that influence selenoproteins levels or functions may affect cancer in concert with or independent of dietary selenium intake. The remainder of this review will summarize issues regarding how different selenoproteins may be involved with metastasis.

4. THIOREDOXIN REDUCTASES AND REDOX TONE

As primary cells of various types transform and develop into aggressive malignant cells, their energy demands and redox tone are altered (Cairns, Harris, & Mak, 2011). As discussed earlier, selenium metabolites may influence redox cycling in different ways, and for the most part the selenoproteins function to tip redox balance toward a reductive state. As described in chapter "Targeting the selenoprotein thioredoxin reductase 1 for anticancer therapy" by Arnér of this series and in other published work (Arner, 2009), one selenoprotein subfamily that plays a key role in redox adjustment during tumor progression is the thioredoxin reductases (TXNRDs). TXNRD1 is localized to the cytosol and nucleus and represents the best-characterized member of the subfamily in regards to cancer. TXNRD2 is localized to the mitochondria in most cells, and TXNRD3 is mainly expressed in sperm during maturation. All three TXNRDs utilize NADPH as an electron donor to regenerate reduced thioredoxin in cells and thus increase reducing capacity (Mustacich & Powis, 2000).

The essential role for TXNRD1 and 2 in development is evident by the fact that deletion of either in mice results in embryonic lethality (Conrad et al., 2004; Jakupoglu et al., 2005). As is the case with many proteins important for development, the TXNRDs are upregulated in many types of cancer. The best example is TXNRD1, which is overexpressed in a variety of malignant cells compared to normal tissue of the same region and with progression toward a metastatic phenotype (Selenius, Rundlof, Olm, Fernandes, & Bjornstedt, 2010). Microarray and immunohistochemical analyses found that increased TXNRD1 expression correlated with regional lymph node metastasis and clinical stage in 50 patients with oral squamous cell carcinoma (Iwasawa et al., 2011). Mice injected with mouse lung carcinoma cells in which RNAi was used to reduce TXNRD1 expression showed a dramatic reduction in tumor progression and metastasis compared with those mice injected with control cells (Yoo, Xu, Carlson, Gladyshev, & Hatfield, 2006). While much has been uncovered regarding the role of

TXNRD1 in steps of carcinogenesis and tumor formation (Schmidt & Arnier, 2016), limited information is available on its specific roles in metastasis. One exception to this is the demonstrated importance of TXNRD1 in angiogenesis during the metastatic process. Reduced expression of TXNRD1 or treatment with reagents that target the TXNRD members exhibit antiangiogenic properties (Xie, Luo, Zhao, & Chen, 2017). A proinvasive role for TXNRD1 was shown in breast cancer studies using MDA-MB-231 cells, and meta-analysis of 25 gene array studies with 5910 patients showed that thioredoxin-1 and TXNRD1 were both associated with a poor patient prognosis in terms of overall survival, distant metastasis-free survival, and disease-free survival (Bhatia et al., 2016). This is consistent with data showing levels of this enzyme in serum as measured by ELISA correlated with prostate cancer metastasis risk (Zhang, Zheng, & Wang, 2015), and TXNRD1 levels measured by histological analyses correlating with gallbladder carcinoma invasiveness risk (Nagano et al., 2012). Also, metastasis of salivary adenoid cystic carcinoma correlated with increased thioredoxin and TXNRD1 and a TXNRD1 inhibitor, BBSKE, reduced metastatic potential by limiting EMT in a xenograft mouse model (Jiang et al., 2015). EMT via anchorage-independent growth properties of mouse lung carcinoma (LLC1) cells in vitro and in mice injected with these cells was reduced using RNAi techniques targeting TXNRD1 (Yoo et al., 2006). In fact, EMT has become a focus of TXNRD1 inhibitors as a potential therapeutic strategy for reducing metastasis (Durand & Storz, 2017). TXNRD2 has been less studied, but this selenoenzyme was upregulated in metastatic osteosarcoma, and the TXNRD inhibitor, auranofin, was found to reduce metastasis to pulmonary tissues (Topkas et al., 2016). Prostate cancer patients carrying a TXNRD2 single-nucleotide polymorphism (rs1005873) had a higher risk of cancer aggressiveness (Xie et al., 2016), although the functional significance of this polymorphism was not determined. Thus, both TXNRD1 and 2 may not only serve as prognostic biomarkers for metastasis but as therapeutic targets for reducing metastatic capacity for a variety of cancers by limiting multiple steps including EMT, angiogenesis, as well as the redox-dependent survival of these cells during the metastatic process.

5. GPX AND ROS REGULATION

While TXNRDs appear to mainly serve a prometastasis role as described earlier, the relationship between the glutathione peroxidases

(GPXs) and metastasis is more dependent on the member of the GPX family being considered. All GPX enzymes utilize glutathione that is oxidized to glutathione disulfide (GSSG) in the process of reducing hydroperoxides (Brigelius-Flohe & Maiorino, 2013). Hydroperoxides are nonradical ROS generated in cells that in some cases can diffuse to extracellular spaces and possibly back into cells (Finkel, 2003). Hydrogen peroxide (H_2O_2) and lipid hydroperoxides are reduced by GPX1–3 and 4, respectively, along with other antioxidant systems such as peroxiredoxins and catalase. This detoxification process serves to reduce damage these ROS may cause in cells and the extracellular matrix but also regulates the signaling functions of these secondary messengers as well. The signaling roles hydroperoxides, particularly H_2O_2, play in noncancer cells can be exploited during the transformation of primary cells to cancer cells and subsequent development into metastatic tumor cells (Liou & Storz, 2010; Storz, 2005; Wu, 2006). In some cases, GPX loss of function may lead to increased ROS that may promote the metastatic potential of tumors. This was clearly demonstrated in studies showing that the treatment of carcinoma cells with H_2O_2 prior to intravenous injection into mice enhanced metastasis (Kundu, Zhang, & Fulton, 1995). This may reflect a particularly important role for extracellular ROS in promoting metastasis, and the enzyme that regulates levels of H_2O_2 in the extracellular space is the secreted GPX3 enzyme. In human thyroid cancer, methylation of the GPX3 gene and reduced GPX3 protein levels was correlated with tumor size and lymph node metastasis (Zhao et al., 2015). Wnt signaling was inhibited by overexpressing GPX3 in TPC-1 and FTC133 cells, suggesting that these cells utilize secreted GPX3 to regulate extracellular H_2O_2 levels and this regulates cellular function. There are other examples of both methylation of the GPX3 gene and reduced GPX3 protein being associated with metastatic potential including studies involving prostate cancer (Yu et al., 2007), hepatocellular carcinoma (Cao et al., 2015), gastric carcinoma (Peng et al., 2012), as well as a cervical cancer study in which downregulation of GPX3 was associated with lymph node metastasis and prognosis (Zhang et al., 2015).

An elevation in bioavailable extracellular H_2O_2 may facilitate cancer cells during various stages of the metastatic process, but since this ROS can diffuse through the plasma membrane, it is likely intracellular GPX enzymes, such as GPX1, 2, and 4, are involved as well. In contrast to secreted GPX3, loss of function of these intracellular GPXs is in some cases associated with decreased metastatic potential. Perhaps the best example is GPX2, which is highly expressed in gastrointestinal tissues and mostly examined in the context of colorectal cancer due to its upregulation in intestinal adenomas and in colon tumors (Florian et al., 2001). In a study using patient-derived colonosphere cultures, GPX2 silencing caused accumulation of ROS, sensitization to H_2O_2-induced apoptosis, and strongly reduced metastases from primary tumors in the spleen to secondary tumors in the liver (Emmink et al., 2014). These cancer cells rely on GPX2 to keep ROS at optimal levels for metastatic capacity. GPX2 expression may be induced through NF-E2-related factor-2 (Nrf2) or through transcription factors of the signal transducers and activators of transcription family (Emmink et al., 2014). Interestingly, GPX2 loss generated a population of stem-like cancer cells, suggesting that GPX2 and ROS play important roles during both early and late stages of cancer development. How these regulatory pathways operate during progression to metastatic stages of cancer cell development are not completely understood, but may offer potential as targets of controlling tumor progression. This is highlighted by studies showing more aggressive tumors have higher levels of Nrf2 compared to less aggressive tumors, and this can upregulate expression of both GPX2 and TXNRD1 for exploitation of their roles in regulating ROS (Brigelius-Flohe, Muller, Lippmann, & Kipp, 2012; Chen et al., 2010). GPX2 was also found to contribute to metastatic potential in hepatocarcinoma (Suzuki et al., 2013). There may be tissue-specific effects of GPX2 regulating metastasis as demonstrated by findings with non-gastrointestinal or liver cancers such as the urothelial carcinoma in which higher risk of metastasis-free survival has been associated with lower levels of GPX2 transcripts (Chang et al., 2015).

Even though ROS-induced damage and apoptosis need to be controlled through the activity of antioxidant enzymes in cancer cells, metastatic cancers still require usable levels of H_2O_2 to migrate and invade. This was shown in the HT-29 human colon cancer cell line in which stable GPX2 knockdown (not complete knockout) cells exhibited a higher capacity to migrate and invade than the GPX2-expressing controls and this was dependent on COX-2 (Yoon, Park, Yoon, Yun, & Chung, 2002). Thus, GPX2 and H_2O_2 levels may need to fluctuate depending on the stage of tumor progression for promoting metastasis. So, how does H_2O_2 promote metastasis? Part of the answer may involve MMPs. As described earlier, prior to migration of cancer cells to distal sites they undergo EMT to release themselves from the restraint of the basal membrane. During this process, MMPs are upregulated to degrade the proteins that compose the basal membrane and are key for the metastatic process (Duffy, Maguire, Hill, McDermott, & O'Higgins, 2000; Nelson, Fingleton, Rothenberg, & Matrisian, 2000), and these enzymes are regulated by ROS at several levels. One example is the regulation of the MMP activation state through redox modulation of the cysteines in their catalytic domains (Koch et al., 2009; Rajagopalan, Meng, Ramasamy, Harrison, & Galis, 1996). In addition, MMPs can be regulated transcriptionally by ROS via Ets- and AP-1-responsive elements present in promoter regions (Nelson et al., 2006). Sustained production of H_2O_2 was also shown to influence expression of MMPs through activation of the phosphatidylinositol 3-kinase/NF-κB pathway, and this affected cell motility and invasion in HT1080 fibrosarcoma cells (Yoon et al., 2002). Also, in esophageal cancer cell lines EC109 and EC9706, enhanced GPX1 activity led to higher capacity for invasion and migration through enzymes MMP-2 and urokinase-type plasminogen activator, which are crucial for tumor formation and metastasis (Qian et al., 2010). Consistent with these results are data showing a role for selenoprotein expression in macrophages for influencing MMP function through transcriptional regulation of inhibitors of ECM proteolysis (Carlson et al., 2009). These inhibitors affected migration of these cells through ECM and may reflect mechanisms by which selenoproteins control ROS regulated migration of cancer cells during metastasis, although specific roles for GPXs were not addressed in the macrophage study.

GPX1 has largely been studied in relation to carcinogenesis, although transcript levels for GPX1 along with other antioxidants have been suggested to have predictive value for metastasis (Giesing, Driesel, Molitor, & Suchy, 2012; Han et al., 2013; Miar et al., 2015; Min et al., 2012). Increased GPX1 activity, but not expression levels, was linked to increased vascular invasion using a hepatocellular carcinoma cell model (Huang et al., 2012). GPX4 has been found to be decreased in many cancer tissues, and GPX4-overexpressing cancer cells have low COX-2 activity, and tumors derived therefrom are smaller than from control cells and do not metastasize (Brigelius-Flohe & Kipp, 2009). Overall, the collective data suggest a dependence on increased extracellular H_2O_2 for metastasis by decreasing GPX3 secretion, but the cancer cells themselves may upregulate other GPXs in a tissue-specific manner to manage intracellular ROS. In this sense, treatments focused on antioxidant drugs that are restricted to the extracellular spaces may prove more effective than those targeting extra- and intracellular spaces.

6. SELENOPROTEINS INVOLVED IN ER STRESS AND CALCIUM FLUX

Seven of the twenty-five members of the selenoprotein family are localized to the ER, including SELENOF, SELENOK, SELENOM, SELENON, SELENOS, SELENOT, and the type-2 iodothyronine deiodinase (Shchedrina, Zhang, Labunskyy, Hatfield, & Gladyshev, 2010). While the latter is an enzyme regulating thyroid hormone metabolism, the other six have been found to be involved in functions such as quality control of protein folding in the ER, retrotranslocation of misfolded proteins from the ER to the cytosol, protein palmitoylation at the ER surface, and regulation of calcium (Ca^{2+}) homeostasis. Ca^{2+} flux from the ER and ER stress is important for regulating metastasis, which implicates the ER-resident selenoproteins involved in these cellular processes.

The importance of ER stress generated through the unfolded protein response in regulating metastasis is well recognized (Hazari, Bashir, Haq, & Fazili, 2016). The ER selenoproteins involved in the unfolded protein response within this organelle include SELENOF, SELENOK, and SELENOS (Kelly, Greene, Carroll, McElvaney, & O'Neill, 2009; Labunskyy, Yoo, Hatfield, & Gladyshev, 2009; Lee et al., 2015). Polymorphisms in SELENOS have been linked to increased risk for different cancers (Hart et al., 2011; Shibata et al., 2009). SELENOF has been shown to have a protumorigenesis and metastasis role, although exactly how it promotes these activities is unclear (Irons et al., 2010; Tsuji et al., 2015). Altogether, little experimental data are available regarding specific mechanisms regulated by these selenoproteins in relation to metastasis. A similar lack of knowledge pertains to roles for the ER selenoproteins involved in Ca^{2+} homeostasis and/or flux, which include SELENOK, SELENOM, SELENON, SELENOT (Appenzeller-Herzog & Simmen, 2016; Grumolato et al., 2008; Reeves, Bellinger, & Berry, 2010; Verma et al., 2011). Direct examination of how these family members may influence metastasis through Ca^{2+} homeostasis or flux from the ER is scant. Ca^{2+} signaling within normal and cancer cells regulates diverse cellular functions including proliferation, migration, differentiation, and cytokine secretion (Berridge, Bootman, & Roderick, 2003). Store-operated Ca^{2+} entry (SOCE) is a mechanism utilized to release Ca^{2+} from ER stores and activate Ca^{2+} signaling, and SOCE has been shown to depend on SELENOK in immune cells (Verma et al., 2011). In particular, SELENOK regulates SOCE by complexing with the

DHHC6 enzyme to promote palmitoylation and stable expression of the inositol 1,4,5-triphosphate receptor for efficient SOCE (Fredericks et al., 2014). There is an emerging recognition of SOCE as playing a major role in metastasis (Chen, Hsu, & Shen, 2016; Umemura et al., 2014; Xia et al., 2016; Yu et al., 2016), suggesting that SELENOK may represent a potential therapeutic target for reducing tumor progression. In fact, ongoing studies in our laboratory support this notion and the fact that SELENOK-deficient mice show no major defects suggests that side effects of targeted SELENOK inhibition for cancer patients may be minimal. The stoichiometry of the SELENOK cofactor and the DHHC6 enzyme may be important for migrating cancer cells in that overexpression of SELENOK was found to inhibit migration of human gastric cancer BGC-823 cells in transwell assays (Ben et al., 2015). This was true for full-length SELENOK and not a C-terminally truncated version, suggesting an overabundance of the selenocysteine containing SELENOK was capable of exerting a dominant-negative effect. Overall, there remains a major gap in knowledge regarding how the ER selenoproteins may be involved in metastasis and further investigation is warranted.

7. OTHER POTENTIAL ROLES FOR SELENOPROTEINS IN METASTASIS

Among all selenoproteins, selenoprotein P (SELENOP) is the only one that contains more than one selenocysteine residue per protein with human and rodent SELENOP containing 10 selenocysteines per molecule. SELENOP is found in the plasma and mainly secreted by the liver, although many cell types are capable of expressing and secreting this selenoprotein (Schweizer, Schomburg, & Kohrle, 2016). A main function of SELENOP is transporting selenium to various tissues, although the selenocysteine most proximal to the N-terminus has antioxidant capacity (Burk & Hill, 2015). In this sense, understanding how SELENOP is related to metastasis parallels the discussion above regarding the relationship of dietary selenium to metastasis since SELENOP delivers bioactive selenium to tissues. However, because different cell types synthesize SELENOP at different levels and its synthesis comes at a potential cost to the synthesis of other selenoproteins, there are some additional issues. For the most part, tumors themselves exhibit lower levels of SELENOP compared to healthy tissues at the same location (Rayman, 2005), but neighbor cells may secrete SELENOP to compensate. For example, a recent study in which tumor cell culture media was used to differentiate macrophages into a protumor phenotype (M2) revealed that SELENOP was upregulated 95-fold at the transcript level (Solinas et al., 2010). Colonocytes from late-stage colorectal cancer (containing both normal and neoplastic human epithelial cells) had increased SELENOP transcript levels compared to early stage and healthy groups (Yajima et al., 2007). Whether this led to more secreted SELENOP for use by the tumor cells was not determined. However, another study looking directly at prostate cells as they transformed into highly metastatic cancer cells found that SELENOP levels decreased (Calvo et al., 2002). There may be a stronger demand of bioavailable selenium for synthesis of other selenoproteins and, since SELENOP consumes 10-fold more selenium than the other members of the selenoprotein family, this may mean that synthesis of SELENOP is sacrificed. In this sense, the increased SELENOP in M2 macrophages may offset the loss of SELENOP in tumor cells and support metastasis by supplying this in a paracrine manner, although this has not been tested experimentally. This may depend also on the location of cancer as a study of gastric adenocarcinoma found that SELENOP expression was low in cancer tissues compared with control tissues and was related to the degree of gastric adenocarcinoma differentiation but not to tumor node metastasis stage (Wang et al., 2009).

The 15-kDa selenoprotein, selenoprotein F (SELENOF; SEP15), has been found to be upregulated in a wide variety of cancers and reduced in others (Carlson, Hartman, & Tsuji, 2016). Regarding metastasis, SELENOF was shown to promote both anchorage-dependent and anchorage-independent growth and formation of experimental metastases of mouse colon carcinoma CT26 cells (Tsuji et al., 2015). In contrast to CT26 cells, SELENOF-targeted downregulation in Lewis lung carcinoma line 1 (LLC1) cells did not affect anchorage-dependent or -independent cell growth (Irons et al., 2010). Thus, the role of SELENOF in facilitating steps of metastasis may be tissue specific and this requires further investigation.

The three selenoenzymes regulating thyroid hormone activity include iodothyronine deiodinases 1, 2, and 3 (DIO1, 2, and 3). DIO1 and 2 convert the prohormone (T4) into active hormone (T3), and DIO3 inactivates T3 (Kohrle, Jakob, Contempre, & Dumont, 2005). Since thyroid hormone regulates a wide range of cellular processes, including the balance between proliferation and differentiation, it is not surprising that these selenoproteins are involved in cancer (Piekielko-Witkowska & Nauman, 2011). DIO3 in particular is reactivated in human neoplasias through MAPK and sonic hedgehog pathways (Romitti et al., 2016). However, specific roles for the DIO enzymes in metastasis are less clear. Some evidence was presented for a potential role of DIO3 in a study involving papillary thyroid carcinoma showing increased mRNA levels and activity for this enzyme was correlated with lymph node or distant metastasis at diagnosis (Romitti et al., 2012). Data from this study also suggested that levels of DIO2 were altered in the opposite direction, although the role in metastasis was not clearly determined. Much of what is known regarding DIOs and metastasis has been garnered through case studies, and more mechanistic studies interrogating the roles they may play in disease progression through metastatic processes are required.

SELENOH is an oxidoreductase localized to the nucleolus in a variety of tissues that has been shown to be involved in p53 pathway activation related to both development and carcinogenesis (Cox et al., 2016). There is evidence using the MDA-MB-231 metastatic breast cancer cell line that SELENOH is transcriptionally upregulated by the antiproliferation and antimigration protein, delta-lactoferrin (Hoedt et al., 2014). These different studies do not clarify how SELENOH may affect metastasis, but redox-directed transcription regulated by SELENOH is likely to both influence and be influenced by the stage of cancer cell progression into aggressive metastatic tumors.

8. CONCLUSIONS

There are several cellular activities regulated by selenoproteins related to metastasis and poor outcome in cancer patients. These include modulation of redox tone and detoxification of ROS by the TXNRDs and GPXs, Ca^{2+} homeostasis and unfolded protein responses regulated by ER selenoproteins, and the multiple physiological responses influenced by other selenoproteins. The tissue-specific expression of different selenoproteins is likely to influence particular roles they play during development of different cancers into metastatic tumors and the establishment of secondary tumors in various tissues. Also, it is worth noting that there are sex-specific mechanisms affecting transcription, alternative splicing, translation, and posttranslational activity of different selenoproteins (Schomburg, 2016). Thus, the effects exerted by selenoproteins on the steps of metastasis may differ between males and females. Acquiring data from studies in humans has proven difficult, although many clues have been obtained through correlative studies. Animal and cell culture model systems have emerged as a source of novel insights, and this is sure to expand in the future as genetic models become more available for the full set of selenoproteins and factors regulating their activity. In fact, the exact biological roles for many selenoproteins themselves are only beginning to emerge. More mechanistic insight is a prerequisite for developing antimetastasis therapies focused on modulating selenoproteins, but progress will continue to be made toward this goal as our understanding of the biological roles of selenoproteins increases.

ACKNOWLEDGMENTS

This work was supported by U.S. National Institutes of Health (NIH) National Institute of Allergy and Infectious Diseases Grant R01AI089999.

REFERENCES

Appenzeller-Herzog, C., & Simmen, T. (2016). ER-luminal thiol/selenol-mediated regulation of Ca2+ signalling. *Biochemical Society Transactions*, *44*, 452–459.

Arner, E. S. (2009). Focus on mammalian thioredoxin reductases—Important selenoproteins with versatile functions. *Biochimica et Biophysica Acta*, *1790*, 495–526.

Ben, S. B., Peng, B., Wang, G. C., Li, C., Gu, H. F., Jiang, H., et al. (2015). Overexpression of selenoprotein SelK in BGC-823 cells inhibits cell adhesion and migration. *Biochemistry Biokhimiia*, *80*, 1344–1353.

Bera, S., De Rosa, V., Rachidi, W., & Diamond, A. M. (2013). Does a role for selenium in DNA damage repair explain apparent controversies in its use in chemoprevention? *Mutagenesis*, *28*, 127–134.

Berridge, M. J., Bootman, M. D., & Roderick, H. L. (2003). Calcium signalling: Dynamics, homeostasis and remodelling. *Nature Reviews. Molecular Cell Biology*, *4*, 517–529.

Bhatia, M., McGrath, K. L., Di Trapani, G., Charoentong, P., Shah, F., King, M. M., et al. (2016). The thioredoxin system in breast cancer cell invasion and migration. *Redox Biology*, *8*, 68–78.

Brigelius-Flohe, R. (2008). Selenium compounds and selenoproteins in cancer. *Chemistry & Biodiversity*, *5*, 389–395.

Brigelius-Flohe, R., & Kipp, A. (2009). Glutathione peroxidases in different stages of carcinogenesis. *Biochimica et Biophysica Acta*, *1790*, 1555–1568.

Brigelius-Flohe, R., & Maiorino, M. (2013). Glutathione peroxidases. *Biochimica et Biophysica Acta*, *1830*, 3289–3303.

Brigelius-Flohe, R., Muller, M., Lippmann, D., & Kipp, A. P. (2012). The yin and yang of nrf2-regulated selenoproteins in carcinogenesis. *International Journal of Cell Biology*, *2012*, 486147.

Burk, R. F., & Hill, K. E. (2015). Regulation of selenium metabolism and transport. *Annual Review of Nutrition*, *35*, 109–134.

Cairns, R. A., Harris, I. S., & Mak, T. W. (2011). Regulation of cancer metabolism. *Nature Reviews. Cancer*, *11*, 85–95.

Calvo, A., Xiao, N., Kang, J., Best, C. J., Leiva, I., Emmert-Buck, M. R., et al. (2002). Alterations in gene expression profiles during prostate cancer progression: Functional correlations to tumorigenicity and down-regulation of selenoprotein-P in mouse and human tumors. *Cancer Research*, *62*, 5325–5335.

Cao, Y., Hoeppner, L. H., Bach, S., Guangqi, E., Guo, Y., Wang, E., et al. (2013). Neuropilin-2 promotes extravasation and metastasis by interacting with endothelial α5 integrin. *Cancer Research*, *73*, 4579–4590.

Cao, S., Yan, B., Lu, Y., Zhang, G., Li, J., Zhai, W., et al. (2015). Methylation of promoter and expression silencing of GPX3 gene in hepatocellular carcinoma tissue. *Clinics and Research in Hepatology and Gastroenterology, 39*, 198–204.

Carlson, B. A., Hartman, J. M., & Tsuji, P. A. (2016). The 15 kDa Selenoprotein: Insights into Its Regulation and Function. In U. Schweizer, P. A. Tsuji, D. L. Hatfield, & V. N. Gladyshev (Eds.), *Selenium: Its molecular biology and role in human health: Vol. 4* (pp. 235–243). New York: Springer.

Carlson, B. A., Yoo, M. H., Sano, Y., Sengupta, A., Kim, J. Y., Irons, R., et al. (2009). Selenoproteins regulate macrophage invasiveness and extracellular matrix-related gene expression. *BMC Immunology, 10*, 57.

Chang, I. W., Lin, V. C., Hung, C. H., Wang, H. P., Lin, Y. Y., Wu, W. J., et al. (2015). GPX2 underexpression indicates poor prognosis in patients with urothelial carcinomas of the upper urinary tract and urinary bladder. *World Journal of Urology, 33*, 1777–1789.

Chen, Y. F., Hsu, K. F., & Shen, M. R. (2016). The store-operated Ca(2+) entry-mediated signaling is important for cancer spread. *Biochimica et Biophysica Acta, 1863*, 1427–1435.

Chen, Y. C., Prabhu, K. S., Das, A., & Mastro, A. M. (2013). Dietary selenium supplementation modifies breast tumor growth and metastasis. *International Journal of Cancer, 133*, 2054–2064.

Chen, Y. C., Prabhu, K. S., & Mastro, A. M. (2013). Is selenium a potential treatment for cancer metastasis? *Nutrients, 5*, 1149–1168.

Chen, N., Yi, X., Abushahin, N., Pang, S., Zhang, D., Kong, B., et al. (2010). Nrf2 expression in endometrial serous carcinomas and its precancers. *International Journal of Clinical and Experimental Pathology, 4*, 85–96.

Conrad, M., Jakupoglu, C., Moreno, S. G., Lippl, S., Banjac, A., Schneider, M., et al. (2004). Essential role for mitochondrial thioredoxin reductase in hematopoiesis, heart development, and heart function. *Molecular and Cellular Biology, 24*, 9414–9423.

Cox, A. G., Tsomides, A., Kim, A. J., Saunders, D., Hwang, K. L., Evason, K. J., et al. (2016). Selenoprotein H is an essential regulator of redox homeostasis that cooperates with p53 in development and tumorigenesis. *Proceedings of the National Academy of Sciences of the United States of America, 113*, E5562–71.

Duffy, M. J., Maguire, T. M., Hill, A., McDermott, E., & O'Higgins, N. (2000). Metalloproteinases: Role in breast carcinogenesis, invasion and metastasis. *Breast Cancer Research, 2*, 252–257.

Durand, N., & Storz, P. (2017). Targeting reactive oxygen species in development and progression of pancreatic cancer. *Expert Review of Anticancer Therapy, 17*, 19–31.

Emmink, B. L., Laoukili, J., Kipp, A. P., Koster, J., Govaert, K. M., Fatrai, S., et al. (2014). GPx2 suppression of H2O2 stress links the formation of differentiated tumor mass to metastatic capacity in colorectal cancer. *Cancer Research, 74*, 6717–6730.

Fernandez, Y., Espana, L., Manas, S., Fabra, A., & Sierra, A. (2000). Bcl-xL promotes metastasis of breast cancer cells by induction of cytokines resistance. *Cell Death and Differentiation, 7*, 350–359.

Finkel, T. (2003). Oxidant signals and oxidative stress. *Current Opinion in Cell Biology, 15*, 247–254.

Florian, S., Wingler, K., Schmehl, K., Jacobasch, G., Kreuzer, O. J., Meyerhof, W., et al. (2001). Cellular and subcellular localization of gastrointestinal glutathione peroxidase in normal and malignant human intestinal tissue. *Free Radical Research, 35*, 655–663.

Fredericks, G. J., Hoffmann, F. W., Rose, A. H., Osterheld, H. J., Hess, F. M., Mercier, F., et al. (2014). Stable expression and function of the inositol 1,4,5-triphosphate receptor requires palmitoylation by a DHHC6/selenoprotein K complex. *Proceedings of the National Academy of Sciences of the United States of America, 111*, 16478–16483.

Giesing, M., Driesel, G., Molitor, D., & Suchy, B. (2012). Molecular phenotyping of circulating tumour cells in patients with prostate cancer: Prediction of distant metastases. *BJU International, 110*, E1202–11.

Gopalakrishna, R., & Gundimeda, U. (2001). Protein kinase C as a molecular target for cancer prevention by selenocompounds. *Nutrition and Cancer, 40*, 55–63.
Grumolato, L., Ghzili, H., Montero-Hadjadje, M., Gasman, S., Lesage, J., Tanguy, Y., et al. (2008). Selenoprotein T is a PACAP-regulated gene involved in intracellular Ca2+ mobilization and neuroendocrine secretion. *The FASEB Journal, 22*, 1756–1768.
Guo, W., & Giancotti, F. G. (2004). Integrin signalling during tumour progression. *Nature Reviews. Molecular Cell Biology, 5*, 816–826.
Gupta, G. P., & Massague, J. (2006). Integrin signalling during tumour progression. *Cell, 127*, 679–695.
Gupta, G. P., Nguyen, D. X., Chiang, A. C., Bos, P. D., Kim, J. Y., Nadal, C., et al. (2007). Mediators of vascular remodelling co-opted for sequential steps in lung metastasis. *Nature, 446*, 765–770.
Han, J. J., Xie de, R., Wang, L. L., Liu, Y. Q., Wu, G. F., Sun, Q., et al. (2013). Significance of glutathione peroxidase 1 and caudal-related homeodomain transcription factor in human gastric adenocarcinoma. *Gastroenterology Research and Practice, 2013*, 380193.
Hart, K., Landvik, N. E., Lind, H., Skaug, V., Haugen, A., & Zienolddiny, S. (2011). A combination of functional polymorphisms in the CASP8, MMP1, IL10 and SEPS1 genes affects risk of non-small cell lung cancer. *Lung Cancer, 71*, 123–129.
Hazari, Y. M., Bashir, A., Haq, E. U., & Fazili, K. M. (2016). Emerging tale of UPR and cancer: An essentiality for malignancy. *Tumour Biology, 37*, 14381–14390.
Hoedt, E., Chaoui, K., Huvent, I., Mariller, C., Monsarrat, B., Burlet-Schiltz, O., et al. (2014). SILAC-based proteomic profiling of the human MDA-MB-231 metastatic breast cancer cell line in response to the two antitumoral lactoferrin isoforms: The secreted lactoferrin and the intracellular delta-lactoferrin. *PLoS One, 9*, e104563.
Huang, C., Ding, G., Gu, C., Zhou, J., Kuang, M., Ji, Y., et al. (2012). Decreased selenium-binding protein 1 enhances glutathione peroxidase 1 activity and downregulates HIF-1α to promote hepatocellular carcinoma invasiveness. *Clinical Cancer Research, 18*, 3042–3053.
Ikeda, K., Iyama, K., Ishikawa, N., Egami, H., Nakao, M., Sado, Y., et al. (2006). Loss of expression of type IV collagen alpha5 and alpha6 chains in colorectal cancer associated with the hypermethylation of their promoter region. *The American Journal of Pathology, 168*, 856–865.
Irons, R., Tsuji, P. A., Carlson, B. A., Ouyang, P., Yoo, M. H., Xu, X. M., et al. (2010). Deficiency in the 15-kDa selenoprotein inhibits tumorigenicity and metastasis of colon cancer cells. *Cancer Prevention Research (Philadelphia, Pa.), 3*, 630–639.
Iwasawa, S., Yamano, Y., Takiguchi, Y., Tanzawa, H., Tatsumi, K., & Uzawa, K. (2011). Upregulation of thioredoxin reductase 1 in human oral squamous cell carcinoma. *Oncology Reports, 25*, 637–644.
Jackson, M. I., & Combs, J. G. F. (2012). *Selenium as a cancer preventive agent*. New York: Springer.
Jakupoglu, C., Przemeck, G. K., Schneider, M., Moreno, S. G., Mayr, N., Hatzopoulos, A. K., et al. (2005). Cytoplasmic thioredoxin reductase is essential for embryogenesis but dispensable for cardiac development. *Molecular and Cellular Biology, 25*, 1980–1988.
Jiang, Y., Feng, X., Zheng, L., Li, S. L., Ge, X. Y., & Zhang, J. G. (2015). Thioredoxin 1 mediates TGF-β-induced epithelial-mesenchymal transition in salivary adenoid cystic carcinoma. *Oncotarget, 6*, 25506–25519.
Kelly, E., Greene, C. M., Carroll, T. P., McElvaney, N. G., & O'Neill, S. J. (2009). Selenoprotein S/SEPS1 modifies endoplasmic reticulum stress in Z variant alpha1-antitrypsin deficiency. *The Journal of Biological Chemistry, 284*, 16891–16897.

Kim, A., Oh, J. H., Park, J. M., & Chung, A. S. (2007). Methylselenol generated from selenomethionine by methioninase downregulates integrin expression and induces caspase-mediated apoptosis of B16F10 melanoma cells. *Journal of Cellular Physiology, 212*, 386–400.

Koch, S., Volkmar, C. M., Kolb-Bachofen, V., Korth, H. G., Kirsch, M., Horn, A. H., et al. (2009). A new redox-dependent mechanism of MMP-1 activity control comprising reduced low-molecular-weight thiols and oxidizing radicals. *Journal of Molecular Medicine, 87*, 261–272.

Kohrle, J., Jakob, F., Contempre, B., & Dumont, J. E. (2005). Selenium, the thyroid, and the endocrine system. *Endocrine Reviews, 26*, 944–984.

Kundu, N., Zhang, S., & Fulton, A. M. (1995). Sublethal oxidative stress inhibits tumor cell adhesion and enhances experimental metastasis of murine mammary carcinoma. *Clinical & Experimental Metastasis, 13*, 16–22.

Labunskyy, V. M., Yoo, M. H., Hatfield, D. L., & Gladyshev, V. N. (2009). Sep15, a thioredoxin-like selenoprotein, is involved in the unfolded protein response and differentially regulated by adaptive and acute ER stresses. *Biochemistry, 48*, 8458–8465.

Lee, J. H., Park, K. J., Jang, J. K., Jeon, Y. H., Ko, K. Y., Kwon, J. H., et al. (2015). Selenoprotein S-dependent selenoprotein K binding to p97(VCP) protein is essential for endoplasmic reticulum-associated degradation. *The Journal of Biological Chemistry, 290*, 29941–29952.

Li, D., Graef, G. L., Yee, J. A., & Yan, L. (2004). Dietary supplementation with high-selenium soy protein reduces pulmonary metastasis of melanoma cells in mice. *The Journal of Nutrition, 134*, 1536–1540.

Liou, G. Y., & Storz, P. (2010). Reactive oxygen species in cancer. *Free Radical Research, 44*, 479–496.

Miar, A., Hevia, D., Munoz-Cimadevilla, H., Astudillo, A., Velasco, J., Sainz, R. M., et al. (2015). Manganese superoxide dismutase (SOD2/MnSOD)/catalase and SOD2/GPx1 ratios as biomarkers for tumor progression and metastasis in prostate, colon, and lung cancer. *Free Radical Biology & Medicine, 85*, 45–55.

Min, S. Y., Kim, H. S., Jung, E. J., Jung, E. J., Jee, C. D., & Kim, W. H. (2012). Prognostic significance of glutathione peroxidase 1 (GPX1) down-regulation and correlation with aberrant promoter methylation in human gastric cancer. *Anticancer Research, 32*, 3169–3175.

Mott, J. D., & Werb, Z. (2004). Regulation of matrix biology by matrix metalloproteinases. *Current Opinion in Cell Biology, 16*, 558–564.

Mustacich, D., & Powis, G. (2000). Thioredoxin reductase. *The Biochemical Journal, 346*(Pt. 1), 1–8.

Nagano, M., Hatakeyama, K., Kai, M., Nakamura, H., Yodoi, J., Asada, Y., et al. (2012). Nuclear expression of thioredoxin-1 in the invasion front is associated with outcome in patients with gallbladder carcinoma. *HPB: The Official Journal of the International Hepato Pancreato Biliary Association, 14*, 573–582.

Nelson, A. R., Fingleton, B., Rothenberg, M. L., & Matrisian, L. M. (2000). Matrix metalloproteinases: Biologic activity and clinical implications. *Journal of Clinical Oncology, 18*, 1135–1149.

Nelson, K. K., Subbaram, S., Connor, K. M., Dasgupta, J., Ha, X. F., Meng, T. C., et al. (2006). Redox-dependent matrix metalloproteinase-1 expression is regulated by JNK through Ets and AP-1 promoter motifs. *The Journal of Biological Chemistry, 281*, 14100–14110.

Park, J. M., Kim, A., Oh, J. H., & Chung, A. S. (2007). Methylseleninic acid inhibits PMA-stimulated pro-MMP-2 activation mediated by MT1-MMP expression and further tumor invasion through suppression of NF-kappaB activation. *Carcinogenesis, 28*, 837–847.

Peng, D. F., Hu, T. L., Schneider, B. G., Chen, Z., Xu, Z. K., & El-Rifai, W. (2012). Silencing of glutathione peroxidase 3 through DNA hypermethylation is associated with lymph node metastasis in gastric carcinomas. *PLoS One*, *7*, e46214.

Piekielko-Witkowska, A., & Nauman, A. (2011). Iodothyronine deiodinases and cancer. *Journal of Endocrinological Investigation*, *34*, 716–728.

Qian, Q., Wang, Q., Zhan, P., Peng, L., Wei, S. Z., Shi, Y., et al. (2010). The role of matrix metalloproteinase 2 on the survival of patients with non-small cell lung cancer: A systematic review with meta-analysis. *Cancer Investigation*, *28*, 661–669.

Rajagopalan, S., Meng, X. P., Ramasamy, S., Harrison, D. G., & Galis, Z. S. (1996). Reactive oxygen species produced by macrophage-derived foam cells regulate the activity of vascular matrix metalloproteinases in vitro. Implications for atherosclerotic plaque stability. *The Journal of Clinical Investigation*, *98*, 2572–2579.

Rayman, M. P. (2005). Selenium in cancer prevention: A review of the evidence and mechanism of action. *The Proceedings of the Nutrition Society*, *64*, 527–542.

Reeves, M. A., Bellinger, F. P., & Berry, M. J. (2010). The neuroprotective functions of selenoprotein M and its role in cytosolic calcium regulation. *Antioxidants & Redox Signaling*, *12*, 809–818.

Reeves, M. A., & Hoffmann, P. R. (2009). The human selenoproteome: Recent insights into functions and regulation. *Cellular and Molecular Life Sciences*, *66*, 2457–2478.

Romitti, M., Wajner, S. M., Ceolin, L., Ferreira, C. V., Ribeiro, R. V., Rohenkohl, H. C., et al. (2016). MAPK and SHH pathways modulate type 3 deiodinase expression in papillary thyroid carcinoma. *Endocrine-Related Cancer*, *23*, 135–146.

Romitti, M., Wajner, S. M., Zennig, N., Goemann, I. M., Bueno, A. L., Meyer, E. L., et al. (2012). Increased type 3 deiodinase expression in papillary thyroid carcinoma. *Thyroid*, *22*(9), 897–904.

Schmidt, E. E., & Arnier, E. S. J. (2016). Thioredoxin Reductase 1 as an Anticancer Drug Target. In D. L. Hatfield, U. Schweizer, P. A. Tsuji, & V. N. Gladyshev (Eds.), *Selenium: Its molecular biology and role in human health: Vol. 4* (pp. 199–210). New York: Springer.

Schomburg, L. (2016). Sex-specific differences in biological effectsand metabolism of selenium. In D. L. Hatfield, U. Schweizer, P. A. Tsuji, & V. N. Gladyshev (Eds.), *Selenium: Its molecular biology and role in human health* (pp. 377–388). New York: Springer.

Schweizer, U., Schomburg, L., & Kohrle, J. (2016). Selenoprotein P and selenium distribution in mammals. In D. L. Hatfield, P. A. Tsuji, U. Schweizer, & V. N. Gladyshev (Eds.), *Selenium: Its molecular biology and role in human health: Vol. 4* (pp. 261–274). New York: Springer.

Selenius, M., Rundlof, A. K., Olm, E., Fernandes, A. P., & Bjornstedt, M. (2010). Selenium and the selenoprotein thioredoxin reductase in the prevention, treatment and diagnostics of cancer. *Antioxidants & Redox Signaling*, *12*, 867–880.

Shchedrina, V. A., Zhang, Y., Labunskyy, V. M., Hatfield, D. L., & Gladyshev, V. N. (2010). Structure-function relations, physiological roles, and evolution of mammalian ER-resident selenoproteins. *Antioxidants & Redox Signaling*, *12*, 839–849.

Shibata, T., Arisawa, T., Tahara, T., Ohkubo, M., Yoshioka, D., Maruyama, N., et al. (2009). Selenoprotein S (SEPS1) gene -105G>A promoter polymorphism influences the susceptibility to gastric cancer in the Japanese population. *BMC Gastroenterology*, *9*, 2.

Singh, B., Berry, J. A., Shoher, A., Ayers, G. D., Wei, C., & Lucci, A. (2007). COX-2 involvement in breast cancer metastasis to bone. *Oncogene*, *26*, 3789–3796.

Solinas, G., Schiarea, S., Liguori, M., Fabbri, M., Pesce, S., Zammataro, L., et al. (2010). Tumor-conditioned macrophages secrete migration-stimulating factor: A new marker for M2-polarization, influencing tumor cell motility. *Journal of Immunology*, *185*, 642–652.

Song, H., Kim, J., Lee, H. K., Park, H. J., Nam, J., Park, G. B., et al. (2011). Selenium inhibits migration of murine melanoma cells via down-modulation of IL-18 expression. *International Immunopharmacology, 11*, 2208–2213.

Stamenkovic, I. (2000). Matrix metalloproteinases in tumor invasion and metastasis. *Seminars in Cancer Biology, 10*, 415–433.

Storz, P. (2005). Reactive oxygen species in tumor progression. *Frontiers in Bioscience, 10*, 1881–1896.

Suzuki, S., Pitchakarn, P., Ogawa, K., Naiki-Ito, A., Chewonarin, T., Punfa, W., et al. (2013). Expression of glutathione peroxidase 2 is associated with not only early hepatocarcinogenesis but also late stage metastasis. *Toxicology, 311*, 115–123.

Tong, R. T., Boucher, Y., Kozin, S. V., Winkler, F., Hicklin, D. J., & Jain, R. K. (2004). Vascular normalization by vascular endothelial growth factor receptor 2 blockade induces a pressure gradient across the vasculature and improves drug penetration in tumors. *Cancer Research, 64*, 3731–3736.

Topkas, E., Cai, N., Cumming, A., Hazar-Rethinam, M., Gannon, O. M., Burgess, M., et al. (2016). Auranofin is a potent suppressor of osteosarcoma metastasis. *Oncotarget, 7*, 831–844.

Tsuji, P. A., Carlson, B. A., Yoo, M. H., Naranjo-Suarez, S., Xu, X. M., He, Y., et al. (2015). The 15kDa selenoprotein and thioredoxin reductase 1 promote colon cancer by different pathways. *PLoS One, 10*, e0124487.

Umemura, M., Baljinnyam, E., Feske, S., De Lorenzo, M. S., Xie, L. H., Feng, X., et al. (2014). Store-operated Ca2+ entry (SOCE) regulates melanoma proliferation and cell migration. *PLoS One, 9*, e89292.

Van Slambrouck, S., Groux-Degroote, S., Krzewinski-Recchi, M. A., Cazet, A., Delannoy, P., & Steelant, W. F. (2014). Carbohydrate-to-carbohydrate interactions between α2,3-linked sialic acids on α2 integrin subunits and asialo-GM1 underlie the bone metastatic behaviour of LNCAP-derivative C4-2B prostate cancer cells. *Bioscience Reports, 34*.

Verma, S., Hoffmann, F. W., Kumar, M., Huang, Z., Roe, K., Nguyen-Wu, E., et al. (2011). Selenoprotein K knockout mice exhibit deficient calcium flux in immune cells and impaired immune responses. *Journal of Immunology, 186*, 2127–2137.

Wang, Q., Gong, L., Dong, R., Qiao, Q., He, X. L., Chu, Y. K., et al. (2009). Tissue microarray assessment of selenoprotein P expression in gastric adenocarcinoma. *The Journal of International Medical Research, 37*, 169–174.

Weber, G. F. (2013). Tissue microarray assessment of selenoprotein P expression in gastric adenocarcinoma. *Cancer Letters, 328*, 207–211.

Wu, W. S. (2006). The signaling mechanism of ROS in tumor progression. *Cancer Metastasis Reviews, 25*, 695–705.

Xia, J., Wang, H., Huang, H., Sun, L., Dong, S., Huang, N., et al. (2016). Elevated Orai1 and STIM1 expressions upregulate MACC1 expression to promote tumor cell proliferation, metabolism, migration, and invasion in human gastric cancer. *Cancer Letters, 381*, 31–40.

Xie, L., Luo, Z., Zhao, Z., & Chen, T. (2017). Anticancer and antiangiogenic iron(II) complexes that target thioredoxin reductase to trigger cancer cell apoptosis. *Journal of Medicinal Chemistry, 60*, 202–214.

Xie, W., Yang, M., Chan, J., Sun, T., Mucci, L. A., Penney, K. L., et al. (2016). Association of genetic variations of selenoprotein genes, plasma selenium levels, and prostate cancer aggressiveness at diagnosis. *Prostate, 76*, 691–699.

Yajima, S., Ishii, M., Matsushita, H., Aoyagi, K., Yoshimatsu, K., Kaneko, H., et al. (2007). Expression profiling of fecal colonocytes for RNA-based screening of colorectal cancer. *International Journal of Oncology, 31*, 1029–1037.

Yan, L., & Combs, G. F., Jr. (2014). Consumption of a high-fat diet abrogates inhibitory effects of methylseleninic acid on spontaneous metastasis of Lewis lung carcinoma in mice. *Carcinogenesis*, *35*, 2308–2313.

Yan, L., & DeMars, L. C. (2012). Dietary supplementation with methylseleninic acid, but not selenomethionine, reduces spontaneous metastasis of Lewis lung carcinoma in mice. *International Journal of Cancer*, *131*, 1260–1266.

Yan, L., Yee, J. A., Li, D., McGuire, M. H., & Graef, G. L. (1999). Dietary supplementation of selenomethionine reduces metastasis of melanoma cells in mice. *Anticancer Research*, *19*, 1337–1342.

Yan, L., Yee, J. A., McGuire, M. H., & Graef, G. L. (1997). Effect of dietary supplementation of selenite on pulmonary metastasis of melanoma cells in mice. *Nutrition and Cancer*, *28*, 165–169.

Yazdi, M. H., Mahdavi, M., Setayesh, N., Esfandyar, M., & Shahverdi, A. R. (2013). Selenium nanoparticle-enriched Lactobacillus brevis causes more efficient immune responses in vivo and reduces the liver metastasis in metastatic form of mouse breast cancer. *Daru*, *21*, 33.

Yoo, M. H., Xu, X. M., Carlson, B. A., Gladyshev, V. N., & Hatfield, D. L. (2006). Thioredoxin reductase 1 deficiency reverses tumor phenotype and tumorigenicity of lung carcinoma cells. *The Journal of Biological Chemistry*, *281*, 13005–13008.

Yoon, S. O., Kim, M. M., & Chung, A. S. (2001). Inhibitory effect of selenite on invasion of HT1080 tumor cells. *The Journal of Biological Chemistry*, *276*, 20085–20092.

Yoon, S. O., Park, S. J., Yoon, S. Y., Yun, C. H., & Chung, A. S. (2002). Sustained production of H(2)O(2) activates pro-matrix metalloproteinase-2 through receptor tyrosine kinases/phosphatidylinositol 3-kinase/NF-kappa B pathway. *The Journal of Biological Chemistry*, *277*, 30271–30282.

Yu, C., Tang, W., Wang, Y., Shen, Q., Wang, B., Cai, C., et al. (2016). Downregulation of ACE2/Ang-(1-7)/Mas axis promotes breast cancer metastasis by enhancing store-operated calcium entry. *Cancer Letters*, *376*, 268–277.

Yu, Y. P., Yu, G., Tseng, G., Cieply, K., Nelson, J., Defrances, M., et al. (2007). Glutathione peroxidase 3, deleted or methylated in prostate cancer, suppresses prostate cancer growth and metastasis. *Cancer Research*, *67*, 8043–8050.

Zeng, H., Briske-Anderson, M., Idso, J. P., & Hunt, C. D. (2006). The selenium metabolite methylselenol inhibits the migration and invasion potential of HT1080 tumor cells. *The Journal of Nutrition*, *136*, 1528–1532.

Zhang, W., Zheng, X., & Wang, X. (2015). Oxidative stress measured by thioredoxin reductase level as potential biomarker for prostate cancer. *American Journal of Cancer Research*, *5*, 2788–2798.

Zhao, H., Li, J., Li, X., Han, C., Zhang, Y., Zheng, L., et al. (2015). Silencing GPX3 expression promotes tumor metastasis in human thyroid cancer. *Current Protein & Peptide Science*, *16*, 316–321.

CHAPTER FOUR

Selenium-Dependent Glutathione Peroxidases During Tumor Development

Anna P. Kipp[1]

Institute of Nutrition, Friedrich Schiller University Jena, Jena, Germany

[1]Corresponding author: e-mail address: anna.kipp@uni-jena.de

Contents

Abstract

Five out of eight human glutathione peroxidases (GPxes) are selenoproteins and thus their expression depends on the selenium (Se) supply. Most Se-dependent GPxes are downregulated in tumor cells, while only GPx2 is considerably upregulated. Whether expression profiles of GPxes predict tumor development and patient survival is controversially discussed. Also, results from in vitro and in vivo studies modulating the expression of GPx isoforms provide evidence for both anti- and procarcinogenic mechanisms. GPxes are able to reduce hydroperoxides, which otherwise would damage DNA, possibly resulting in DNA mutations, modulate redox-sensitive signaling pathways affecting proliferation, differentiation, and cellular metabolism or initiate cell death. Considering these different processes, the role and functions of individual Se-dependent GPx isoforms will be discussed herein in the context of tumorigenesis.

Advances in Cancer Research, Volume 136
ISSN 0065-230X
http://dx.doi.org/10.1016/bs.acr.2017.07.004

ABBREVIATIONS

ACSL4 acyl-CoA synthetase long-chain family member 4
AOM azoxymethane
CCS colon cancer subtypes
DCF dichlorodihydrofluorescein
EGFR epidermal growth factor receptor
GSH glutathione
GPx glutathione peroxidase
HSPA5 heat shock 70 kDa protein 5
HCC hepatocellular carcinoma
iPSCs induced pluripotent stem cells
LSCs leukemia stem cells
LOX lipoxygenase
MAPK MAP kinase
MMP matrix metalloproteinase
hucMSC-Ex mesenchymal stem cells
NOX NADPH-dependent oxidases
Prdx peroxiredoxin
PI3K phosphoinositide 3-kinase
RTKs receptor tyrosine kinases
Se selenium
SOD2 superoxide dismutase 2

1. INTRODUCTION

The family of glutathione peroxidases (GPxes) comprises eight members (GPx1–8) in mammals, five of which are selenoproteins in humans (GPx1–4 and 6). The other three isoforms contain a cysteine instead of a selenocysteine (Sec) moiety. This chapter will focus on the selenium-dependent GPxes only. GPxes reduce hydroperoxides to their corresponding alcohols (water in case of hydrogen peroxide (H_2O_2)) mainly using glutathione (GSH) as reductant (Ursini et al., 1995). A common feature of all of them is a conserved tetrade with Sec as one of the catalytically active amino acids. Furthermore, the tetrade consists of glutamine, tryptophan, and an asparagine (Tosatto et al., 2008). In GPxes, the Sec residue is present as selenol, which reacts as a selenolate with H_2O_2 or other hydroperoxides to selenenic acid. The reduction back to selenol in turn uses two molecules of GSH forming glutathione disulfide (GSSG) and water (for more details on mechanistical aspects, see Brigelius-Flohé & Maiorino, 2013). GPx1, 2, 3, and 6 are homotetramers, while GPx4 acts as a monomer. These structural prerequisites appear to affect substrate specificity: while the smaller GPx4 is able to reduce complex fatty acid, phospholipid, and cholesterol hydroperoxides, GPx1, 2, and 6 are supposed to react with less complex, soluble but also small fatty acid hydroperoxides. However, systematic studies on substrate specificity have only been carried out on GPx1 so far and further studies are required for GPx2 and 6. The extracellular isoform GPx3 has an intermediate substrate spectrum ranging from small synthetic hydroperoxides to phospholipid hydroperoxides (overview in Brigelius-Flohé & Maiorino, 2013).

Besides their substrate specificity, GPx isoforms differ in their localization both intracellularly but also with respect to different cell types and tissues. GPx1 is ubiquitously expressed. Its mRNA is detectable in almost all tissues with highest levels in liver, kidney, and erythrocytes. Within cells, GPx1 is localized in the cytosol as well as in mitochondria (overview in Lubos, Loscalzo, & Handy, 2011). In contrast, the expression pattern of GPx2 is restricted to epithelial cells, in particular of the complete gastrointestinal tract. In addition, GPx2 is also localized in the epithelium lining the lung, bladder, gall bladder, and breast (Kipp & Müller, 2015). Until now, GPx2 has only been detected in the cytosol and not within cellular organelles. *GPx3* transcripts are expressed in almost all cells with highest levels in kidney, liver, and the thyroid. GPx3 is the only extracellular isoform of this family and thus GPx3 protein is detectable in the plasma but also in other extracellular body fluids such as chamber water of the eye, thyroid colloid lumen, or amniotic fluid (reviewed in Brigelius-Flohé & Maiorino, 2013). Together with selenoprotein P, GPx3 accounts for more than 97% of all plasma selenium in mice. However, its function as selenium transport protein has not been clearly deciphered yet (Olson et al., 2010). Recently, it has been shown that GPx3 is secreted by bronchial epithelial cells and is bound to the extracellular matrix (Schamberger et al., 2016). In addition, circulating GPx3 binds to basement membranes of epithelial cells such as those of the gastrointestinal tract (Burk, Olson, Winfrey, Hill, & Yin, 2011), cortex tubule cells of the kidney (Malinouski et al., 2012; Olson et al., 2010), and the epididymis (Burk et al., 2011). GPx4 exists as three different isoforms which are synthetized by the use of individual promoters. These isoforms are located in the cytosol, mitochondria, nucleus, or at the plasma membrane (recently reviewed in Cardoso, Hare, Bush, & Roberts, 2017). cGPx4 is not only located in the cytosol but also in membrane compartments in a broad spectrum of cell types but predominantly in neurons. Mitochondrial (mGPx4) and sperm-nuclear (snGPx4) GPx4 are almost exclusively expressed in testes. So far, GPx6 has only been described to be localized in the olfactory epithelium (Dear, Campbell, & Rabbitts, 1991).

As they are selenoproteins, the translation of the mentioned GPxes essentially depends on the availability of selenium. Based on the so-called concept of "selenoprotein hierarchy" GPx isoforms are differentially sensitive to a limited selenium supply. GPx4 and GPx2 are both retained for a relatively long period of time when selenium becomes limiting, while GPx1 mRNA and protein levels decline very fast. GPx3 comprises an intermediate position within the selenoprotein hierarchy, while this has not been studied for GPx6, yet (Brigelius-Flohé & Maiorino, 2013). Until now, different mechanisms have been discussed for this phenomenon, which modulate, e.g., selenoprotein mRNA stability, translation efficiency, and incorporation of Sec in the growing peptide chain. For a recent review on selenoprotein synthesis and hierarchy, see Bulteau and Chavatte (2015) and Labunskyy, Hatfield, and Gladyshev (2014). Because of their inherent Se sensitivity, GPx1 activity in erythrocytes and GPx3 activity in plasma are widely used as biomarkers for the Se status (Ashton et al., 2009). Vice versa, expression levels of these two GPxes might not always be consistent between different studies because they highly depend on the Se status of the individual person or experimental setup.

Previously, several overviews on the role of GPxes in health and disease (Brigelius-Flohé & Maiorino, 2013; Lubos, Loscalzo, et al., 2011) and during carcinogenesis (Brigelius-Flohé & Kipp, 2009) have been published, which provide the basis of the current review. Beyond that, this review focusses on recent findings on the role of Se-dependent GPxes in redox signaling and cellular metabolism in the context of tumor development.

2. EXPRESSION PROFILE OF SE-DEPENDENT GPxes IN DIFFERENT TYPES OF TUMORS

Recently published data from prospective cohort studies, such as the EPIC study, show that it appears to be favorable to have an adequate Se status because this reduces the risk to develop different types of tumors such as colorectal and hepatobiliary cancer (Hughes et al., 2016, 2015). The proposed mechanisms range from direct effects of Se and specific Se species to only selenoprotein-mediated effects. The prevailing idea is that physiological doses of Se itself and/or as integral part of (several) selenoproteins modulate the cellular redox status and thus prevent oxidative DNA damage, which would otherwise have a high probability to result in DNA mutations and tumor initiation. In addition, the cellular redox status substantially modulates signaling pathways involved in proliferation, differentiation, and further processes of tumorigenesis (as discussed recently Brigelius-Flohé & Flohé, 2017). Lower levels of plasma GPx3 as biomarker for the Se status significantly correlated with a larger tumor size and a higher recurrence rate of hepatocellular carcinoma (HCC) (Qi et al., 2014). Also in lung cancer patients, plasma GPx3 levels were lower than in healthy controls (Oh et al., 2014). However, Se intervention studies conducted so far provided inconsistent results indicating that not all cancer patients profit from increasing the Se status (Rayman, 2012). Many open questions remain with regard to the right dosage and form of Se, the responsive tumor type and tumor stage. Besides the effects of Se on limiting tumor initiation, established tumor cells may benefit from an enhanced Se supply and thus upregulation of selenoproteins as they rely on antioxidant systems more than normal cells (Trachootham, Alexandre, & Huang, 2009). Thus, Table 1 provides an overview of the expression profiles of Se-dependent GPxes in different types of tumors including cancer cell lines on both mRNA and protein levels. In general, GPx2 is upregulated in most tumors in comparison to healthy tissue, while the four other Se-dependent GPxes are downregulated. For example, in breast invasive ductal carcinoma, GPx4 levels decreased gradually with tumor grade (Cejas et al., 2007). Also, GPx2 upregulation appears to depend on the tumor stage. Colorectal tumors have been recently categorized into three colon cancer subtypes (CCS). CCS1 and CCS2 comprise chromosomal-instable and microsatellite-instable cancers, while CCS3 tumors are poorly differentiated, have a high recurrence rate, and express genes involved in matrix remodeling and epithelial–mesenchymal transition (De Sousa et al., 2013). *GPx2* mRNA expression was the highest in CCS1 in comparison to CCS2 and CCS3 (Emmink et al., 2014). A low tumor-resident *GPx2* expression was significantly correlated with a longer relapse-free survival time of colon cancer patients with the aggressive subtype CCS3 (Emmink et al., 2014). Also in castration-resistant prostate cancer, low or even undetectable GPx2 levels in biopsies predicted a better survival probability than high levels (Naiki et al., 2014). In contrast, loss of GPx2 expression predicted a poor prognosis in patients with esophageal squamous cell carcinoma (Lei, Tian, Zhang, Zhao, & Su, 2016).

Table 1 GPx Profiles in a Selection of Different Types of Tumors

Tumor Type	GPx Profile	References
Esophagus cancer	GPx2 is upregulated, GPx3 is downregulated	Mörk et al. (2003)
Gastric cancer	*GPx1* and *GPx4* mRNA is decreased in gastric tumor and *para*-carcinoma tissues in the gastric cancer group compared to the control group	Lan et al. (2016)
	GPx3 is downregulated	Zhang et al. (2010)
Colorectal cancer	GPx2 expression is upregulated. GPx3 is downregulated	Murawaki et al. (2008)
	GPx1 mRNA is downregulated	Nalkiran et al. (2015)
	GPx1 and GPx4 were upregulated in carcinoma tissues at both protein and mRNA levels	Yagublu et al. (2011)
Liver cancer	In TGFα/c-myc transgenic mice GPx1 and GPx4 protein was downregulated in tumor compared to normal tissue	Gladyshev, Factor, Housseau, and Hatfield (1998)
	GPx2 protein expression is upregulated	Suzuki et al. (2013)
	GPx3 protein is downregulated	Qi et al. (2014)
	GPx6 mRNA is downregulated upon c-myc induction in liver stem cells	Aravalli, Talbot, and Steer (2015)
Renal cell carcinoma	mRNA expression of *GPx1*, *GPx3*, and *GPx4* genes was decreased in comparison to healthy individuals	Rudenko et al. (2015)
	GPx activity was reduced in tumor compared to noncancerous adjacent tissues	Durak et al. (1997)
Lung cancer	GPx2 expression is upregulated	Wönckhaus et al. (2006)
Prostate cancer	In prostate cancer cells derived from matched normal and malignant tissues, GPx1 protein expression was reduced	Gladyshev et al. (1998)
	GPx3 is downregulated	Yu et al. (2007)
	GPx2 expression is upregulated	Naiki et al. (2014)
Breast cancer	In estrogen-negative breast cancer cell lines GPx1 levels were downregulated	Esworthy, Baker, and Chu (1995)
	GPx2 expression is upregulated	Naiki-Ito et al. (2007)

The downregulation of *GPx3* mRNA in HCC in comparison to the surrounding tissue significantly correlated with advanced tumor stage and poor prognosis of HCC patients (Qi et al., 2014). The same inter-dependency has been described for other tumors as well (overview in Brigelius-Flohé & Maiorino, 2013). In addition, loss of tumor-resident GPx1 expression was associated with a higher aggressiveness and poor survival of patients with gastric cancer (Min et al., 2012). Several genetic polymorphisms have been described for *GPx1*. The resulting proteins differ in their responsiveness to Se and in their cellular distribution, thus also affecting their role in cancer (Brigelius-Flohé & Kipp, 2009). Individual polymorphisms most frequently associated with cancer have been described to correlate with the risk to develop certain types of cancer and with the survival rate.

3. HYDROPEROXIDES AND REDOX SIGNALING DURING CANCER DEVELOPMENT

Cancer is a multifactorial disease driven by acquired mutations that escape the cellular repair machinery. Both the original and the recently extended concept of "The Hallmarks of Cancer" are based on these mutations and their modulatory effects on signal transduction pathways resulting in unrestricted proliferation and invasion of tumor cells (Hanahan & Weinberg, 2000, 2011). Redox signaling processes have recently been discussed to play essential roles in every hallmark of cancer (Hornsveld & Dansen, 2016). The term redox signaling comprises a form of signal transduction that proceeds through the reversible oxidation of cysteine residues in proteins by mainly using H_2O_2 as a second messenger (Fig. 1; overview in Brigelius-Flohé & Flohé, 2011; Holmström & Finkel, 2014). In that sense, H_2O_2 is not merely a damaging byproduct of cellular metabolism anymore, but is specifically produced as signaling molecule, e.g., by NADPH-dependent oxidases (NOX). NOX enzymes rather synthetize superoxide to the extracellular space where it becomes dismutated by superoxide dismutatse 3 (SOD3) to H_2O_2. H_2O_2 is shuttled back into the cell by aquaporins. Besides that, the other relevant source of H_2O_2 in terms of signaling is the mitochondrial respiratory chain, which directly correlates with the metabolic activity of cells. Next to H_2O_2, especially products of 12,15-lipoxygenase (12,15-LOX) have been shown to be relevant modulators of redox signaling resulting, e.g., in efficient inactivation of protein

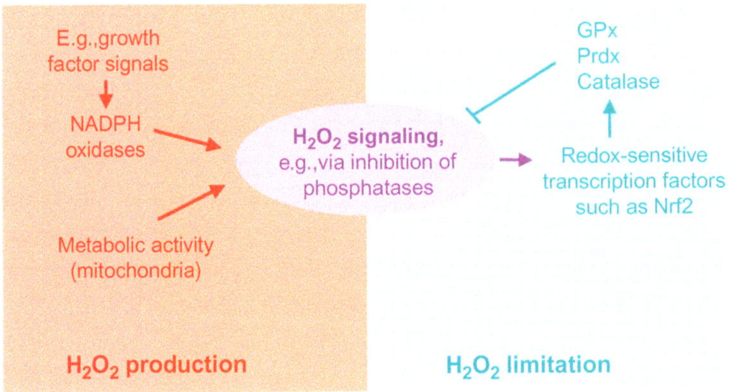

Fig. 1 H_2O_2 as signaling molecule in cancer cells. For details, see description in Section 3.

tyrosine phosphatases (Conrad et al., 2010). An unbalanced LOX activity might, therefore, lead to dysregulation of many signaling cascades. Because of the fact that GPxes are able to reduce these second messengers and thus limit the signal transduction, they are important modulators of redox signaling processes. However, besides Se-dependent GPxes other hydroperoxide-reducing enzymes are present in cells such as Cys-dependent GPxes, peroxiredoxins (PRDXes), and catalase. In most cell types, only peroxisomes contain relevant amounts of catalase taking care for the locally produced H_2O_2. However, different compartmentalization is not obvious for GPxes and PRDXes, which both colocalize in the cytosol and in many organelles. Nevertheless, it is extensively discussed that H_2O_2-reducing enzymes need to be closely connected to the H_2O_2 source, e.g., at membranes in case of NOX enzymes, to efficiently impact on H_2O_2 signaling. Thus, small differences in localization can have a big impact on signal transduction.

PRDXes are a family of six isoforms, which are classified based on having either one (1-Cys) or two (2-Cys) conserved cysteine residues that form inter- or intramolecular disulfide bounds upon reduction of hydroperoxides (Rhee, Kang, Chang, Jeong, & Kim, 2001). Notably, both GPxes and PRDXes ultimately depend on NADPH to maintain their catalytic activity, making NADPH regeneration a key feature of redox regulation. Enzymatic characterizations have indicated that PRDXes can efficiently compete with GPxes, especially because of their high expression levels in most cell types (discussed in Winterbourn, 2013). Under which conditions these individual thiol-peroxidases come into play has not been fully elucidated, yet, but will be of relevance to understand how each single enzyme affects cellular signaling of healthy and cancer cells.

The main cancer-relevant, redox-sensitive pathways modulate growth factor signals. Upon binding of extracellular growth factors to receptor tyrosine kinases (RTKs), NOX enzymes become activated and increase the local H_2O_2 concentration. Thus, protein tyrosine phosphatases are inhibited, which otherwise would dephosphorylate and inhibit RTKs. The transfer of the RTK signal to downstream MAP kinase (MAPK) cascades can only proceed, if an H_2O_2 signal prevails for a certain window of time. In consequence, modulation of H_2O_2 levels, e.g., by GPxes, is essential for the fine-tuning and limitation of growth factor signaling. As explained earlier, cellular metabolism is central to redox signaling, as oxidative phosphorylation in the mitochondria is a sources of H_2O_2. In turn, H_2O_2 itself is able to limit mitochondrial respiration by inhibiting glycolysis through oxidative inhibition of glycolytic enzymes such as glyceraldehyde-3-phosphate dehydrogenase (GAPDH) and the cancer cell-specific pyruvate kinase isozyme M2 (PKM2) (Anastasiou et al., 2011; Peralta et al., 2015). Besides that, transcription factors or their binding proteins are most often modified by oxidation (overview in Brigelius-Flohé & Flohé, 2011), which include transcription factors such as Nrf2, NF-kB, P53, FOXOs, Wnt, and many more. This way, the cell is able to adapt to changes in the redox balance by upregulating antioxidant enzymes including GPxes. Under conditions of high levels of H_2O_2 or lipid peroxidation, induction of cell death takes place, which is also triggered by redox signals (overview in Galadari, Rahman, Pallichankandy, & Thayyullathil, 2017; Gaschler & Stockwell, 2017).

4. IMPACT OF INDIVIDUAL GPxes ON TUMOR DEVELOPMENT

4.1 GPx1

GPx1 was identified in 1973 as the first Se-dependent enzyme (Flohé, Günzler, & Schock, 1973; Rotruck et al., 1973). Based on its hydroperoxide-reducing capacity and its broad abundance it has been widely considered a major antioxidant enzyme. Mainly because of the high dependency of GPx1 activity on the Se status, antioxidant characteristics were also ascribed to Se itself. Surprisingly, *GPx1* knockout mice display no obvious phenotype under unstressed conditions (de Haan et al., 1998; Ho et al., 1997). However, when these mice are exposed to redox cyclers or bacterial endotoxins to trigger H_2O_2 production they are severely impaired (Cheng et al., 1998; Lei & Cheng, 2005). This indicates that GPx1 is dispensable under basal conditions, but is required to protect the cell under stressed conditions. Thus, it appeared to be obvious that GPx1 is able to protect cells from oxidative damage, resulting DNA mutations and thus tumor cell initiation. In vitro cell culture models revealed that a knockdown of *GPx1* increased UV-induced DNA damage and micronuclei formation in response to UV irradiation (Baliga et al., 2008), while overexpression had the opposite effect (Baliga, Wang, Zhuo, Schwartz, & Diamond, 2007). In line with this, *GPx1* overexpression in Ras-positive pancreatic cancer cells reduced cell growth and tumor size when injected into nude mice (Liu et al., 2004). However, overexpression of superoxide dismutase 2 (SOD2) and thus higher H_2O_2 levels inhibited proliferation and tumorigenic properties of human glioma cells and these characteristics could be reversed by additionally overexpressing *GPx1* (Li, Yan, Yang, Oberley, & Oberley, 2000). Also in a chemically induced skin tumor model, mice overexpressing *GPx1* developed more tumors than WT controls (Lu et al., 1997). Vice versa, cell proliferation was reduced in *GPx1* knockdown breast cancer cells together with increased H_2O_2 levels (Jin et al., 2015). This indicates that GPx1 cannot only limit but obviously also enhance tumor growth under specific circumstances, which appears to involve the specific amount and source of H_2O_2. As described earlier, physiological H_2O_2 levels are not necessarily harmful for cells but are essentially involved in modulating signaling pathways out of which several have been shown to depend on GPx1 levels.

4.1.1 Cell Death

Overexpression of *GPx1* in ECV304 cells reduced the amount of apoptotic cells by lowering the ratio of proapoptotic Bax to antiapoptotic Bcl-2 (Faucher et al., 2005). Also in transgenic mice overexpressing *GPx1*, renal toxicity of cocaine treatment was attenuated via the induction of antiapoptotic factors such as Bcl-2 and Bcl-xl and concomitant decrease in Bax and cleaved-caspase 3 (Mai et al., 2016). In line with that, *GPx1* knockout mice had an increased number of TUNEL-positive neuronal cells in response to cold-induced injury (Flentjar et al., 2002). After treatment with exogenous H_2O_2, more cell death was detected in primary neurons isolated from GPx1 knockout mice together with activation of the phosphoinositide 3-kinase (PI3K)-AKT survival pathway (see Section 4.1.2) (Taylor, Ali, Iannello, Hertzog, & Crack, 2005). In ATDC5 chondrocyte cells, a *GPx1* knockdown impaired chondrogenic differentiation and induced cell apoptosis (Yan, Guo et al., 2017). Exosomes, which contain functional mRNA, microRNA, and proteins, are secreted by mesenchymal stem cells (hucMSC-Ex) to enhance tissue repair, e.g., in liver. Those hucMSC-Ex also contain GPx1, which is involved in modulating antiapoptotic abilities of hucMSC-Ex and thus supports their hepatoprotective effects (Yan, Jiang et al., 2017). GPx1 activity is regulated by posttranslational modifications such as phosphorylation catalyzed by the nonreceptor tyrosine kinases c-Abl and Arg (Cao, Leng, Huang, Liu, & Kufe, 2003). These two kinases are essential for protecting cells from apoptosis during oxidative stress. Thus, independent of the model system studied GPx1 appears to block apoptosis induction consistently. In the context of cancer, tumor cells with high GPx1 expression might evade from apoptosis induction, increase their survival rate, and thus increase tumor growth.

4.1.2 Redox Signaling

TNFα is a pleiotropic cytokine that induces either cell proliferation or cell death by differentially modulating pathways such as ERK1/2, JNK, and NF-kB involving H_2O_2 as signaling molecule (Kamata et al., 2005). Loss of GPx1 increases the activity of these pathways in response to TNFα (Lubos, Kelly, et al., 2011), while excess GPx1 is able to suppress the activation of proinflammatory signaling responses (Garg & Aggarwal, 2002; Kretz-Remy, Mehlen, Mirault, & Arrigo, 1996) by modulating intracellular H_2O_2 levels. In *GPx1* knockdown cells, DUSP4, a dual specificity phosphatase which is known to be an antagonist of MAPK signaling, was significantly downregulated and discussed to act downstream of GPx1 (Lubos, Kelly, et al., 2011). Following cerebral ischemia–reperfusion injury, a more pronounced NF-kB activation was detectable in neurons of *GPx1* knockout mice contributing to an increased neuronal cell death (Crack, Taylor, Ali, Mansell, & Hertzog, 2006). However, under inflammatory conditions such as LPS treatment, protein levels of proinflammatory mediators and the number of macrophages were lower in broncho-alveolar lavage fluid of *GPx1* knockout than in WT mice (Bozinovski, Seow, Crack, Anderson, & Vlahos, 2012). In the hypothalamus, inhibition of GPx1 expression was associated with enhanced JAK2/STAT3 signaling and nuclear translocation of p65 (Tran et al., 2016).

Cells overexpressing *GPx1* have diminished epidermal growth factor receptor (EGFR)-mediated AKT activation in response to H_2O_2 or EGF while a *GPx1* knockdown had the opposite effect (Handy et al., 2009). Enhanced H_2O_2 levels, PTEN oxidation, and PI3K/AKT signaling were observed in muscle (Loh et al., 2009), liver, and pancreatic β cells (Merry et al., 2014), respectively, of whole-body *GPx1* knockout mice. Recently, *GPx1* has been also conditionally deleted in hepatocytes to study its in vivo effects on AKT signaling locally in those cells. AKT signaling was enhanced in Alb-Cre;*Gpx1*$^{lox/lox}$ mice which was modulated by increased oxidation and thus inactivation of protein tyrosine phosphatases such as TCPTP, PTP1B, and SHP1 (Merry et al., 2016). The activation status of other signaling pathways, e.g., those mediated by STAT3, ERK1/2, and JNK were not altered in this model system. MAPK signaling pathways, represented by the three main pathways ERK, JNK, and p38, are essentially involved in balancing proliferation and differentiation processes and are often deregulated in cancer cells (Dhillon, Hagan, Rath, & Kolch, 2007; Zhang & Liu, 2002). Recently, it has been shown that cancer-associated fibroblasts have an impaired TGFβ signaling which mediates their loss of GPx1 expression and activity. Consequently, a more oxidized tumor microenvironment was established which enhanced cancer cell growth by activating AKT-ERK1/2-linked signaling (Chan et al., 2017).

4.1.3 Metabolism

In the murine heart, GPx1 protects mitochondria against reoxygenation damage caused by H_2O_2 (Thu et al., 2010). Besides that, GPx1 was shown to decreased EGFR signaling by decreasing mitochondria-derived oxidants. In parallel, *GPx1* overexpressing cells are characterized by an overall mitochondrial dysfunction with a decrease in mitochondrial potential and a decrease in ATP production (Handy et al., 2009). To assess whether the distribution of GPx1 between the cytoplasm and mitochondria had different biological consequences, GPx1 proteins were either targeted to the mitochondria or located in the cytosol. Interestingly only the mitochondrial GPx1 was able to increase the oxygen consumption rate in relation to glycolysis (Bera et al., 2014). Vice versa, *SOD2* overexpression shifted cellular adaptation toward glycolytic metabolism (Hart et al., 2015). Thus, H_2O_2 generated from the dismutation of O_2^- by SOD2 appears to be more efficiently reduced by mitochondrially located GPx1 than by the cytosolic form. H_2O_2 released from mitochondria is an important signal that activates AMPK to enhance glycolysis and thus support cancer cell survival. Interestingly, also hearts of *ErbB2*^{tg} mice display increased GPx1 protein and activity in isolated mitochondria (Belmonte et al., 2015). This appears to be modulated by two nonreceptor tyrosine kinases c-Abl and Arg, which previously have been shown to activate GPx1 by posttranslational modification (Cao et al., 2003). Also in the rat heart, the Abl kinase inhibitor imatinib mesylate decreased GPx activity (Saad, Alkharfy, & Arafah, 2006). Hyperglycemia increases H_2O_2 levels primarily through a higher mitochondrial oxidative phosphorylation. Accordingly, hyperglycemia upregulated GPx1 activity, which was modulated by cellular *O*-GlcNAcylation (Yang et al., 2010). Furthermore, a recent publication shows that glutamate dehydrogenase 1 contributes to redox homeostasis by increasing intracellular fumarate levels, which bind to and activate GPx1 in both breast and lung cancer cell lines (Jin et al., 2015). Thus, the metabolic intermediate fumarate functions as a signaling molecule providing a direct link between metabolism and redox signaling.

4.2 GPx2

As described earlier, GPx2 is not ubiquitously detectable but rather specifically localizes to epithelial cells of different tissues (for recent reviews, see Brigelius-Flohé & Kipp, 2016; Kipp & Müller, 2015). GPx2 is part of a stem cell signature of human-induced pluripotent stem cells (iPSCs) (Dannenmann et al., 2015) and highly expressed in mouse embryonic stem cells (Blaschke et al., 2013). In the gastrointestinal tract, GPx2 is localized at the crypt base, where stem cells and transit amplifying cells reside. Thus, the location of GPx2 appears to be connected to stem cell-like cells. Those cells are pluripotent, give rise to all differentiated epithelial cells and thus need an efficient protection to avoid the mutation of their DNA. Similar to GPx1, GPx2 was proposed to reduce oxidative DNA damage by reducing hydroperoxides. Indeed, iPSCs appear to ensure their genomic integrity by upregulation of GPx2 expression (Dannenmann et al., 2015). In contrast to the other GPxes, GPx2 is mostly upregulated in tumor tissue indicating that GPx2 expression is colocalized with high proliferation.

In line with *GPx1* knockout mice, also *GPx2* knockout mice develop normally (Esworthy, Mann, Sam, & Chu, 2000). However, they have some impairments in the gastrointestinal epithelium including increased numbers of apoptotic cells at the crypt base of the intestinal epithelium (Florian et al., 2010), longer crypts with an enlarged proliferation zone, and a low-grade inflammation with higher numbers of intraepithelial leukocytes (Müller et al., 2013). Interestingly, these morphological changes prevail even though an upregulation of GPx1 and thus total GPx activity takes place (Florian et al., 2010). As these changes in GPx1 expression were not detectable on mRNA level, it is discussed that loss of GPx2 results in enhanced translation of GPx1 mediated by a redistribution of Se. Concomitantly, a double knockout of *GPx1* and *GPx2* (*GPx1*/*GPx2*-DKO) has stronger effects on intestinal integrity resulting in the spontaneous development of an ileocolitis (Esworthy et al., 2001) and colon cancer (Chu, Esworthy, Chu, et al., 2004). To identify the source of H_2O_2 responsible for the observed effects, *GPx1*/*GPx2*-DKO mice were crossbred with knockout mice for either *Nox1* or *Duox2* both being highly expressed in the intestinal epithelium. Almost all characteristics of DKO mice could be reversed by loss of Nox1 (Esworthy et al., 2014), while loss of Duox2 rescued only the inflammatory phenotype but not the higher number of apoptotic cells (Chu et al., 2016). Both ileocolitis and colon cancer could be also efficiently rescued by bringing back one allele of *GPx2* but not *GPx1* (Esworthy, Yang, Frankel, & Chu, 2005). Altogether, these results indicate that intestinal GPx2 rather than GPx1 is important for the reduction of mainly Nox1-derived H_2O_2 and thus protection from intestinal inflammation and colorectal cancer. Single *GPx2* knockout mice indeed develop more squamous cell carcinoma in response to UV radiation with concomitant increase in H_2O_2 levels (Walshe et al., 2007). H_2O_2 levels are also increased during inflammation. Thus, *GPx2* knockout mice develop a more severe inflammation using a mouse model for allergic airway inflammation (Dittrich et al., 2010) or a chemically induced colitis model, which later on resulted in a higher number of inflammation-driven colonic adenomas (Krehl et al., 2012). Much to our surprise, *GPx2* knockout mice, however, developed fewer adenomas and preneoplastic lesions in response to the carcinogen azoxymethane (AOM) alone (Müller et al., 2013). The underlying mechanisms might involve the high number of basal and AOM-induced apoptotic cells in *GPx2* knockout mice, which might have contributed to the efficient elimination of initiated cells (see Section 4.2.1). In line with this finding, different *GPx2* knockdown cell lines have been characterized by a lower proliferation rate and a lower ability to form tumors in xenograft models, while cell lines overexpressing *GPx2* show the opposite effect (Banning, Kipp, et al., 2008; Emmink et al., 2014; Naiki et al., 2014). Thus, GPx2 appears to have anticarcinogenic properties in the context of inflammation-driven carcinogenesis while it can also contribute to the survival of already established tumor cells.

4.2.1 Cell Death

Upon overexpression, GPx2 decreased the sensitivity of breast and colorectal cancer cells to H_2O_2-induced cell death, while the opposite was observed with *GPx2* knockdown cells (Emmink et al., 2014; Yan & Chen, 2006). In both rat and human mammary carcinoma cell lines as well as in hepatocellular and breast carcinoma cells, a *GPx2* knockdown caused the inhibition of proliferation and enhanced apoptosis but only in cells with wild type p53 (Naiki-Ito et al., 2007; Suzuki et al., 2013). Human HT-29 colorectal cancer cells have an activating mutation in the p53 gene (Rodrigues et al., 1990). Thus, a *GPx2* knockdown did not affect proliferation and apoptosis rates in those cells (Banning, Kipp, et al., 2008). However, when HT-29 cells were injected into the spleen of immune incompetent mice to form metastasis in the liver, GPx2 expression supported the survival of floating cells resulting in more metastatic nodules. Obviously, *GPx2* knockdown cells are more prone to anoikis, a type of apoptosis induced by loss of contact to the basement membrane and associated with high H_2O_2 levels (Emmink et al., 2014). Together with the high apoptosis rate found at the crypt base of the intestinal epithelium this points to a specific function of GPx2 in inhibiting apoptotic processes, which could not be compensated by GPx1 at least in the intestinal epithelium. It appears that the GPx2 effect on apoptosis involves p53, but until now, mechanistic details are still missing.

4.2.2 Redox Signaling

Prostate cancer cell lines with *GPx2* knockdown have a higher dichlorodihydrofluorescein (DCF) fluorescence indicating higher H_2O_2 levels than GPx2 competent control cells (Naiki et al., 2014). When patient-derived colorectal cancer cells were FACS-sorted based on their CM-H_2DCFDA fluorescence, GPx2 expression was far lower in those cells with high H_2O_2 levels (Emmink et al., 2014). This indicates that GPx2 has a relevant impact on regulating cellular H_2O_2 levels. Under conditions of intestinal inflammation, GPx2 is generally upregulated both in mice and humans (Banning, Florian, et al., 2008; Hiller, Besselt, Deubel, Brigelius-Flohé, & Kipp, 2015). During acute chemically induced colitis, GPx2 expression is detectable in whole crypts while in the chronic phase GPx2 specifically resides next to ulcerations (Hiller et al., 2015). Also under these circumstances GPx2 is supposed to modulate cellular H_2O_2 levels and thus redox-sensitive pathways, which would contribute to, e.g., wound healing processes including proliferation (recently reviewed in Pérez, Talens-Visconti, Rius-Perez, Finamor, & Sastre, 2017). These functions could be inferred from observations in HT-29 cells with *GPx2* knockdown. Those cells show enhanced COX2 expression both under basal and IL-1 stimulated conditions, as well as COX2-mediated migration and invasion (Banning,

Florian, et al., 2008; Banning, Kipp, et al., 2008). Preliminary data show that GPx2 modulates NF-kB activity as has been described for GPx1 as well. Besides that, GPx2 is supposed to modulate the activity of redox-sensitive matrix metalloproteinase-7 (MMP-7), which is required to cleave and activate microbicidal defensins secreted by Paneth cells (Chu, Esworthy, & Doroshow, 2004). In hepatocarcinoma cells, GPx2 modified the secretion of MMP-9 (Suzuki et al., 2013).

Besides inflammation, GPx2 has recently been shown to also affect differentiation processes. Together with p63, FOXM1 strongly contributes to the maintenance of a high proliferative potential in keratinocytes. Depletion of FOXM1 was sufficient to induce keratinocyte senescence which was associated with a strong suppression of GPx2 expression (Smirnov et al., 2016). Also differentiation of embryonic stem cells (Saretzki et al., 2008) and Caco2 cells (Speckmann et al., 2011) led to a reduced mRNA expression of *GPx2*. Growing patient-derived colorectal cancer cells as 3D spheroids revealed that a *GPx2* knockdown favors the development of stem-like cancer cells that lack multilineage differentiation capacity. However, overexpression of *GPx2* caused an increase in the expression of markers associated with various differentiation lineages present in the colon, including MUC2 (goblet cells), FABP1 (enterocytes), and chromogranin A (enteroendocrine cells) (Emmink et al., 2014). The underlying pathways have not been identified yet, but could potentially involve redox-sensitive Notch and Wnt signaling, which are very important for the maintenance of the intestinal epithelium but also during cancer development (Pérez et al., 2017).

4.3 GPx3

GPx3 is mainly synthesized in kidney, secreted into the plasma, and bound to basement membranes of several tissues. It is likely that other tissues besides kidney contribute to plasma GPx3 as nephrectomized rats retain 30% of their plasma GPx activity (Avissar et al., 1994). In most tumors, *GPx3* expression is downregulated, which is mainly mediated via epigenetic regulation (see also Chapter "Selenium and epigenetics in cancer–Focus on DNA methylation" by Ewa Jabłońska and Edyta Reszka). As described for *GPx1* and *GPx2* knockout mice, also *GPx3* knockout mice are viable and have no obvious phenotype (Olson et al., 2010). In prostate cancer cell lines, *GPx3* overexpression suppressed colony formation and tumor size after injection of these cells into a xenograft model (Yu et al., 2007). Also in TRAMP mice, a model for prostate tumorigenesis, GPx3 inhibits cancer growth (Chang, Lee, Oh, & Park, 2016). *GPx3* overexpression suppressed migration and invasion of HCC cells. In addition, those cells show less intrahepatic tumor spread in orthotopic nude mice and less lung metastasis of HCC cells injected through the tail vein of nude mice (Qi et al., 2016). Reduced GPx3 expression was associated with increased tumor size and lymph node metastasis in human thyroid cancer (Zhao et al., 2015). GPx3 has been furthermore described to have a tumor suppressor function in a model of colitis-associated colorectal cancer (Barrett et al., 2013) similar to GPx2. Tumors with *GPx3* knockout showed increased DNA damage, more infiltration of M2 macrophages, and more proliferation. In different colorectal cancer samples and cell lines, reduced GPx3 expression was associated with increased oxaliplatin and cisplatin sensitivity involving high H_2O_2 levels (Pelosof et al., 2016). However, a *GPx3* knockdown in Caco2 cells led to decreased colony formation, indicating that under specific circumstances GPx3 could also support tumor growth as also described for GPx2 (Barrett et al., 2013). Specifically in leukemia stem cells (LSCs), the GPx3 activity was increased resulting in decreased H_2O_2 levels. However, proliferation and differentiation of progenitor LSCs were not affected by *GPx3* shRNA (Herault et al., 2012). Altogether, the tumor suppressor function of GPx3 appears to prevail, but it needs to be clarified under which circumstances GPx3 could also have protumorigenic effects.

4.3.1 Cell Death

In response to H_2O_2 treatment, a knockdown of *GPx3* in Caco2 cells led to increased apoptosis most probably due to increased H_2O_2-induced DNA damage (Barrett et al., 2013). Similar results were obtained in human myoblasts with *GPx3* knockdown, showing enhanced basal rates of cell death (El Haddad et al., 2012). In prostate cancer, however, GPx3-induced cell death by stimulation of p53-induced gene 3 (PIG3), a NADPH-dependent reductase, which is essential for p53-mediated apoptosis. Overexpressed *GPx3* directly interacted with PIG3 independent of its peroxidase activity. Within this intracellular complex, GPx3 rather produced H_2O_2, which resulted in the activation of caspases (Wang et al., 2012). This mechanism is obviously mostly relevant for UV-mediated cell death.

4.3.2 Redox Signaling

Tumor suppressing effects of GPx3 correlated with a downregulation of c-Met, a procarcinogenic tyrosine kinase receptor for the hepatic growth factor in epithelial cells (Yu et al., 2007). However, the underlying mechanisms are not resolved yet but might involve the oxidative modification of protein phosphatases (as discussed in Brigelius-Flohé & Maiorino, 2013). In skeletal muscle cells, *GPx3* overexpression decreased extracellular H_2O_2 concentrations and concomitantly inhibited activation of JNK, p38, and ERK1/2 by glucose oxidase (Chung et al., 2009). Among the three MMP subtypes (MMP-2, MMP-7, and MMP-9) expressed in liver cancer, only MMP-2 was suppressed upon *GPx3* overexpression, which was accompanied by less nuclear translocation of JNK upstream of MMP-2. Thus, the tumor suppressive function of GPx3 may be mediated through inhibition of the JNK-MMP-2 signaling pathway (Qi et al., 2016). In addition, GPx3 inhibited Wnt signaling in human thyroid cancer cell lines (TPC-1 and FTC133 cells) (Zhao et al., 2015). In line with this, *GPx3* knockout intestinal tumors had increased levels of intratumoral nuclear β-catenin (Barrett et al., 2013).

4.4 GPx4

GPx4 had initially been characterized as lipid peroxidation inhibiting protein (Ursini, Maiorino, Valente, Ferri, & Gregolin, 1982) due to its unique ability to reduce complex lipid hydroperoxides in free and membrane-bound forms (Thomas, Geiger, Maiorino, Ursini, & Girotti, 1990). Accordingly, a *GPx4* knockdown in NIH3T3 cells increased lipid peroxidation in membranes (Yoo et al., 2010). GPx4 exists in three different isoforms, a cytosolic (cGPx4) and ubiquitously expressed form and a mitochondrial (mGPx4) and sperm nuclear GPx4 (snGPx4) form both mainly restricted to testis (for further details on the latter two isoforms, see Brigelius-Flohé & Maiorino, 2013). A systemic knockout of the entire *GPx4* gene (Yant et al., 2003) as well as a knockout of cytosolic *GPx4* (Liang et al., 2009) are embryonically lethal. In 2008, the first mouse carrying a conditional *GPx4* allele was generated (Seiler et al., 2008). Studies using this model organism revealed that GPx4 is essential for maintaining tissue homeostasis in several organs including brain, skin, and endothelium (Seiler et al., 2008; Wortmann et al., 2013). GPx4 function has extensively been studied in healthy brain and in the context of neurodegenerative diseases (overview in Cardoso, Hare, Bush, & Roberts, 2017). Induction of lipid peroxidation by treatment with docosahexaenoic acid killed cancer cells with *GPx4* knockdown more efficiently than control cells (Ding & Lind, 2007), while *GPx4* overexpression protected dermal fibroblasts from UV-induced damage (Wenk et al., 2004). Accordingly, *GPx4* overexpression reduced growth properties in different tumor models (Heirman et al., 2006; Liu et al., 2006).

4.4.1 Cell Death

Upon GPx4 deletion, mouse embryonic fibroblasts have been initially described to undergo a caspase-independent form of apoptotic cell death, which is modulated by lipid peroxidation resulting from 12/15-LOX that is otherwise inhibited by GPx4 (Seiler et al., 2008). Lipid peroxidation than results in translocation of apoptosis inducing factor from the mitochondria to the nucleus, leading to large-scale DNA fragmentation and cell death. Another recently described form of cell death, called ferroptosis, is characterized by the iron-dependent inhibition of the xc^- transport system. This transporter is responsible for cysteine import and its inhibition limits GSH biosynthesis within cells, which efficiently inhibits GPx4. The resulting increase in lipid peroxidation including cardiolipin oxidation triggers ferroptosis induction (overview in Yang & Stockwell, 2016; Yu, Guo, Xie, Wang, & Chen, 2017). Interestingly, p53 is able to repress *Slc7A11*, the gene encoding for the key component of xc^-. Most tumor cells establish a mechanism to evade from ferroptosis by inactivating mutations of p53 and high xc^- levels (Jiang et al., 2015). Testing 177 cancer cell lines revealed that diffuse large B cell lymphomas and renal cell carcinomas are particularly susceptible to GPx4-regulated ferroptosis and thus cancer cell death (Yang et al., 2014). In mice with inducible deletion of *GPx4* massive renal tubule cell death and acute kidney failure became obvious within 2 weeks, indicating that the kidney is most sensitive to ferroptotic cell death (Friedmann Angeli et al., 2014). Besides its role in kidney, GPx4 is critical for hepatocyte survival and proper liver function. Interestingly, vitamin E can compensate for the loss of GPx4 by protecting cells against deleterious lipid peroxidation (Carlson et al., 2016). T cells with inducible *GPx4* knockout rapidly accumulated membrane lipid peroxides and underwent cell death driven by ferroptosis (Matsushita et al., 2015). Ferroptosis inducers inhibit GPx4 by covalently targeting the active site selenocysteine (Yang et al., 2016). In human pancreatic ductal adenocarcinoma cells, the heat shock 70 kDa protein 5 (HSPA5) binds and inactivates GPx4 (Zhu et al., 2017). Recently, acyl-CoA synthetase long-chain family member 4 (ACSL4) was identified as an essential component for ferroptosis execution. ACSL4 enriches cellular membranes with long polyunsaturated ω6 fatty acids. In a subset of basal-like breast cancer cell lines, ACSL4 predicts sensitivity to ferroptosis (Doll et al., 2017). In mouse erythroid precursors, GPx4 prevents necroptosis, another form of cell death with concomitant increase in lipid peroxidation, resulting in the inactivation of caspase 8 via glutathionylation and subsequent formation of the RIP1/RIP3-containing necrosome (Canli et al., 2016).

4.4.2 Redox Signaling

Microarray and subsequent pathway analysis of Caco2 cells with *GPx4* knockdown identified oxidative phosphorylation, ubiquinone biosynthesis, and mitochondrial dysfunction as affected processes thus mainly influencing mitochondrial function and integrity (Cole-Ezea, Swan, Shanley, & Hesketh, 2012). In human dermal fibroblasts, overexpression of *GPx4* resulted in the suppression of phosphatidylcholine hydroperoxide-induced NF-κB and limited induction of MMP-1 (Wenk et al., 2004). Inhibitory effects of GPx4 on NF-kB activity have been described in additional cell types such as endothelial cells (Brigelius-Flohé, Friedrichs, Maurer, Schultz, & Streicher, 1997). Suppression of COX and LOX enzymatic activity clearly is an antiinflammatory function also of GPx4 and might occur downstream of NF-kB inhibition, but could also involve a direct modulation of enzyme activity by reducing lipid peroxides needed to activate COX/LOX enzymes (Heirman et al., 2006). In addition, a *GPx4* knockout reduced T cell survival, which can be prevented by inhibiting MEK1/2 and subsequent ERK1/2 activation, indicating that deregulated ERK activation is caused by enhanced lipid peroxidation (Matsushita et al., 2015). Recently, it has been shown that hepatitis C virus (HCV) induces oxidative stress, but controls it tightly by inducing antioxidant enzymes, especially GPx4. Thus, GPx4 plays an essential role in the HCV life cycle, supports its survival, and finally increases the risk of hepatitis-induced heptocarcinoma (Brault et al., 2016).

5. CONCLUSIONS

Overall, Se-dependent GPxes are relevant enzymes for limiting hydroperoxide signals, which affect redox signaling pathways. Thus, the Se supply and subsequently GPx expression profiles essentially modulate processes involved in the carcinogenic process. While prevention of oxidative DNA damage clearly has anticarcinogenic effects, modulation of redox signaling differs depending on the specific circumstances, such as tumor stage or inflammatory conditions. Inhibition of cell death might be a procarcinogenic feature, because cancer cells profit from this otherwise protective effect. Further studies are clearly needed to decipher in more detail, which

GPx isoform reduces which type of hydroperoxide under which specific condition and how their expression in the next step could be modulated to efficiently combat carcinogenesis.

ACKNOWLEDGMENT

The support of Dr. Solveigh Koeberle in proofreading the manuscript is highly acknowledged.

REFERENCES

Anastasiou, D., Poulogiannis, G., Asara, J. M., Boxer, M. B., Jiang, J. K., Shen, M., et al. (2011). Inhibition of pyruvate kinase M2 by reactive oxygen species contributes to cellular antioxidant responses. *Science*, *334*, 1278–1283.

Aravalli, R. N., Talbot, N. C., & Steer, C. J. (2015). Gene expression profiling of MYC-driven tumor signatures in porcine liver stem cells by transcriptome sequencing. *World Journal of Gastroenterology*, *21*, 2011–2029.

Ashton, K., Hooper, L., Harvey, L. J., Hurst, R., Casgrain, A., & Fairweather-Tait, S. J. (2009). Methods of assessment of selenium status in humans: A systematic review. *The American Journal of Clinical Nutrition*, *89*, 2025S–2039S.

Avissar, N., Ornt, D. B., Yagil, Y., Horowitz, S., Watkins, R. H., Kerl, E. A., et al. (1994). Human kidney proximal tubules are the main source of plasma glutathione peroxidase. *The American Journal of Physiology*, *266*, C367–C375.

Baliga, M. S., Diwadkar-Navsariwala, V., Koh, T., Fayad, R., Fantuzzi, G., & Diamond, A. M. (2008). Selenoprotein deficiency enhances radiation-induced micronuclei formation. *Molecular Nutrition & Food Research*, *52*, 1300–1304.

Baliga, M. S., Wang, H., Zhuo, P., Schwartz, J. L., & Diamond, A. M. (2007). Selenium and GPx-1 overexpression protect mammalian cells against UV-induced DNA damage. *Biological Trace Element Research*, *115*, 227–242.

Banning, A., Florian, S., Deubel, S., Thalmann, S., Müller-Schmehl, K., Jacobasch, G., et al. (2008). GPx2 counteracts PGE2 production by dampening COX-2 and mPGES-1 expression in human colon cancer cells. *Antioxidants & Redox Signaling*, *10*, 1491–1500.

Banning, A., Kipp, A., Schmitmeier, S., Löwinger, M., Florian, S., Krehl, S., et al. (2008). Glutathione peroxidase 2 inhibits cyclooxygenase-2-mediated migration and invasion of HT-29 adenocarcinoma cells but supports their growth as tumors in nude mice. *Cancer Research*, *68*, 9746–9753.

Barrett, C. W., Ning, W., Chen, X., Smith, J. J., Washington, M. K., Hill, K. E., et al. (2013). Tumor suppressor function of the plasma glutathione peroxidase gpx3 in colitis-associated carcinoma. *Cancer Research*, *73*, 1245–1255.

Belmonte, F., Das, S., Sysa-Shah, P., Sivakumaran, V., Stanley, B., Guo, X., et al. (2015). ErbB2 overexpression upregulates antioxidant enzymes, reduces basal levels of reactive oxygen species, and protects against doxorubicin cardiotoxicity. *American Journal of Physiology Heart and Circulatory Physiology*, *309*, H1271–H1280.

Bera, S., Weinberg, F., Ekoue, D. N., Ansenberger-Fricano, K., Mao, M., Bonini, M. G., et al. (2014). Natural allelic variations in glutathione peroxidase-1 affect its subcellular localization and function. *Cancer Research*, *74*, 5118–5126.

Blaschke, K., Ebata, K. T., Karimi, M. M., Zepeda-Martinez, J. A., Goyal, P., Mahapatra, S., et al. (2013). Vitamin C induces Tet-dependent DNA demethylation and a blastocyst-like state in ES cells. *Nature*, *500*, 222–226.

Bozinovski, S., Seow, H. J., Crack, P. J., Anderson, G. P., & Vlahos, R. (2012). Glutathione peroxidase-1 primes pro-inflammatory cytokine production after LPS challenge in vivo. *PloS One*, *7*, e33172.
Brault, C., Levy, P., Duponchel, S., Michelet, M., Salle, A., Pecheur, E. I., et al. (2016). Glutathione peroxidase 4 is reversibly induced by HCV to control lipid peroxidation and to increase virion infectivity. *Gut*, *65*, 144–154.
Brigelius-Flohé, R., & Flohé, L. (2011). Basic principles and emerging concepts in the redox control of transcription factors. *Antioxidants & Redox Signaling*, *15*, 2335–2381.
Brigelius-Flohé, R., & Flohé, L. (2017). Selenium and redox signaling. *Archives of Biochemistry and Biophysics*, *617*, 48–59.
Brigelius-Flohé, R., Friedrichs, B., Maurer, S., Schultz, M., & Streicher, R. (1997). Interleukin-1-induced nuclear factor kappa B activation is inhibited by overexpression of phospholipid hydroperoxide glutathione peroxidase in a human endothelial cell line. *The Biochemical Journal*, *328*(Pt. 1), 199–203.
Brigelius-Flohé, R., & Kipp, A. (2009). Glutathione peroxidases in different stages of carcinogenesis. *Biochimica et Biophysica Acta*, *1790*, 1555–1568.
Brigelius-Flohé, R., & Kipp, A. P. (2016). Glutathione peroxidase 2, a selenoprotein exhibiting a dual personality in preventing and promoting cancer. In D. L. Hatfield, M. J. Berry, & V. N. Gladyshev (Eds.), *Selenium. Its molecular biology and role in human health* (pp. 451–462). New York: Springer Science and Business Media.
Brigelius-Flohé, R., & Maiorino, M. (2013). Glutathione peroxidases. *Biochimica et Biophysica Acta*, *1830*, 3289–3303.
Bulteau, A. L., & Chavatte, L. (2015). Update on selenoprotein biosynthesis. *Antioxidants & Redox Signaling*, *23*, 775–794.
Burk, R. F., Olson, G. E., Winfrey, V. P., Hill, K. E., & Yin, D. (2011). Glutathione peroxidase-3 produced by the kidney binds to a population of basement membranes in the gastrointestinal tract and in other tissues. *American Journal of Physiology Gastrointestinal and Liver Physiology*, *301*, G32–G38.
Canli, O., Alankus, Y. B., Grootjans, S., Vegi, N., Hultner, L., Hoppe, P. S., et al. (2016). Glutathione peroxidase 4 prevents necroptosis in mouse erythroid precursors. *Blood*, *127*, 139–148.
Cao, C., Leng, Y., Huang, W., Liu, X., & Kufe, D. (2003). Glutathione peroxidase 1 is regulated by the c-Abl and Arg tyrosine kinases. *The Journal of Biological Chemistry*, *278*, 39609–39614.
Cardoso, B. R., Hare, D. J., Bush, A. I., & Roberts, B. R. (2017). Glutathione peroxidase 4: A new player in neurodegeneration? *Molecular Psychiatry*, *22*, 328–335.
Carlson, B. A., Tobe, R., Yefremova, E., Tsuji, P. A., Hoffmann, V. J., Schweizer, U., et al. (2016). Glutathione peroxidase 4 and vitamin E cooperatively prevent hepatocellular degeneration. *Redox Biology*, *9*, 22–31.
Cejas, P., Garcia-Cabezas, M. A., Casado, E., Belda-Iniesta, C., De Castro, J., Fresno, J. A., et al. (2007). Phospholipid hydroperoxide glutathione peroxidase (PHGPx) expression is downregulated in poorly differentiated breast invasive ductal carcinoma. *Free Radical Research*, *41*, 681–687.
Chan, J. S., Tan, M. J., Sng, M. K., Teo, Z., Phua, T., Choo, C. C., et al. (2017). Cancer-associated fibroblasts enact field cancerization by promoting extratumoral oxidative stress. *Cell Death & Disease*, *8*, e2562.
Chang, S. N., Lee, J. M., Oh, H., & Park, J. H. (2016). Glutathione peroxidase 3 inhibits prostate tumorigenesis in TRAMP mice. *Prostate*, *76*, 1387–1398.
Cheng, W. H., Ho, Y. S., Valentine, B. A., Ross, D. A., Combs, G. F., Jr., & Lei, X. G. (1998). Cellular glutathione peroxidase is the mediator of body selenium to protect against paraquat lethality in transgenic mice. *The Journal of Nutrition*, *128*, 1070–1076.

Chu, F. F., Esworthy, R. S., Chu, P. G., Longmate, J. A., Huycke, M. M., Wilczynski, S., et al. (2004). Bacteria-induced intestinal cancer in mice with disrupted Gpx1 and Gpx2 genes. *Cancer Research, 64,* 962–968.

Chu, F. F., Esworthy, R. S., & Doroshow, J. H. (2004). Role of Se-dependent glutathione peroxidases in gastrointestinal inflammation and cancer. *Free Radical Biology & Medicine, 36,* 1481–1495.

Chu, F. F., Esworthy, R. S., Doroshow, J. H., Grasberger, H., Donko, A., Leto, T. L., et al. (2016). Deficiency in Duox2 activity alleviates ileitis in GPx1- and GPx2-knockout mice without affecting apoptosis incidence in the crypt epithelium. *Redox Biology, 11,* 144–156.

Chung, S. S., Kim, M., Youn, B. S., Lee, N. S., Park, J. W., Lee, I. K., et al. (2009). Glutathione peroxidase 3 mediates the antioxidant effect of peroxisome proliferator-activated receptor gamma in human skeletal muscle cells. *Molecular and Cellular Biology, 29,* 20–30.

Cole-Ezea, P., Swan, D., Shanley, D., & Hesketh, J. (2012). Glutathione peroxidase 4 has a major role in protecting mitochondria from oxidative damage and maintaining oxidative phosphorylation complexes in gut epithelial cells. *Free Radical Biology & Medicine, 53,* 488–497.

Conrad, M., Sandin, A., Forster, H., Seiler, A., Frijhoff, J., Dagnell, M., et al. (2010). 12/15-lipoxygenase-derived lipid peroxides control receptor tyrosine kinase signaling through oxidation of protein tyrosine phosphatases. *Proceedings of the National Academy of Sciences of the United States of America, 107,* 15774–15779.

Crack, P. J., Taylor, J. M., Ali, U., Mansell, A., & Hertzog, P. J. (2006). Potential contribution of NF-kappaB in neuronal cell death in the glutathione peroxidase-1 knockout mouse in response to ischemia-reperfusion injury. *Stroke, 37,* 1533–1538.

Dannenmann, B., Lehle, S., Hildebrand, D. G., Kubler, A., Grondona, P., Schmid, V., et al. (2015). High glutathione and glutathione peroxidase-2 levels mediate cell-type-specific DNA damage protection in human induced pluripotent stem cells. *Stem Cell Reports, 4,* 886–898.

de Haan, J. B., Bladier, C., Griffiths, P., Kelner, M., O'Shea, R. D., Cheung, N. S., et al. (1998). Mice with a homozygous null mutation for the most abundant glutathione peroxidase, Gpx1, show increased susceptibility to the oxidative stress-inducing agents paraquat and hydrogen peroxide. *The Journal of Biological Chemistry, 273,* 22528–22536.

De Sousa, E. M. F., Wang, X., Jansen, M., Fessler, E., Trinh, A., de Rooij, L. P., et al. (2013). Poor-prognosis colon cancer is defined by a molecularly distinct subtype and develops from serrated precursor lesions. *Nature Medicine, 19,* 614–618.

Dear, T. N., Campbell, K., & Rabbitts, T. H. (1991). Molecular cloning of putative odorant-binding and odorant-metabolizing proteins. *Biochemistry, 30,* 10376–10382.

Dhillon, A. S., Hagan, S., Rath, O., & Kolch, W. (2007). MAP kinase signalling pathways in cancer. *Oncogene, 26,* 3279–3290.

Ding, W. Q., & Lind, S. E. (2007). Phospholipid hydroperoxide glutathione peroxidase plays a role in protecting cancer cells from docosahexaenoic acid-induced cytotoxicity. *Molecular Cancer Therapeutics, 6,* 1467–1474.

Dittrich, A. M., Meyer, H. A., Krokowski, M., Quarcoo, D., Ahrens, B., Kube, S. M., et al. (2010). Glutathione peroxidase-2 protects from allergen-induced airway inflammation in mice. *The European Respiratory Journal, 35,* 1148–1154.

Doll, S., Proneth, B., Tyurina, Y. Y., Panzilius, E., Kobayashi, S., Ingold, I., et al. (2017). ACSL4 dictates ferroptosis sensitivity by shaping cellular lipid composition. *Nature Chemical Biology, 13,* 91–98.

Durak, I., Beduk, Y., Kavutcu, M., Ozturk, S., Canbolat, O., & Ulutepe, S. (1997). Activities of superoxide dismutase and glutathione peroxidase enzymes in cancerous and noncancerous human kidney tissues. *International Urology and Nephrology, 29,* 5–11.

El Haddad, M., Jean, E., Turki, A., Hugon, G., Vernus, B., Bonnieu, A., et al. (2012). Glutathione peroxidase 3, a new retinoid target gene, is crucial for human skeletal muscle precursor cell survival. *Journal of Cell Science, 125*, 6147–6156.

Emmink, B. L., Laoukili, J., Kipp, A. P., Koster, J., Govaert, K. M., Fatrai, S., et al. (2014). GPx2 suppression of H2O2 stress links the formation of differentiated tumor mass to metastatic capacity in colorectal cancer. *Cancer Research, 74*, 6717–6730.

Esworthy, R. S., Aranda, R., Martin, M. G., Doroshow, J. H., Binder, S. W., & Chu, F. F. (2001). Mice with combined disruption of Gpx1 and Gpx2 genes have colitis. *American Journal of Physiology. Gastrointestinal and Liver Physiology, 281*, G848–G855.

Esworthy, R. S., Baker, M. A., & Chu, F. F. (1995). Expression of selenium-dependent glutathione peroxidase in human breast tumor cell lines. *Cancer Research, 55*, 957–962.

Esworthy, R. S., Kim, B. W., Chow, J., Shen, B., Doroshow, J. H., & Chu, F. F. (2014). Nox1 causes ileocolitis in mice deficient in glutathione peroxidase-1 and -2. *Free Radical Biology & Medicine, 68*, 315–325.

Esworthy, R. S., Mann, J. R., Sam, M., & Chu, F. F. (2000). Low glutathione peroxidase activity in Gpx1 knockout mice protects jejunum crypts from gamma-irradiation damage. *American Journal of Physiology. Gastrointestinal and Liver Physiology, 279*, G426–G436.

Esworthy, R. S., Yang, L., Frankel, P. H., & Chu, F. F. (2005). Epithelium-specific glutathione peroxidase, Gpx2, is involved in the prevention of intestinal inflammation in selenium-deficient mice. *The Journal of Nutrition, 135*, 740–745.

Faucher, K., Rabinovitch-Chable, H., Cook-Moreau, J., Barriere, G., Sturtz, F., & Rigaud, M. (2005). Overexpression of human GPX1 modifies Bax to Bcl-2 apoptotic ratio in human endothelial cells. *Molecular and Cellular Biochemistry, 277*, 81–87.

Flentjar, N. J., Crack, P. J., Boyd, R., Malin, M., de Haan, J. B., Hertzog, P., et al. (2002). Mice lacking glutathione peroxidase-1 activity show increased TUNEL staining and an accelerated inflammatory response in brain following a cold-induced injury. *Experimental Neurology, 177*, 9–20.

Flohé, L., Günzler, W. A., & Schock, H. H. (1973). Glutathione peroxidase: A selenoenzyme. *FEBS Letters, 32*, 132–134.

Florian, S., Krehl, S., Loewinger, M., Kipp, A., Banning, A., Esworthy, S., et al. (2010). Loss of GPx2 increases apoptosis, mitosis, and GPx1 expression in the intestine of mice. *Free Radical Biology & Medicine, 49*, 1694–1702.

Friedmann Angeli, J. P., Schneider, M., Proneth, B., Tyurina, Y. Y., Tyurin, V. A., Hammond, V. J., et al. (2014). Inactivation of the ferroptosis regulator Gpx4 triggers acute renal failure in mice. *Nature Cell Biology, 16*, 1180–1191.

Galadari, S., Rahman, A., Pallichankandy, S., & Thayyullathil, F. (2017). Reactive oxygen species and cancer paradox: To promote or to suppress? *Free Radical Biology & Medicine, 104*, 144–164.

Garg, A. K., & Aggarwal, B. B. (2002). Reactive oxygen intermediates in TNF signaling. *Molecular Immunology, 39*, 509–517.

Gaschler, M. M., & Stockwell, B. R. (2017). Lipid peroxidation in cell death. *Biochemical and Biophysical Research Communications, 482*, 419–425.

Gladyshev, V. N., Factor, V. M., Housseau, F., & Hatfield, D. L. (1998). Contrasting patterns of regulation of the antioxidant selenoproteins, thioredoxin reductase, and glutathione peroxidase, in cancer cells. *Biochemical and Biophysical Research Communications, 251*, 488–493.

Hanahan, D., & Weinberg, R. A. (2000). The hallmarks of cancer. *Cell, 100*, 57–70.

Hanahan, D., & Weinberg, R. A. (2011). Hallmarks of cancer: The next generation. *Cell, 144*, 646–674.

Handy, D. E., Lubos, E., Yang, Y., Galbraith, J. D., Kelly, N., Zhang, Y. Y., et al. (2009). Glutathione peroxidase-1 regulates mitochondrial function to modulate redox-dependent cellular responses. *The Journal of Biological Chemistry, 284*, 11913–11921.

Hart, P. C., Mao, M., de Abreu, A. L., Ansenberger-Fricano, K., Ekoue, D. N., Ganini, D., et al. (2015). MnSOD upregulation sustains the Warburg effect via mitochondrial ROS and AMPK-dependent signalling in cancer. *Nature Communications*, *6*, 6053.

Heirman, I., Ginneberge, D., Brigelius-Flohé, R., Hendrickx, N., Agostinis, P., Brouckaert, P., et al. (2006). Blocking tumor cell eicosanoid synthesis by GPx 4 impedes tumor growth and malignancy. *Free Radical Biology & Medicine*, *40*, 285–294.

Herault, O., Hope, K. J., Deneault, E., Mayotte, N., Chagraoui, J., Wilhelm, B. T., et al. (2012). A role for GPx3 in activity of normal and leukemia stem cells. *The Journal of Experimental Medicine*, *209*, 895–901.

Hiller, F., Besselt, K., Deubel, S., Brigelius-Flohé, R., & Kipp, A. P. (2015). GPx2 induction is mediated through STAT transcription factors during acute colitis. *Inflammatory Bowel Diseases*, *21*, 2078–2089.

Ho, Y. S., Magnenat, J. L., Bronson, R. T., Cao, J., Gargano, M., Sugawara, M., et al. (1997). Mice deficient in cellular glutathione peroxidase develop normally and show no increased sensitivity to hyperoxia. *The Journal of Biological Chemistry*, *272*, 16644–16651.

Holmström, K. M., & Finkel, T. (2014). Cellular mechanisms and physiological consequences of redox-dependent signalling. *Nature Reviews. Molecular Cell Biology*, *15*, 411–421.

Hornsveld, M., & Dansen, T. B. (2016). The hallmarks of cancer from a redox perspective. *Antioxidants & Redox Signaling*, *25*, 300–325.

Hughes, D. J., Duarte-Salles, T., Hybsier, S., Trichopoulou, A., Stepien, M., Aleksandrova, K., et al. (2016). Prediagnostic selenium status and hepatobiliary cancer risk in the European prospective investigation into cancer and nutrition cohort. *The American Journal of Clinical Nutrition*, *104*, 406–414.

Hughes, D. J., Fedirko, V., Jenab, M., Schomburg, L., Meplan, C., Freisling, H., et al. (2015). Selenium status is associated with colorectal cancer risk in the European prospective investigation of cancer and nutrition cohort. *International Journal of Cancer*, *136*, 1149–1161.

Jiang, L., Kon, N., Li, T., Wang, S. J., Su, T., Hibshoosh, H., et al. (2015). Ferroptosis as a p53-mediated activity during tumour suppression. *Nature*, *520*, 57–62.

Jin, L., Li, D., Alesi, G. N., Fan, J., Kang, H. B., Lu, Z., et al. (2015). Glutamate dehydrogenase 1 signals through antioxidant glutathione peroxidase 1 to regulate redox homeostasis and tumor growth. *Cancer Cell*, *27*, 257–270.

Kamata, H., Honda, S., Maeda, S., Chang, L., Hirata, H., & Karin, M. (2005). Reactive oxygen species promote TNFalpha-induced death and sustained JNK activation by inhibiting MAP kinase phosphatases. *Cell*, *120*, 649–661.

Kipp, A. P., & Müller, M. F. (2015). Dual functions of selenoproteins in cancer—Glutathione peroxidase 2. In R. Brigelius-Flohé & H. Sies (Eds.), *Diversity of selenium functions in health and disease* (pp. 189–203): Abingdon, USA: CRC Press, Taylor & Francis Group.

Krehl, S., Loewinger, M., Florian, S., Kipp, A., Banning, A., Wessjohann, L., et al. (2012). Glutathione peroxidase-2 and selenium decreased inflammation and tumors in a mouse model of inflammation-associated carcinogenesis whereas sulforaphane effects differed with selenium supply. *Carcinogenesis*, *33*, 620–628.

Kretz-Remy, C., Mehlen, P., Mirault, M. E., & Arrigo, A. P. (1996). Inhibition of I kappa B-alpha phosphorylation and degradation and subsequent NF-kappa B activation by glutathione peroxidase overexpression. *The Journal of Cell Biology*, *133*, 1083–1093.

Labunskyy, V. M., Hatfield, D. L., & Gladyshev, V. N. (2014). Selenoproteins: Molecular pathways and physiological roles. *Physiological Reviews*, *94*, 739–777.

Lan, X., Xing, J., Gao, H., Li, S., Quan, L., Jiang, Y., et al. (2017). Decreased expression of selenoproteins as a poor prognosticator of gastric cancer in humans. *Biological Trace Element Research*, *178*, 22–28.

Lei, X. G., & Cheng, W. H. (2005). New roles for an old selenoenzyme: Evidence from glutathione peroxidase-1 null and overexpressing mice. *The Journal of Nutrition*, *135*, 2295–2298.

Lei, Z., Tian, D., Zhang, C., Zhao, S., & Su, M. (2016). Clinicopathological and prognostic significance of GPX2 protein expression in esophageal squamous cell carcinoma. *BMC Cancer*, *16*, 410.

Li, S., Yan, T., Yang, J. Q., Oberley, T. D., & Oberley, L. W. (2000). The role of cellular glutathione peroxidase redox regulation in the suppression of tumor cell growth by manganese superoxide dismutase. *Cancer Research*, *60*, 3927–3939.

Liang, H., Yoo, S. E., Na, R., Walter, C. A., Richardson, A., & Ran, Q. (2009). Short form glutathione peroxidase 4 is the essential isoform required for survival and somatic mitochondrial functions. *The Journal of Biological Chemistry*, *284*, 30836–30844.

Liu, J., Du, J., Zhang, Y., Sun, W., Smith, B. J., Oberley, L. W., et al. (2006). Suppression of the malignant phenotype in pancreatic cancer by overexpression of phospholipid hydroperoxide glutathione peroxidase. *Human Gene Therapy*, *17*, 105–116.

Liu, J., Hinkhouse, M. M., Sun, W., Weydert, C. J., Ritchie, J. M., Oberley, L. W., et al. (2004). Redox regulation of pancreatic cancer cell growth: Role of glutathione peroxidase in the suppression of the malignant phenotype. *Human Gene Therapy*, *15*, 239–250.

Loh, K., Deng, H., Fukushima, A., Cai, X., Boivin, B., Galic, S., et al. (2009). Reactive oxygen species enhance insulin sensitivity. *Cell Metabolism*, *10*, 260–272.

Lu, Y. P., Lou, Y. R., Yen, P., Newmark, H. L., Mirochnitchenko, O. I., Inouye, M., et al. (1997). Enhanced skin carcinogenesis in transgenic mice with high expression of glutathione peroxidase or both glutathione peroxidase and superoxide dismutase. *Cancer Research*, *57*, 1468–1474.

Lubos, E., Kelly, N. J., Oldebeken, S. R., Leopold, J. A., Zhang, Y. Y., Loscalzo, J., et al. (2011). Glutathione peroxidase-1 deficiency augments proinflammatory cytokine-induced redox signaling and human endothelial cell activation. *The Journal of Biological Chemistry*, *286*, 35407–35417.

Lubos, E., Loscalzo, J., & Handy, D. E. (2011). Glutathione peroxidase-1 in health and disease: From molecular mechanisms to therapeutic opportunities. *Antioxidants & Redox Signaling*, *15*, 1957–1997.

Mai, H. N., Jeong, J. H., Kim, D. J., Chung, Y. H., Shin, E. J., Nguyen, L. T., et al. (2016). Genetic overexpressing of GPx-1 attenuates cocaine-induced renal toxicity via induction of anti-apoptotic factors. *Clinical and Experimental Pharmacology & Physiology*, *43*, 428–437.

Malinouski, M., Kehr, S., Finney, L., Vogt, S., Carlson, B. A., Seravalli, J., et al. (2012). High-resolution imaging of selenium in kidneys: A localized selenium pool associated with glutathione peroxidase 3. *Antioxidants & Redox Signaling*, *16*, 185–192.

Matsushita, M., Freigang, S., Schneider, C., Conrad, M., Bornkamm, G. W., & Kopf, M. (2015). T cell lipid peroxidation induces ferroptosis and prevents immunity to infection. *The Journal of Experimental Medicine*, *212*, 555–568.

Merry, T. L., Tran, M., Dodd, G. T., Mangiafico, S. P., Wiede, F., Kaur, S., et al. (2016). Hepatocyte glutathione peroxidase-1 deficiency improves hepatic glucose metabolism and decreases steatohepatitis in mice. *Diabetologia*, *59*, 2632–2644.

Merry, T. L., Tran, M., Stathopoulos, M., Wiede, F., Fam, B. C., Dodd, G. T., et al. (2014). High-fat-fed obese glutathione peroxidase 1-deficient mice exhibit defective insulin secretion but protection from hepatic steatosis and liver damage. *Antioxidants & Redox Signaling*, *20*, 2114–2129.

Min, S. Y., Kim, H. S., Jung, E. J., Jung, E. J., Jee, C. D., & Kim, W. H. (2012). Prognostic significance of glutathione peroxidase 1 (GPX1) down-regulation and correlation with aberrant promoter methylation in human gastric cancer. *Anticancer Research*, *32*, 3169–3175.

Mörk, H., Scheurlen, M., Al-Taie, O., Zierer, A., Kraus, M., Schottker, K., et al. (2003). Glutathione peroxidase isoforms as part of the local antioxidative defense system in normal and Barrett's esophagus. *International Journal of Cancer*, *105*, 300–304.

Müller, M. F., Florian, S., Pommer, S., Osterhoff, M., Esworthy, R. S., Chu, F. F., et al. (2013). Deletion of glutathione peroxidase-2 inhibits azoxymethane-induced colon cancer development. *PloS One*, *8*, e72055.

Murawaki, Y., Tsuchiya, H., Kanbe, T., Harada, K., Yashima, K., Nozaka, K., et al. (2008). Aberrant expression of selenoproteins in the progression of colorectal cancer. *Cancer Letters*, *259*, 218–230.

Naiki, T., Naiki-Ito, A., Asamoto, M., Kawai, N., Tozawa, K., Etani, T., et al. (2014). GPX2 overexpression is involved in cell proliferation and prognosis of castration-resistant prostate cancer. *Carcinogenesis*, *35*, 1962–1967.

Naiki-Ito, A., Asamoto, M., Hokaiwado, N., Takahashi, S., Yamashita, H., Tsuda, H., et al. (2007). Gpx2 is an overexpressed gene in rat breast cancers induced by three different chemical carcinogens. *Cancer Research*, *67*, 11353–11358.

Nalkiran, I., Turan, S., Arikan, S., Kahraman, O. T., Acar, L., Yaylim, I., et al. (2015). Determination of gene expression and serum levels of MnSOD and GPX1 in colorectal cancer. *Anticancer Research*, *35*, 255–259.

Oh, I. J., Kim, H. E., Song, S. Y., Na, K. J., Kim, K. S., Kim, Y. C., et al. (2014). Diagnostic value of serum glutathione peroxidase 3 levels in patients with lung cancer. *Thoracic Cancer*, *5*, 425–430.

Olson, G. E., Whitin, J. C., Hill, K. E., Winfrey, V. P., Motley, A. K., Austin, L. M., et al. (2010). Extracellular glutathione peroxidase (Gpx3) binds specifically to basement membranes of mouse renal cortex tubule cells. *American Journal of Physiology. Renal Physiology*, *298*, F1244–F1253.

Pelosof, L., Yerram, S., Armstrong, T., Chu, N., Danilova, L., Yanagisawa, B., et al. (2016). GPX3 promoter methylation predicts platinum sensitivity in colorectal cancer. *Epigenetics*, 1–11. http://dx.doi.org/10.1080/15592294.2016.1265711.

Peralta, D., Bronowska, A. K., Morgan, B., Doka, E., Van Laer, K., Nagy, P., et al. (2015). A proton relay enhances H2O2 sensitivity of GAPDH to facilitate metabolic adaptation. *Nature Chemical Biology*, *11*, 156–163.

Pérez, S., Talens-Visconti, R., Rius-Perez, S., Finamor, I., & Sastre, J. (2017). Redox signaling in the gastrointestinal tract. *Free Radical Biology & Medicine*, *104*, 75–103.

Qi, X., Ng, K. T., Lian, Q. Z., Liu, X. B., Li, C. X., Geng, W., et al. (2014). Clinical significance and therapeutic value of glutathione peroxidase 3 (GPx3) in hepatocellular carcinoma. *Oncotarget*, *5*, 11103–11120.

Qi, X., Ng, K. T., Shao, Y., Li, C. X., Geng, W., Ling, C. C., et al. (2016). The clinical significance and potential therapeutic role of GPx3 in tumor recurrence after liver transplantation. *Theranostics*, *6*, 1934–1946.

Rayman, M. P. (2012). Selenium and human health. *Lancet*, *379*, 1256–1268.

Rhee, S. G., Kang, S. W., Chang, T. S., Jeong, W., & Kim, K. (2001). Peroxiredoxin, a novel family of peroxidases. *IUBMB Life*, *52*, 35–41.

Rodrigues, N. R., Rowan, A., Smith, M. E., Kerr, I. B., Bodmer, W. F., Gannon, J. V., et al. (1990). p53 mutations in colorectal cancer. *Proceedings of the National Academy of Sciences of the United States of America*, *87*, 7555–7559.

Rotruck, J. T., Pope, A. L., Ganther, H. E., Swanson, A. B., Hafeman, D. G., & Hoekstra, W. G. (1973). Selenium: Biochemical role as a component of glutathione peroxidase. *Science*, *179*, 588–590.

Rudenko, E., Kondratov, O., Gerashchenko, G., Lapska, Y., Kravchenko, S., Koliada, O., et al. (2015). Aberrant expression of selenium-containing glutathione peroxidases in clear cell renal cell carcinomas. *Experimental Oncology*, *37*, 105–110.

Saad, S. Y., Alkharfy, K. M., & Arafah, M. M. (2006). Cardiotoxic effects of arsenic trioxide/imatinib mesilate combination in rats. *The Journal of Pharmacy and Pharmacology*, *58*, 567–573.
Saretzki, G., Walter, T., Atkinson, S., Passos, J. F., Bareth, B., Keith, W. N., et al. (2008). Downregulation of multiple stress defense mechanisms during differentiation of human embryonic stem cells. *Stem Cells*, *26*, 455–464.
Schamberger, A. C., Schiller, H. B., Fernandez, I. E., Sterclova, M., Heinzelmann, K., Hennen, E., et al. (2016). Glutathione peroxidase 3 localizes to the epithelial lining fluid and the extracellular matrix in interstitial lung disease. *Scientific Reports*, *6*, 29952.
Seiler, A., Schneider, M., Forster, H., Roth, S., Wirth, E. K., Culmsee, C., et al. (2008). Glutathione peroxidase 4 senses and translates oxidative stress into 12/15-lipoxygenase dependent- and AIF-mediated cell death. *Cell Metabolism*, *8*, 237–248.
Smirnov, A., Panatta, E., Lena, A., Castiglia, D., Di Daniele, N., Melino, G., et al. (2016). FOXM1 regulates proliferation, senescence and oxidative stress in keratinocytes and cancer cells. *Aging (Albany NY)*, *8*, 1384–1397.
Speckmann, B., Bidmon, H. J., Pinto, A., Anlauf, M., Sies, H., & Steinbrenner, H. (2011). Induction of glutathione peroxidase 4 expression during enterocytic cell differentiation. *The Journal of Biological Chemistry*, *286*, 10764–10772.
Suzuki, S., Pitchakarn, P., Ogawa, K., Naiki-Ito, A., Chewonarin, T., Punfa, W., et al. (2013). Expression of glutathione peroxidase 2 is associated with not only early hepatocarcinogenesis but also late stage metastasis. *Toxicology*, *311*, 115–123.
Taylor, J. M., Ali, U., Iannello, R. C., Hertzog, P., & Crack, P. J. (2005). Diminished Akt phosphorylation in neurons lacking glutathione peroxidase-1 (Gpx1) leads to increased susceptibility to oxidative stress-induced cell death. *Journal of Neurochemistry*, *92*, 283–293.
Thomas, J. P., Geiger, P. G., Maiorino, M., Ursini, F., & Girotti, A. W. (1990). Enzymatic reduction of phospholipid and cholesterol hydroperoxides in artificial bilayers and lipoproteins. *Biochimica et Biophysica Acta*, *1045*, 252–260.
Thu, V. T., Kim, H. K., Ha, S. H., Yoo, J. Y., Park, W. S., Kim, N., et al. (2010). Glutathione peroxidase 1 protects mitochondria against hypoxia/reoxygenation damage in mouse hearts. *Pflügers Archiv*, *460*, 55–68.
Tosatto, S. C., Bosello, V., Fogolari, F., Mauri, P., Roveri, A., Toppo, S., et al. (2008). The catalytic site of glutathione peroxidases. *Antioxidants & Redox Signaling*, *10*, 1515–1526.
Trachootham, D., Alexandre, J., & Huang, P. (2009). Targeting cancer cells by ROS-mediated mechanisms: A radical therapeutic approach? *Nature Reviews. Drug Discovery*, *8*, 579–591.
Tran, T. H., Mai, H. N., Shin, E. J., Nam, Y., Nguyen, B. T., Lee, Y. J., et al. (2016). Repeated exposure to far infrared ray attenuates acute restraint stress in mice via inhibition of JAK2/STAT3 signaling pathway by induction of glutathione peroxidase-1. *Neurochemistry International*, *94*, 9–22.
Ursini, F., Maiorino, M., Brigelius-Flohé, R., Aumann, K. D., Roveri, A., Schomburg, D., et al. (1995). Diversity of glutathione peroxidases. *Methods in Enzymology*, *252*, 38–53.
Ursini, F., Maiorino, M., Valente, M., Ferri, L., & Gregolin, C. (1982). Purification from pig liver of a protein which protects liposomes and biomembranes from peroxidative degradation and exhibits glutathione peroxidase activity on phosphatidylcholine hydroperoxides. *Biochimica et Biophysica Acta*, *710*, 197–211.
Walshe, J., Serewko-Auret, M. M., Teakle, N., Cameron, S., Minto, K., Smith, L., et al. (2007). Inactivation of glutathione peroxidase activity contributes to UV-induced squamous cell carcinoma formation. *Cancer Research*, *67*, 4751–4758.
Wang, H., Luo, K., Tan, L. Z., Ren, B. G., Gu, L. Q., Michalopoulos, G., et al. (2012). p53-induced gene 3 mediates cell death induced by glutathione peroxidase 3. *The Journal of Biological Chemistry*, *287*, 16890–16902.

Wenk, J., Schüller, J., Hinrichs, C., Syrovets, T., Azoitei, N., Podda, M., et al. (2004). Overexpression of phospholipid-hydroperoxide glutathione peroxidase in human dermal fibroblasts abrogates UVA irradiation-induced expression of interstitial collagenase/matrix metalloproteinase-1 by suppression of phosphatidylcholine hydroperoxide-mediated NFkappaB activation and interleukin-6 release. *The Journal of Biological Chemistry, 279*, 45634–45642.

Winterbourn, C. C. (2013). The biological chemistry of hydrogen peroxide. *Methods in Enzymology, 528*, 3–25.

Wönckhaus, M., Klein-Hitpass, L., Grepmeier, U., Merk, J., Pfeifer, M., Wild, P., et al. (2006). Smoking and cancer-related gene expression in bronchial epithelium and non-small-cell lung cancers. *The Journal of Pathology, 210*, 192–204.

Wortmann, M., Schneider, M., Pircher, J., Hellfritsch, J., Aichler, M., Vegi, N., et al. (2013). Combined deficiency in glutathione peroxidase 4 and vitamin E causes multiorgan thrombus formation and early death in mice. *Circulation Research, 113*, 408–417.

Yagublu, V., Arthur, J. R., Babayeva, S. N., Nicol, F., Post, S., & Keese, M. (2011). Expression of selenium-containing proteins in human colon carcinoma tissue. *Anticancer Research, 31*, 2693–2698.

Yan, W., & Chen, X. (2006). GPX2, a direct target of p63, inhibits oxidative stress-induced apoptosis in a p53-dependent manner. *The Journal of Biological Chemistry, 281*, 7856–7862.

Yan, J., Guo, Y., Fei, Y., Zhang, R., Han, Y., & Lu, S. (2017). GPx1 knockdown suppresses chondrogenic differentiation of ATDC5 cells through induction of reductive stress. *Acta Biochimica et Biophysica Sinica, 49*, 110–118.

Yan, Y., Jiang, W., Tan, Y., Zou, S., Zhang, H., Mao, F., et al. (2017). hucMSC exosome-derived GPX1 is required for the recovery of hepatic oxidant injury. *Molecular Therapy, 25*, 465–479.

Yang, W. S., Kim, K. J., Gaschler, M. M., Patel, M., Shchepinov, M. S., & Stockwell, B. R. (2016). Peroxidation of polyunsaturated fatty acids by lipoxygenases drives ferroptosis. *Proceedings of the National Academy of Sciences of the United States of America, 113*, E4966–E4975.

Yang, W. H., Park, S. Y., Ji, S., Kang, J. G., Kim, J. E., Song, H., et al. (2010). O-GlcNAcylation regulates hyperglycemia-induced GPX1 activation. *Biochemical and Biophysical Research Communications, 391*, 756–761.

Yang, W. S., SriRamaratnam, R., Welsch, M. E., Shimada, K., Skouta, R., Viswanathan, V. S., et al. (2014). Regulation of ferroptotic cancer cell death by GPX4. *Cell, 156*, 317–331.

Yang, W. S., & Stockwell, B. R. (2016). Ferroptosis: Death by lipid peroxidation. *Trends in Cell Biology, 26*, 165–176.

Yant, L. J., Ran, Q., Rao, L., Van Remmen, H., Shibatani, T., Belter, J. G., et al. (2003). The selenoprotein GPX4 is essential for mouse development and protects from radiation and oxidative damage insults. *Free Radical Biology & Medicine, 34*, 496–502.

Yoo, M. H., Gu, X., Xu, X. M., Kim, J. Y., Carlson, B. A., Patterson, A. D., et al. (2010). Delineating the role of glutathione peroxidase 4 in protecting cells against lipid hydroperoxide damage and in Alzheimer's disease. *Antioxidants & Redox Signaling, 12*, 819–827.

Yu, H., Guo, P., Xie, X., Wang, Y., & Chen, G. (2017). Ferroptosis, a new form of cell death, and its relationships with tumourous diseases. *Journal of Cellular and Molecular Medicine, 21*, 648–657.

Yu, Y. P., Yu, G., Tseng, G., Cieply, K., Nelson, J., Defrances, M., et al. (2007). Glutathione peroxidase 3, deleted or methylated in prostate cancer, suppresses prostate cancer growth and metastasis. *Cancer Research, 67*, 8043–8050.

Zhang, W., & Liu, H. T. (2002). MAPK signal pathways in the regulation of cell proliferation in mammalian cells. *Cell Research*, *12*, 9–18.

Zhang, X., Yang, J. J., Kim, Y. S., Kim, K. Y., Ahn, W. S., & Yang, S. (2010). An 8-gene signature, including methylated and down-regulated glutathione peroxidase 3, of gastric cancer. *International Journal of Oncology*, *36*, 405–414.

Zhao, H., Li, J., Li, X., Han, C., Zhang, Y., Zheng, L., et al. (2015). Silencing GPX3 expression promotes tumor metastasis in human thyroid cancer. *Current Protein & Peptide Science*, *16*, 316–321.

Zhu, S., Zhang, Q., Sun, X., Zeh, H. J., Lotze, M. T., Kang, R., et al. (2017). HSPA5 regulates ferroptotic cell death in cancer cells. *Cancer Research*, *77*, 2064–2077.

CHAPTER FIVE

Targeting the Selenoprotein Thioredoxin Reductase 1 for Anticancer Therapy

Elias S.J. Arnér[1]
Karolinska Institutet, Stockholm, Sweden
[1]Corresponding author: e-mail address: elias.arner@ki.se

Contents

Abstract

The cytosolic selenoprotein thioredoxin reductase 1 (TrxR1, encoded in human by *TXNRD1*) is implied to have several different roles in relation to cancer. Its physiologic functions may protect normal cells from carcinogenesis, but may also promote cancer progression if carcinogenesis nonetheless occurs. With distinct links to Nrf2 signaling, ribonucleotide reductase-dependent production of deoxyribonucleotides and its support of several antioxidant systems counteracting oxidative stress, the metabolic pathways regulated, and affected by TrxR1, are altogether of crucial importance in cancer. These pathways and causal relationships are at the same time highly intricate. In spite of the complexity in the cellular redox networks, several observations discussed in this chapter suggest that specific targeting of TrxR1 may be promising as a mechanistic principle for anticancer therapy.

1. INTRODUCTION

1.1 The Thioredoxin System

The thioredoxin (Trx) system is an important reductive enzyme system in cells acting in conjunction with the glutathione (GSH) system (Arner & Holmgren, 2000; Becker, Gromer, Schirmer, & Müller, 2000; Gromer,

Advances in Cancer Research, Volume 136
ISSN 0065-230X
http://dx.doi.org/10.1016/bs.acr.2017.07.005

Urig, & Becker, 2004; Nordberg & Arner, 2001). The Trx system encompasses Trx1 (encoded in human by *TXN*) and several additional Trx-fold enzymes, which are kept redox active through the activities of thioredoxin reductases (TrxRs) that use NADPH to reduce the active site disulfides of Trxs to dithiols. The Trx-fold proteins can subsequently act upon several downstream targets involved in reductive cellular pathways and redox regulation of cellular functions. The human genome encodes for three specific TrxR isoenzymes, namely cytosolic TrxR1 (encoded by *TXNRD1*), mitochondrial TrxR2 (encoded by *TXNRD2*), and testis-specific TGR (encoded by *TXNRD3*), which all are selenoproteins (Arnér, 2009; Arner & Holmgren, 2000; Gromer et al., 2004; Martin, 1995; Miranda-Vizuete et al., 2004; Nordberg & Arner, 2001). The reader is referred to some of these prior review articles for further general information on these enzyme systems. In this chapter, the specific targeting of cytosolic TrxR1 will be discussed in relation to anticancer therapy.

1.2 Thioredoxin Reductase 1

The TrxR1 (also named TXNRD1) enzyme is mainly cytosolic. However, it can also be found in the nucleus, in form of the splice variant TXNRD1_v2, where it can modulate the activities of several transcription factors including estrogen receptors (Arnér, 2009; Damdimopoulos, Miranda-Vizuete, Treuter, Gustafsson, & Spyrou, 2004). A third splice variant of the enzyme, TXNRD1_v3, is targeted to the cell membrane where it colocalizes with membrane rafts and can promote formation of filopodia upon overexpression (Arnér, 2009; Cebula, Moolla, Capovilla, & Arner, 2013; Damdimopoulou, Miranda-Vizuete, Arner, Gustafsson, & Damdimopoulos, 2009; Dammeyer et al., 2008). These different splice variants of TrxR1 have not yet been examined at depth in relation to anticancer therapy, but probably need to be considered for a full understanding of the therapeutic effects of targeting the enzyme. However, the cytosolic splice variant (TXNRD1_v1) is considered the main form of the enzyme and this is the form that has been mostly studied in relation to anticancer therapy.

The primary substrate of TrxR1 is Trx1 and most cellular actions of TrxR1 are likely to be carried out by Trx1 (Arnér, 2009; Arner & Holmgren, 2000; Gromer et al., 2004; Nishinaka, Masutani, Nakamura, & Yodoi, 2001; Nishiyama, Masutani, Nakamura, Nishinaka, & Yodoi, 2001; Nordberg & Arner, 2001; Sengupta & Holmgren, 2013). However, additional substrate proteins for TrxR1, such as TRP14, are also likely to have importance for the effects of TrxR1 targeting (Dagnell et al., 2013; Doka et al., 2016; Jeong, Chang, Boja, Fales, & Rhee, 2004; Jeong, Yoon, Lee, & Rhee, 2004; Pader et al., 2014; Woo et al., 2004). Reduced Trx1 supports the activities of enzymes such as ribonucleotide reductase (RNR), peroxiredoxins (Prxs), methionine sulfoxide reductases (Msrs), and many more important redox-active proteins that need to be reduced by Trx1 to sustain their functions (Arnér, 2009; Arner & Holmgren, 2000; Gromer et al., 2004; Lillig & Holmgren, 2007). Additional protein motifs beyond disulfides that either Trx1 or TRP14 can reduce include S-nitrosylated (Benhar, Forrester, Hess, & Stamler, 2008; Hashemy & Holmgren, 2008; Pader et al., 2014; Sengupta & Holmgren, 2013; Shen et al., 2016; Tannenbaum & White, 2006) and persulfidated (Doka et al., 2016; Wedmann et al., 2016) Cys residues in target proteins, and a sulfenylamide in oxidized PTP1B (Dagnell et al., 2013; Salmeen et al., 2003; Schwertassek et al., 2014). Reduced Trx1 also inhibits apoptosis signaling kinase ASK1 by directly binding the protein, which thus is activated upon oxidation of Trx1 (Lim, Liu, Go, & Boelsterli, 2008; Matsuzawa & Ichijo, 2008; Nishiyama et al., 2001; Saitoh et al., 1998; Tobiume et al., 2001). All of these activities, as schematically summarized in Fig. 1, are likely to be affected or modulated upon inhibition of TrxR1. It should, however, be emphasized that after a genetic

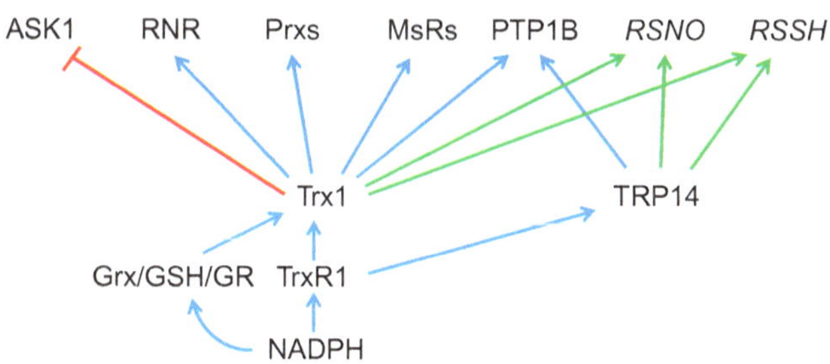

Fig. 1 *Activities of the cytosolic Trx system.* This figure schematically summarizes activities of the cytosolic Trx system, as mentioned in the text. In short, TrxR1 is an NADPH-dependent reductase that supports the activities of Trx1 and TRP14, which in turn reduce and thereby support the activities of a number of downstream enzymes (*top row, blue arrows*). Trx1 and TRP14 can also reduce a number of diverse oxidized post-translational modifications in proteins or similar motifs in small compounds (*top row, green arrows*). Reduced Trx1 furthermore inhibits apoptosis signaling kinase 1 (ASK1, *red arrow*). The glutathione system can aid in keeping Trx1 reduced, as also indicated in the scheme. For further information, see the text and cited references. *GR*, glutathione reductase; *Grx*, glutaredoxin; *GSH*, glutathione; *MsRs*, methionine sulfoxide reductases; *Prxs*, peroxiredoxins; *PTP1B*, phosphoprotein tyrosine phosphatase 1B; *RNR*, ribonucleotide reductase; *RSNO*, *S*-nitrosylated motifs in proteins or compounds; *RSSH*, *S*-persulfide motifs in proteins or compounds.

deletion of TrxR1, Trx1 can still be kept reduced, at least in normal cells, due to significant cross talk in the cellular redox pathways and reduction of Trx1 through GSH- and glutaredoxin (Grx)-dependent pathways (Du, Zhang, Zhang, Lu, & Holmgren, 2013; Iverson et al., 2013; Lei et al., 2016; Peng et al., 2016; Prigge et al., 2012; Rollins et al., 2010). Thus, any effects on cells of inhibited TrxR1 may perhaps only in part be due to downstream effects of impaired Trx1 or TRP14 functions, and should likely also involve more direct effects of the inhibited TrxR1 protein.

2. TrxR1 AS AN IMPORTANT ENZYME IN CANCER

Many studies have suggested that the whole Trx system is important in carcinogenesis and in cancer progression (Arnér, 2009; Arnér & Holmgren, 2006; Bhatia et al., 2016; Cadenas et al., 2010; Cebula, Schmidt, & Arner, 2015; Gorrini, Harris, & Mak, 2013; Gromer et al., 2004; Harris et al., 2015; Lincoln, Ali Emadi, Tonissen, & Clarke, 2003; Manda et al., 2015; Singh, Li, et al., 2008; Wang et al., 2012; Watson, 2013; Yamada, Tomida, Yoshikawa, Taketani, & Tsuruo, 1996; Yoo, Xu, Carlson, Gladyshev, & Hatfield, 2006; Yoo et al., 2007; Zhang, Wang, Li, Zhi, & Lu, 2016; Zhang, Zhang, et al., 2016). It is less clear, however, which, if any, of the functions of the Trx system that are *required* for cancer development and, conversely, which of these functions that can perhaps be selectively targeted for anticancer therapy. It is also not certain if any particular Trx-fold enzyme, or TrxR isoenzyme, is of more importance than other components of the Trx system for a therapeutic effect. Still it is likely, and proposed here, that specific targeting of cytosolic TrxR1 can yield anticancer efficacy.

Due to its inherent chemical reactivity, TrxR1 is easily inhibited by a wide range of small compounds, especially electrophiles such as metal compounds or Michael acceptors (Zhang, Zhang, et al., 2016). Typically, TrxR1 is inhibited by electrophiles through the targeting of its highly accessible and very reactive nucleophilic Sec residue (Cheng, Sandalova, Lindqvist, & Arnér, 2009) that thereby becomes covalently derivatized. Inhibition of TrxR1 has been shown to occur with many different electrophiles that furthermore have anticancer effects (Arnér, 2009; Arnér & Holmgren, 2006; Becker et al., 2000; Cai et al., 2012; Carvalho, Chew, Hashemy, Lu, & Holmgren, 2008; Casini et al., 2008; Cebula et al., 2015; Chew, Lu, Bradshaw, & Holmgren, 2008; Eriksson, Prast-Nielsen, Flaberg, Szekely, & Arner, 2009; Fang, Lu, & Holmgren, 2005; Gromer et al., 2004; Hashemy, Ungerstedt, Zahedi Avval, & Holmgren, 2006;

Hedstrom, Eriksson, Zawacka-Pankau, Arner, & Selivanova, 2009; Krishnamurthy et al., 2008; Lincoln et al., 2003; Liu, Du, Huang, Lee, & Gu, 2008; Liu, Huang, et al., 2008; Lu, Chew, & Holmgren, 2007; Lu et al., 2006; Marzano et al., 2007; Peng et al., 2013; Prast-Nielsen, Cebula, Pader, & Arner, 2010; Prast-Nielsen et al., 2011; Shi et al., 2014; Urig & Becker, 2006; Wang, Zhang, & Xu, 2008; Wipf et al., 2004; Witte, Anestål, Jerremalm, Ehrsson, & Arnér, 2005). For recent comprehensive reviews on the different classes of TrxR1 inhibitors, see Cai et al. (2012) and Zhang, Zhang, et al. (2016). It is not clear, however, whether an efficient anticancer therapy can be obtained by TrxR1 targeting alone, or whether the inhibition of this enzyme by such different anticancer compounds is merely coincidental, and not of causal importance with regards to the therapeutic efficacy. As discussed elsewhere, the targeting of TrxR1 may have either beneficial or detrimental effects with regards to physiological functions, depending upon context in which the enzyme is inhibited (Arnér, 2009; Iverson et al., 2013; Lei et al., 2016; Locy et al., 2012). It shall be emphasized, however, that it seems clear that normal adult cells and tissues survive well without TrxR1, at least in genetic mouse models (Eriksson, Prigge, Talago, Arner, & Schmidt, 2015; Iverson et al., 2013; Prigge et al., 2012; Rollins et al., 2010). If the enzyme is nonetheless crucial for support of cancer progression, which has been implied by a number of studies (Bhatia et al., 2016; Cadenas et al., 2010; Harris et al., 2015; Mandal et al., 2010; Wang et al., 2012; Yoo et al., 2006, 2007; Zhang, Zhang, et al., 2016), then its dispensability for normal cells and tissues indeed provides a promising foundation for anticancer therapy. The notion would hence be that normal cells can handle and survive TrxR1 inhibition, while cancer cells can perhaps not.

It needs to be emphasized that therapeutically beneficial effects of TrxR1 targeting in cancer treatment remain to be confirmed in a clinical setting, although this possibility has been implied by many studies as discussed and cited in this chapter. The notion that cancer cells have inherently increased oxidative stress as a result of oncogenic signaling, distorted metabolism, impaired mitochondrial function, and other mechanisms, is rapidly gaining wider recognition (Galluzzi, Kepp, Vander Heiden, & Kroemer, 2013; Harris et al., 2015; Luo, Solimini, & Elledge, 2009; Manda et al., 2015; Shi et al., 2014; Trachootham, Alexandre, & Huang, 2009; Trachootham et al., 2006). Their enzymatic systems that counteract oxidative stress are thereby pushed to the limits of their capacity, which typically include overexpression of different enzymes of the thioredoxin system, including TrxR1 (Bhatia et al., 2016; Lincoln et al., 2003). This renders tumor cells more vulnerable than normal cells to agents that enhance oxidative stress, or that lower their capacity to cope with such stress, as tumor cells typically have less reserve capacity to additional oxidative challenges. This fact may likely be utilized to selectively kill tumor cells (Gorrini et al., 2013; Wondrak, 2009). Indeed, many classical anticancer treatments—at least in part—are today known to act by further increasing the oxidative stress in cancer cells (Gorrini et al., 2013; Wondrak, 2009). Metabolic perturbations resulting from this increased stress will at the same time activate the major reductive Nrf2-driven Trx and GSH systems, which support cell survival under challenges of increased endogenous oxidative stress (Brigelius-Flohe & Flohe, 2011; Higgins & Hayes, 2011; Mitsuishi, Motohashi, & Yamamoto, 2012; Singh, Boldin-Adamsky, et al., 2008; Zhang, 2006). It is thus not far-fetched to propose that TrxR1 inhibition may be a potential mechanism of action for anticancer drugs. TrxR1 inhibition can hereby lead to an impairment of the required antioxidant capacity particularly in cancer cells, while normal cells can survive a loss of TrxR1 activity (Arnér, 2009; Arnér & Holmgren, 2006; Chew et al., 2010; Harris et al., 2015; Shi et al., 2014; Trachootham et al., 2009). This should indeed be a major mechanism by which TrxR1 inhibition by small-molecule inhibitors may yield anticancer efficacy, which is further corroborated by the fact that genetically lowered TrxR1 expression in cancer cells impairs tumor development (Hatfield, Yoo, Carlson, & Gladyshev, 2009; Mandal et al., 2010; Yoo et al., 2006, 2007). In addition, drug targeting of TrxR1 can convert the enzyme to toxic prooxidant SecTRAPs (*Se*lenium *c*ompromised *t*hioredoxin *r*eductase-derived *a*poptotic *p*rotein*s*) that can further aggravate the oxidative stress, i.e., compounds targeting TrxR1 may not only yield loss of TrxR1 activity but also enzymatically *increased* production of reactive oxygen species (Anestål & Arnér, 2003; Anestål, Prast-Nielsen, Cenas, & Arnér, 2008; Cai et al., 2012; Cebula et al., 2015; Hashemy et al., 2006).

3. CONCLUSIONS

The pathways in both normal and cancer cells that are modulated or regulated by the Trx system are many, and are intricately linked with each other. It is thereby very difficult, if not impossible, to fully understand the impact of drug targeting of the cytosolic selenoprotein TrxR1. Nonetheless, based upon observations from genetic mouse models as well as biochemical studies with small molecules, the picture emerges that this enzyme may provide a promising anticancer drug target. The major effects that constitute the arguments for this notion are schematically summarized in Fig. 2. The events include the higher endogenous oxidative stress and higher dependence upon TrxR1 for cancer cells compared to normal cells, the fact that TrxR1

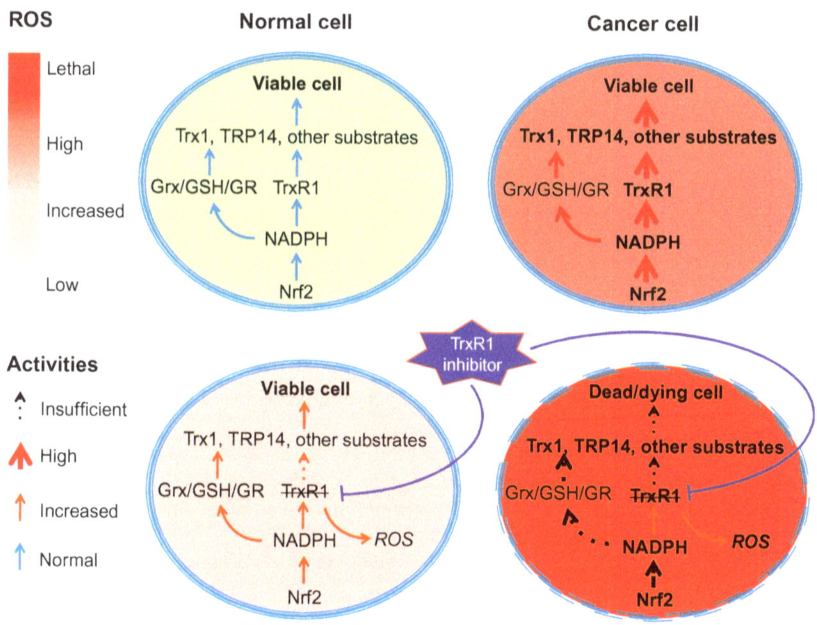

Fig. 2 *Potential mechanistic principle of TrxR1 targeting for anticancer therapy*. This figure schematically summarizes mechanisms of specific drug targeting of TrxR1 that may yield anticancer effects without overt toxicity to normal cells and tissues. All cells typically drive their reductive cytosolic Trx system using NADPH channeled either through the GSH system or through TrxR1, with key players of both systems being Nrf2 targets, as summarized in these schematically drawn cells. Due to their distorted metabolism, cancer cells (*right*) have higher endogenous levels of reactive oxygen species (ROS) than normal cells (*left*), with the color of each cell here reflecting the steady-state ROS level using the indicator at left. The activities of all reductive pathways are typically increased in cancer cells compared to normal cells, in order to maintain viability. Upon exposure to a TrxR1 inhibitor (*blue star*), this may inhibit normal TrxR1 activities, while also increasing NADPH-dependent production of ROS, as TrxR1 can be converted to SecTRAPs (see text for details). This is indicated in the two lower schemes. Normal cells can still maintain viability upon exposure to a TrxR1 inhibitor by further activating Nrf2 and thus maintain reductive pathways by utilizing increased activities of the GSH system. Cancer cells, however, die upon such treatment because they reach lethal levels of ROS, as their reductive systems cannot be further increased to sufficient activities for maintained cell viability.

because of its structure and chemical reactivity is easily targeted by electrophilic compounds that have anticancer efficacy, and the fact that the inhibited enzyme can be converted to SecTRAPs that produce additional reactive oxygen species to provide a further toxic burden particularly to cancer cells. If future studies can demonstrate that novel specific inhibitors of TrxR1 will have anticancer properties without toxic effects on normal cells and tissues, then this mechanistic principle for anticancer therapy will be further strengthened.

ACKNOWLEDGMENTS

Work of the author has support from The National Institute of Health (NIH), Swedish Cancer Society, Swedish Research Council (Medicine and Natural Sciences sections), Knut and Alice Wallenberg Foundation, and Karolinska Institutet.

REFERENCES

Anestål, K., & Arnér, E. S. J. (2003). Rapid induction of cell death by selenium-compromised thioredoxin reductase 1 but not by the fully active enzyme containing selenocysteine. *The Journal of Biological Chemistry*, *278*, 15966–15972.

Anestål, K., Prast-Nielsen, S., Cenas, N., & Arnér, E. S. J. (2008). Cell death by SecTRAPs—Thioredoxin reductase as a prooxidant killer of cells. *PloS One*, *3*, e1846.

Arnér, E. S. J. (2009). Focus on mammalian thioredoxin reductases—Important selenoproteins with versatile functions. *Biochimica et Biophysica Acta*, *1790*, 495–526.

Arner, E. S., & Holmgren, A. (2000). Physiological functions of thioredoxin and thioredoxin reductase. *European Journal of Biochemistry*, *267*, 6102–6109.

Arnér, E. S. J., & Holmgren, A. (2006). The thioredoxin system in cancer. *Seminars in Cancer Biology*, *16*, 420–426.

Becker, K., Gromer, S., Schirmer, R. H., & Müller, S. (2000). Thioredoxin reductase as a pathophysiological factor and drug target. *European Journal of Biochemistry*, *267*, 6118–6125.

Benhar, M., Forrester, M. T., Hess, D. T., & Stamler, J. S. (2008). Regulated protein denitrosylation by cytosolic and mitochondrial thioredoxins. *Science*, *320*, 1050–1054.

Bhatia, M., McGrath, K. L., Di Trapani, G., Charoentong, P., Shah, F., King, M. M., et al. (2016). The thioredoxin system in breast cancer cell invasion and migration. *Redox Biology*, *8*, 68–78.

Brigelius-Flohe, R., & Flohe, L. (2011). Basic principles and emerging concepts in the redox control of transcription factors. *Antioxidants & Redox Signaling*, *15*, 2335–2381.

Cadenas, C., Franckenstein, D., Schmidt, M., Gehrmann, M., Hermes, M., Geppert, B., et al. (2010). Role of thioredoxin reductase 1 and thioredoxin interacting protein in prognosis of breast cancer. *Breast Cancer Research*, *12*, R44.

Cai, W., Zhang, L., Song, Y., Wang, B., Zhang, B., Cui, X., et al. (2012). Small molecule inhibitors of mammalian thioredoxin reductase. *Free Radical Biology & Medicine*, *52*, 257–265.

Carvalho, C. M., Chew, E. H., Hashemy, S. I., Lu, J., & Holmgren, A. (2008). Inhibition of the human thioredoxin system. A molecular mechanism of mercury toxicity. *The Journal of Biological Chemistry*, *283*, 11913–11923.

Casini, A., Gabbiani, C., Sorrentino, F., Rigobello, M. P., Bindoli, A., Geldbach, T. J., et al. (2008). Emerging protein targets for anticancer metallodrugs: inhibition of thioredoxin reductase and cathepsin B by antitumor ruthenium(II)-arene compounds. *Journal of Medicinal Chemistry*, *51*, 6773–6781.

Cebula, M., Moolla, N., Capovilla, A., & Arner, E. S. (2013). The rare TXNRD1_v3 ("v3") splice variant of human thioredoxin reductase 1 protein is targeted to membrane rafts by N-acylation and induces filopodia independently of its redox active site integrity. *The Journal of Biological Chemistry*, *288*, 10002–10011.

Cebula, M., Schmidt, E. E., & Arner, E. S. (2015). TrxR1 as a potent regulator of the Nrf2-Keap1 response system. *Antioxidants & Redox Signaling*, *23*, 823–853.

Cheng, Q., Sandalova, T., Lindqvist, Y., & Arnér, E. S. J. (2009). Crystal structure and catalysis of the selenoprotein thioredoxin reductase 1. *The Journal of Biological Chemistry*, *284*, 3998–4008.

Chew, E. H., Lu, J., Bradshaw, T. D., & Holmgren, A. (2008). Thioredoxin reductase inhibition by antitumor quinols: A quinol pharmacophore effect correlating to antiproliferative activity. *The FASEB Journal*, *22*, 2072–2083.

Chew, E. H., Nagle, A. A., Zhang, Y., Scarmagnani, S., Palaniappan, P., Bradshaw, T. D., et al. (2010). Cinnamaldehydes inhibit thioredoxin reductase and induce Nrf2: Potential candidates for cancer therapy and chemoprevention. *Free Radical Biology & Medicine*, *48*, 98–111.

Dagnell, M., Frijhoff, J., Pader, I., Augsten, M., Boivin, B., Xu, J., et al. (2013). Selective activation of oxidized PTP1B by the thioredoxin system modulates PDGF-beta receptor tyrosine kinase signaling. *Proceedings of the National Academy of Sciences of the United States of America*, *110*, 13398–13403.

Damdimopoulos, A. E., Miranda-Vizuete, A., Treuter, E., Gustafsson, J. Å., & Spyrou, G. (2004). An alternative splicing variant of the selenoprotein thioredoxin reductase is a modulator of estrogen signaling. *The Journal of Biological Chemistry*, *279*, 38721–38729.

Damdimopoulou, P. E., Miranda-Vizuete, A., Arner, E. S., Gustafsson, J. A., & Damdimopoulos, A. E. (2009). The human thioredoxin reductase-1 splice variant TXNRD1_v3 is an atypical inducer of cytoplasmic filaments and cell membrane filopodia. *Biochimica et Biophysica Acta*, *1793*, 1588–1596.

Dammeyer, P., Damdimopoulos, A. E., Nordman, T., Jimenez, A., Miranda-Vizuete, A., & Arner, E. S. (2008). Induction of cell membrane protrusions by the N-terminal glutaredoxin domain of a rare splice variant of human thioredoxin reductase 1. *The Journal of Biological Chemistry*, *283*, 2814–2821.

Doka, E., Pader, I., Biro, A., Johansson, K., Cheng, Q., Ballago, K., et al. (2016). A novel persulfide detection method reveals protein persulfide- and polysulfide-reducing functions of thioredoxin and glutathione systems. *Science Advances*, *2*, e1500968.

Du, Y., Zhang, H., Zhang, X., Lu, J., & Holmgren, A. (2013). Thioredoxin 1 is inactivated due to oxidation induced by peroxiredoxin under oxidative stress and reactivated by the glutaredoxin system. *The Journal of Biological Chemistry*, *288*, 32241–32247.

Eriksson, S. E., Prast-Nielsen, S., Flaberg, E., Szekely, L., & Arner, E. S. (2009). High levels of thioredoxin reductase 1 modulate drug-specific cytotoxic efficacy. *Free Radical Biology & Medicine*, *47*, 1661–1671.

Eriksson, S., Prigge, J. R., Talago, E. A., Arner, E. S., & Schmidt, E. E. (2015). Dietary methionine can sustain cytosolic redox homeostasis in the mouse liver. *Nature Communications*, *6*, 6479.

Fang, J., Lu, J., & Holmgren, A. (2005). Thioredoxin reductase is irreversibly modified by curcumin: A novel molecular mechanism for its anticancer activity. *The Journal of Biological Chemistry*, *280*, 25284–25290.

Galluzzi, L., Kepp, O., Vander Heiden, M. G., & Kroemer, G. (2013). Metabolic targets for cancer therapy. *Nature Reviews Drug Discovery*, *12*, 829–846.

Gorrini, C., Harris, I. S., & Mak, T. W. (2013). Modulation of oxidative stress as an anticancer strategy. *Nature Reviews. Drug Discovery*, *12*, 931–947.

Gromer, S., Urig, S., & Becker, K. (2004). The thioredoxin system—From science to clinic. *Medicinal Research. Reviews*, *24*, 40–89.
Harris, I. S., Treloar, A. E., Inoue, S., Sasaki, M., Gorrini, C., Lee, K. C., et al. (2015). Glutathione and thioredoxin antioxidant pathways synergize to drive cancer initiation and progression. *Cancer Cell*, *27*, 211–222.
Hashemy, S. I., & Holmgren, A. (2008). Regulation of the catalytic activity and structure of human thioredoxin 1 via oxidation and S-nitrosylation of cysteine residues. *The Journal of Biological Chemistry*, *283*, 21890–21898.
Hashemy, S. I., Ungerstedt, J. S., Zahedi Avval, F., & Holmgren, A. (2006). Motexafin gadolinium, a tumor-selective drug targeting thioredoxin reductase and ribonucleotide reductase. *The Journal of Biological Chemistry*, *281*, 10691–10697.
Hatfield, D. L., Yoo, M. H., Carlson, B. A., & Gladyshev, V. N. (2009). Selenoproteins that function in cancer prevention and promotion. *Biochimica et Biophysica Acta*, *1790*(11), 1541–1545.
Hedstrom, E., Eriksson, S., Zawacka-Pankau, J., Arner, E. S., & Selivanova, G. (2009). p53-dependent inhibition of TrxR1 contributes to the tumor-specific induction of apoptosis by RITA. *Cell Cycle*, *8*, 3576–3583.
Higgins, L. G., & Hayes, J. D. (2011). The cap'n'collar transcription factor Nrf2 mediates both intrinsic resistance to environmental stressors and an adaptive response elicited by chemopreventive agents that determines susceptibility to electrophilic xenobiotics. *Chemico-Biological Interactions*, *192*, 37–45.
Iverson, S. V., Eriksson, S., Xu, J., Prigge, J. R., Talago, E. A., Meade, T. A., et al. (2013). A Txnrd1-dependent metabolic switch alters hepatic lipogenesis, glycogen storage, and detoxification. *Free Radical Biology & Medicine*, *63*, 369–380.
Jeong, W., Chang, T. S., Boja, E. S., Fales, H. M., & Rhee, S. G. (2004). Roles of TRP14, a thioredoxin-related protein in tumor necrosis factor-alpha signaling pathways. *The Journal of Biological Chemistry*, *279*, 3151–3159.
Jeong, W., Yoon, H. W., Lee, S. R., & Rhee, S. G. (2004). Identification and characterization of TRP14, a thioredoxin-related protein of 14 kDa. New insights into the specificity of thioredoxin function. *The Journal of Biological Chemistry*, *279*, 3142–3150.
Krishnamurthy, D., Karver, M. R., Fiorillo, E., Orru, V., Stanford, S. M., Bottini, N., et al. (2008). Gold(I)-mediated inhibition of protein tyrosine phosphatases: A detailed in vitro and cellular study. *Journal of Medicinal Chemistry*, *51*, 4790–4795.
Lei, X. G., Zhu, J. H., Cheng, W. H., Bao, Y., Ho, Y. S., Reddi, A. R., et al. (2016). Paradoxical roles of antioxidant enzymes: Basic mechanisms and health implications. *Physiological Reviews*, *96*, 307–364.
Lillig, C. H., & Holmgren, A. (2007). Thioredoxin and related molecules-from biology to health and disease. *Antioxidants & Redox Signaling*, *9*, 25–47.
Lim, P. L., Liu, J., Go, M. L., & Boelsterli, U. A. (2008). The mitochondrial superoxide/thioredoxin-2/Ask1 signaling pathway is critically involved in troglitazone-induced cell injury to human hepatocytes. *Toxicological Sciences*, *101*, 341–349.
Lincoln, D. T., Ali Emadi, E. M., Tonissen, K. F., & Clarke, F. M. (2003). The thioredoxin-thioredoxin reductase system: Over-expression in human cancer. *Anticancer Research*, *23*, 2425–2433.
Liu, Z., Du, Z. Y., Huang, Z. S., Lee, K. S., & Gu, L. Q. (2008). Inhibition of thioredoxin reductase by curcumin analogs. *Bioscience, Biotechnology, and Biochemistry*, *72*, 2214–2218.
Liu, Z., Huang, S., Li, M., Huang, Z., Lee, K. S., & Gu, L. (2008). Inhibition of thioredoxin reductase by mansonone F analogues: Implications for anticancer activity. *Chemico-Biological Interactions*, *177*, 48–57.
Locy, M. L., Rogers, L. K., Prigge, J. R., Schmidt, E. E., Arner, E. S., & Tipple, T. E. (2012). Thioredoxin reductase inhibition elicits Nrf2-mediated responses in Clara cells: Implications for oxidant-induced lung injury. *Antioxidants & Redox Signaling*, *17*, 1407–1416.

Lu, J., Chew, E. H., & Holmgren, A. (2007). Targeting thioredoxin reductase is a basis for cancer therapy by arsenic trioxide. *Proceedings of the National Academy of Sciences of the United States of America, 104*, 12288–12293.

Lu, J., Papp, L. V., Fang, J., Rodriguez-Nieto, S., Zhivotovsky, B., & Holmgren, A. (2006). Inhibition of mammalian thioredoxin reductase by some flavonoids: Implications for myricetin and quercetin anticancer activity. *Cancer Research, 66*, 4410–4418.

Luo, J., Solimini, N. L., & Elledge, S. J. (2009). Principles of cancer therapy: Oncogene and non-oncogene addiction. *Cell, 136*, 823–837.

Manda, G., Isvoranu, G., Comanescu, M. V., Manea, A., Debelec Butuner, B., & Korkmaz, K. S. (2015). The redox biology network in cancer pathophysiology and therapeutics. *Redox Biology, 5*, 347–357.

Mandal, P. K., Schneider, M., Kolle, P., Kuhlencordt, P., Forster, H., Beck, H., et al. (2010). Loss of thioredoxin reductase 1 renders tumors highly susceptible to pharmacologic glutathione deprivation. *Cancer Research, 70*, 9505–9514.

Martin, J. L. (1995). Thioredoxin—A fold for all reasons. *Structure, 3*, 245–250.

Marzano, C., Gandin, V., Folda, A., Scutari, G., Bindoli, A., & Rigobello, M. P. (2007). Inhibition of thioredoxin reductase by auranofin induces apoptosis in cisplatin-resistant human ovarian cancer cells. *Free Radical Biology & Medicine, 42*, 872–881.

Matsuzawa, A., & Ichijo, H. (2008). Redox control of cell fate by MAP kinase: Physiological roles of ASK1-MAP kinase pathway in stress signaling. *Biochimica et Biophysica Acta, 1780*, 1325–1336.

Miranda-Vizuete, A., Sadek, C. M., Jimenez, A., Krause, W. J., Sutovsky, P., & Oko, R. (2004). The mammalian testis-specific thioredoxin system. *Antioxidants & Redox Signaling, 6*, 25–40.

Mitsuishi, Y., Motohashi, H., & Yamamoto, M. (2012). The Keap1-Nrf2 system in cancers: Stress response and anabolic metabolism. *Frontiers in Oncology, 2*, 200.

Nishinaka, Y., Masutani, H., Nakamura, H., & Yodoi, J. (2001). Regulatory roles of thioredoxin in oxidative stress-induced cellular responses. *Redox Report, 6*, 289–295.

Nishiyama, A., Masutani, H., Nakamura, H., Nishinaka, Y., & Yodoi, J. (2001). Redox regulation by thioredoxin and thioredoxin-binding proteins. *IUBMB Life, 52*, 29–33.

Nordberg, J., & Arner, E. S. (2001). Reactive oxygen species, antioxidants, and the mammalian thioredoxin system. *Free Radical Biology & Medicine, 31*, 1287–1312.

Pader, I., Sengupta, R., Cebula, M., Xu, J., Lundberg, J. O., Holmgren, A., et al. (2014). Thioredoxin-related protein of 14 kDa is an efficient L-cystine reductase and S-denitrosylase. *Proceedings of the National Academy of Sciences of the United States of America, 111*, 6964 6969.

Peng, X., Gimenez-Cassina, A., Petrus, P., Conrad, M., Ryden, M., & Arner, E. S. (2016). Thioredoxin reductase 1 suppresses adipocyte differentiation and insulin responsiveness. *Scientific Reports, 6*, 28080.

Peng, X., Zhang, M. Q., Conserva, F., Hosny, G., Selivanova, G., Bykov, V. J., et al. (2013). APR-246/PRIMA-1MET inhibits thioredoxin reductase 1 and converts the enzyme to a dedicated NADPH oxidase. *Cell Death & Disease, 4*, e881.

Prast-Nielsen, S., Cebula, M., Pader, I., & Arner, E. S. (2010). Noble metal targeting of thioredoxin reductase—Covalent complexes with thioredoxin and thioredoxin-related protein of 14 kDa triggered by cisplatin. *Free Radical Biology & Medicine, 49*, 1765–1778.

Prast-Nielsen, S., Dexheimer, T. S., Schultz, L., Stafford, W. C., Cheng, Q., Xu, J., et al. (2011). Inhibition of thioredoxin reductase 1 by porphyrins and other small molecules identified by a high-throughput screening assay. *Free Radical Biology & Medicine, 50*, 1114–1123.

Prigge, J. R., Eriksson, S., Iverson, S. V., Meade, T. A., Capecchi, M. R., Arnér, E. S. J., et al. (2012). Hepatocyte DNA replication in growing liver requires either glutathione or a single allele of txnrd1. *Free Radical Biology & Medicine, 52*, 803–810.

Rollins, M. F., van der Heide, D. M., Weisend, C. M., Kundert, J. A., Comstock, K. M., Suvorova, E. S., et al. (2010). Hepatocytes lacking thioredoxin reductase 1 have normal replicative potential during development and regeneration. *Journal of Cell Science, 123*, 2402–2412.
Saitoh, M., Nishitoh, H., Fujii, M., Takeda, K., Tobiume, K., Sawada, Y., et al. (1998). Mammalian thioredoxin is a direct inhibitor of apoptosis signal-regulating kinase (ASK) 1. *The EMBO Journal, 17*, 2596–2606.
Salmeen, A., Andersen, J. N., Myers, M. P., Meng, T. C., Hinks, J. A., Tonks, N. K., et al. (2003). Redox regulation of protein tyrosine phosphatase 1B involves a sulphenyl-amide intermediate. *Nature, 423*, 769–773.
Schwertassek, U., Haque, A., Krishnan, N., Greiner, R., Weingarten, L., Dick, T. P., et al. (2014). Reactivation of oxidized PTP1B and PTEN by thioredoxin 1. *The FEBS Journal, 281*, 3545–3558.
Sengupta, R., & Holmgren, A. (2013). Thioredoxin and thioredoxin reductase in relation to reversible S-nitrosylation. *Antioxidants & Redox Signaling, 18*, 259–269.
Shen, X., Burguillos, M. A., Osman, A. M., Frijhoff, J., Carrillo-Jimenez, A., Kanatani, S., et al. (2016). Glioma-induced inhibition of caspase-3 in microglia promotes a tumor-supportive phenotype. *Nature Immunology, 17*, 1282–1290.
Shi, Y., Nikulenkov, F., Zawacka-Pankau, J., Li, H., Gabdoulline, R., Xu, J., et al. (2014). ROS-dependent activation of JNK converts p53 into an efficient inhibitor of oncogenes leading to robust apoptosis. *Cell Death and Differentiation, 21*, 612–623.
Singh, A., Boldin-Adamsky, S., Thimmulappa, R. K., Rath, S. K., Ashush, H., Coulter, J., et al. (2008). RNAi-mediated silencing of nuclear factor erythroid-2-related factor 2 gene expression in non-small cell lung cancer inhibits tumor growth and increases efficacy of chemotherapy. *Cancer Research, 68*, 7975–7984.
Singh, S. S., Li, Y., Ford, O. H., Wrzosek, C. S., Mehedint, D. C., Titus, M. A., et al. (2008). Thioredoxin reductase 1 expression and castration-recurrent growth of prostate cancer. *Translational Oncology, 1*, 153–157.
Tannenbaum, S. R., & White, F. M. (2006). Regulation and specificity of S-nitrosylation and denitrosylation. *ACS Chemical Biology, 1*, 615–618.
Tobiume, K., Matsuzawa, A., Takahashi, T., Nishitoh, H., Morita, K., Takeda, K., et al. (2001). ASK1 is required for sustained activations of JNK/p38 MAP kinases and apoptosis. *EMBO Reports, 2*, 222–228.
Trachootham, D., Alexandre, J., & Huang, P. (2009). Targeting cancer cells by ROS-mediated mechanisms: A radical therapeutic approach? *Nature Reviews. Drug Discovery, 8*, 579–591.
Trachootham, D., Zhou, Y., Zhang, H., Demizu, Y., Chen, Z., Pelicano, H., et al. (2006). Selective killing of oncogenically transformed cells through a ROS-mediated mechanism by beta-phenylethyl isothiocyanate. *Cancer Cell, 10*, 241–252.
Urig, S., & Becker, K. (2006). On the potential of thioredoxin reductase inhibitors for cancer therapy. *Seminars in Cancer Biology, 16*, 452–465.
Wang, L., Yang, Z., Fu, J., Yin, H., Xiong, K., Tan, Q., et al. (2012). Ethaselen: A potent mammalian thioredoxin reductase 1 inhibitor and novel organoselenium anticancer agent. *Free Radical Biology & Medicine, 52*, 898–908.
Wang, X., Zhang, J., & Xu, T. (2008). Thioredoxin reductase inactivation as a pivotal mechanism of ifosfamide in cancer therapy. *European Journal of Pharmacology, 579*, 66–73.
Watson, J. (2013). Oxidants, antioxidants and the current incurability of metastatic cancers. *Open Biology, 3*, 120144.
Wedmann, R., Onderka, C., Wei, S., Szijarto, I. A., Miljkovic, J. L., Mitrovic, A., et al. (2016). Improved tag-switch method reveals that thioredoxin acts as depersulfidase and controls the intracellular levels of protein persulfidation. *Chemical Science, 7*, 3414–3426.

Wipf, P., Lynch, S. M., Birmingham, A., Tamayo, G., Jimenez, A., Campos, N., et al. (2004). Natural product based inhibitors of the thioredoxin-thioredoxin reductase system. *Organic & Biomolecular Chemistry*, *2*, 1651–1658.

Witte, A. B., Anestål, K., Jerremalm, E., Ehrsson, H., & Arnér, E. S. J. (2005). Inhibition of thioredoxin reductase but not of glutathione reductase by the major classes of alkylating and platinum-containing anticancer compounds. *Free Radical Biology & Medicine*, *39*, 696–703.

Wondrak, G. T. (2009). Redox-directed cancer therapeutics: Molecular mechanisms and opportunities. *Antioxidants & Redox Signaling*, *11*, 3013–3069.

Woo, J. R., Kim, S. J., Jeong, W., Cho, Y. H., Lee, S. C., Chung, Y. J., et al. (2004). Structural basis of cellular redox regulation by human TRP14. *The Journal of Biological Chemistry*, *279*, 48120–48125.

Yamada, M., Tomida, A., Yoshikawa, H., Taketani, Y., & Tsuruo, T. (1996). Increased expression of thioredoxin/adult T-cell leukemia-derived factor in cisplatin-resistant human cancer cell lines. *Clinical Cancer Research*, *2*, 427–432.

Yoo, M. H., Xu, X. M., Carlson, B. A., Gladyshev, V. N., & Hatfield, D. L. (2006). Thioredoxin reductase 1 deficiency reverses tumor phenotype and tumorigenicity of lung carcinoma cells. *The Journal of Biological Chemistry*, *281*, 13005–13008.

Yoo, M. H., Xu, X. M., Carlson, B. A., Patterson, A. D., Gladyshev, V. N., & Hatfield, D. L. (2007). Targeting thioredoxin reductase 1 reduction in cancer cells inhibits self-sufficient growth and DNA replication. *PloS One*, *2*, e1112.

Zhang, D. D. (2006). Mechanistic studies of the Nrf2-Keap1 signaling pathway. *Drug Metabolism Reviews*, *38*, 769–789.

Zhang, Z., Wang, A., Li, H., Zhi, H., & Lu, F. (2016). STAT3-dependent TXNDC17 expression mediates Taxol resistance through inducing autophagy in human colorectal cancer cells. *Gene*, *584*, 75–82.

Zhang, B., Zhang, J., Peng, S., Liu, R., Li, X., Hou, Y., et al. (2016). Thioredoxin reductase inhibitors: A patent review. *Expert Opinion on Therapeutic Patents*, 1–10.

CHAPTER SIX

The Regulation of Pathways of Inflammation and Resolution in Immune Cells and Cancer Stem Cells by Selenium

Bastihalli T. Diwakar, Arvind M. Korwar, Robert F. Paulson, K. Sandeep Prabhu[1]
Center for Molecular Immunology and Infectious Disease and Center for Molecular Toxicology and Carcinogenesis, The Pennsylvania State University, University Park, PA, United States
[1]Corresponding author: e-mail address: ksprabhu@psu.edu

Contents

Abstract

Cancer is a complex disease where cancer stem cells (CSCs) maintain unlimited replicative potential, but evade chemotherapy drugs through cellular quiescence. CSCs are able to give rise to bulk tumor cells that have the capability to override antiproliferative signals and evade apoptosis. Numerous pathways are dysregulated in tumor cells, where increased levels of prooxidant reactive oxygen and nitrogen species can lead to localized inflammation to exacerbate all three stages of tumorigenesis: initiation, progression, and metastasis. Modulation of cellular metabolism in tumor cells as well as immune cells in the tumor microenvironment (TME) can impact inflammatory networks. Altering these pathways can potentially serve as a portal for therapy. It is well known that selenium, through selenoproteins, modulates inflammatory pathways in addition to regulating redox homeostasis in cells. Therefore, selenium has the potential to impact the interaction between tumor cells, CSCs, and immune cells. In the sections later, we review the current status of knowledge regarding this interaction, with reference to leukemia stem cells, and the importance of selenium-dependent regulation of inflammation as a potential mechanism to affect the TME and tumor cell survival.

Advances in Cancer Research, Volume 136
ISSN 0065-230X
http://dx.doi.org/10.1016/bs.acr.2017.07.003

1. TUMOR MICROENVIRONMENT: INTERPLAY WITH IMMUNE CELLS AND PATHWAYS OF INFLAMMATION

The tumor tissue consists of parenchyma that includes malignant cells and the stromal compartment. This heterogeneous milieu comprises various cell types that express adhesion molecules, which provide functional activity and structural support to the tumor microenvironment (TME). The TME is also highly vascularized, which allows infiltrating inflammatory cells and a variety of associated tissue-resident cells entrance into the tumor. Most importantly, the infiltrates of inflammatory cells in TME are enriched with different T cell populations, B cells, and innate immune cells such as macrophages, polymorphonuclear leukocytes (neutrophils), and natural killer (NK cells) (Whiteside, 2008). Among these, the presence of cytotoxic $CD8^+$ memory T cells ($CD8^+$ $CD45RO^+$) and their supporter $CD4^+$ T helper 1 (Th1) cells with signature cytokines such as interleukin-2 (IL-2) and interferon-γ (IFN-γ) in TME correlate with better prognosis (Fridman, Pagès, Sautès-Fridman, & Galon, 2012). Other $CD4^+$ cells such as Th2 cells producing IL-4, IL-5, and IL-13 that support B cell responses, macrophages, or $CD4^+$ Th17 T cells that produce IL-17A, IL-17F, IL-21, and IL-22, all of which favor antimicrobial tissue inflammation, are generally thought to promote tumor growth (Fridman et al., 2012). The immunosuppressive T regulatory cells (Tregs; $CD4^+CD25^{hi}FOXP3^+$) exert tumor-promoting function through production of IL-10 and transforming growth factor-β (TGF-β) inhibiting recognition and clearance of tumor cells. Furthermore, high numbers of Tregs in TME correlate with poor prognosis in many types of cancer (Bates et al., 2006; Curiel et al., 2004; Hiraoka, Onozato, Kosuge, & Hirohashi, 2006). Studies in rodent models have shown that selenium synergizes with fish oil to affect antitumor immunity through a decrease in Tregs that is also associated with a decrease in cancer cachexia (Wang et al., 2013). However, in an iodine-induced autoimmune thyroiditis (AIT) model in mice, selenite administration in drinking water (0.3 mg/L) actually increased Tregs with a concomitant decrease in lymphocyte infiltration and antithyroglobulin antibody titres leading to decreased AIT (Xue et al., 2010). This clearly suggests divergent effects of selenium supplementation with regard to the regulation of Tregs in immune suppression that adds another layer of complexity to the Yin–Yang functions of these cells in immune regulation.

NK cells spontaneously kill cells deemed dangerous to the host. This process is regulated by complex integration of signals from receptors expressed on them including inhibitory receptors (Ly49 in mouse and killer immunoglobulin receptor in humans) that recognize the inhibitory signals and activate receptors that recognize stress-induced ligands on target cells such as antibodies and soluble factors like cytokines (Guillerey, Huntington, & Smyth, 2016). In the TME, tumor cells, tumor-associated fibroblasts, and other tumor-induced aberrant infiltrates including tolerogenic macrophages, dendritic cells, and T cells can either secrete immunosuppressive products or interfere with complex receptor array that regulates the activation and antitumor activity of NK cells (Vitale, Cantoni, Pietra, Mingari, & Moretta, 2014). Additionally, activated platelets can directly inhibit NK cells and platelet cloak formation seems to be critical for inhibition of NK cell-mediated killing of circulating tumor cells (Krasnova, Putz, Smyth, & Souza-Fonseca-Guimaraes, 2017). Selenium potentiates NK cell activity in two ways (Fig. 1). First, it prevents the non-enzymatic formation of parafibrin that surrounds tumor cells rendering the tumor vulnerable to immune surveillance (Lipinski, 2016). Second, selenium activates the NK cell population in TME (Kiremidjian-Schumacher, Roy, Wishe, Cohen, & Stotzky, 1996). It is shown that

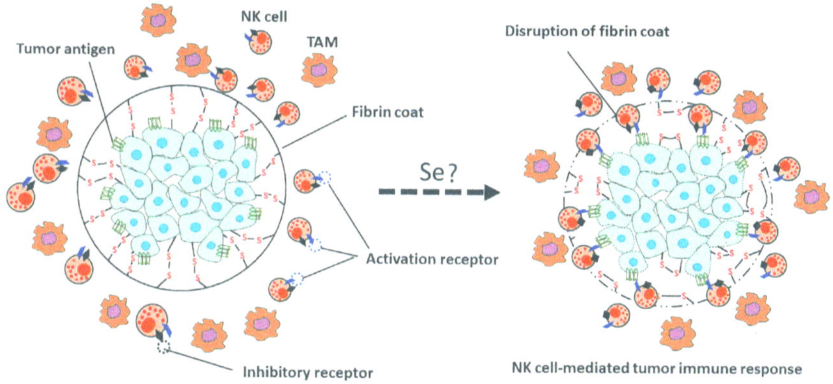

Fig. 1 Potentiating NK cell function in TME by selenium. Hypoxic environment within the TME can shift the redox equilibrium resulting in an increased interaction between tumor cell membrane and fibrinogen or plasma proteins through disulfide bond formation. As a result, a protective fibrin coat facilitating evasion of immune response in TME is formed. Treatment with selenium could reduce the disulfides within the protective fibrin coat to expose neoantigens on tumor cells to NK cells eventually initiating an immune response against the tumor.

selenite has a property of oxidizing polythiols but not monothiols such as cysteine or glutathione (Frenkel, Falvey, & MacVicar, 1991). The oxidant property of selenite is seen only in the presence of polythiols where selenite oxidizes thiol groups to form selenopersulfides at the surface of tumor cells that abolishes proteolytically resistant fibrin coat formed around the tumor (Lipinski, 2005). These changes provide NK cells an opportunity to recognize tumor cells as foreign and mount an immune response (Lipinski, 2016). Selenite (0.5 μM and higher)-induced oxidation of polythiols is implicated in death of prostate cancer cells (Zhong & Oberley, 2001) and hepatoma cells (Shen, Yang, & Ong, 1999).

Selenium also exhibits cancer preventive properties as a regulator of cellular redox via selenoproteins at low concentrations, whereas at higher concentrations selenite is shown to induce apoptosis in malignant cells through oxidative stress (Nilsonne et al., 2006). Further, Se-organic compounds (diselenides) act as glutathione peroxidase (GPx) mimetic drugs in cells with high thiol content such as cells exhibiting a high rate of metabolism and proliferation, all characteristics of tumor (Bartolini, Piroddi, et al., 2015; Frenkel et al., 1991). Furthermore, in malignant cells, toxicity to selenium is reached at considerably lower concentrations compared to nonmalignant cells, suggesting the cytotoxic effects are specific for tumor cells (Husbeck, Nonn, Peehl, & Knox, 2006). While the exact mechanisms are unclear, it is plausible that redox modulation of signaling pathways through changes in cellular oxidant tone combined with changes in redox status of protein thiols, selenium could interfere with the premetastatic niche created by platelet-mediated fibrin deposition on tumor cells (Bartolini, Commodi, et al., 2015; Bartolini, Piroddi, et al., 2015; Quail & Joyce, 2013).

Inflammatory responses play decisive roles at different stages of tumor development, including initiation, promotion, malignant conversion, invasion, and metastasis. Immune cells that infiltrate the tumor engage in an extensive and dynamic cross talk with cancer cells. The key molecular pathways involved are activation of transcription factors, mainly nuclear factor-κ B (NF-κB), signal transducer and activator of transcription 3 (STAT3), and hypoxia-inducible factor-1α (HIF-1α), in tumor cells. These transcription factors coordinate the production of inflammatory mediators, including cytokines and chemokines, as well as the production of cyclooxygenase-2 (COX-2) leading to production of prostaglandins (PGs). These factors recruit and activate various leukocytes, most notably cells of the myelomonocytic lineage. One of the key leukocyte infiltrate includes the tumor-associated macrophages (TAMs) and related cell types that are involved in carcinogenesis and/or tumor invasion and metastasis (Bunt et al., 2007; Coussens, Tinkle, Hanahan, & Werb, 2000; Mantovani, Bottazzi, Colotta, Sozzani, & Ruco, 1992). The tumor cells derived factors activate the same key transcription factors in inflammatory cells, stromal cells, and tumor cells, resulting in more inflammatory mediators being produced and a cancer-related inflammatory microenvironment is generated. Therefore, targeting of inflammatory mediators (chemokines and TNF-α and IL-1β), key transcription factors involved in inflammation (such as NF-κB and STAT3) or inflammatory cells decreases the incidence and spread of cancer (Mantovani, Allavena, Sica, & Balkwill, 2008).

TAMs are a major component of tumor infiltrates that are reprogrammed to be protumorigenic. TAMs often exhibit an alternatively activated (M2)-like phenotype, which inhibits lymphocyte functions through release of antiinflammatory cytokines such as IL-10 to negatively impact antitumor immune surveillance (Mantovani, Sozzani, Locati, Allavena, & Sica, 2002; Martinez, Sica, Mantovani, & Locati, 2008). Furthermore, TAMs are major contributors of tumor angiogenesis. The comparative analysis of TAM transcriptomes with available clinical databases shows a transcriptional signature predictive of tumor survival (Ojalvo, Whittaker, Condeelis, & Pollard, 2010).

Myeloid suppressor cells (MDSCs; $CD34^{+}CD33^{+}CD13^{+}CD15^{-}$) are a heterogeneous population of bone marrow-derived early progenitors that resemble immature dendritic cells and granulocytes. These cells are characterized by the increased production of extracellular degradative enzymes, cytokines, and RONS (Draghiciu, Lubbers, Nijman, & Daemen, 2015). MDSCs promote tumor growth and suppress immune cell functions by blunting leukocyte responses (Ochoa, Zea, Hernandez, & Rodriguez, 2007). Given their plasticity and their ability to exhibit variable phenotypes, MDSCs differentiate into Tregs. While there are no reports on the role of selenium in MDSCs per se, it is known that antioxidant vitamin E treatment can significantly reduce the number of MDSCs in a HPV16 E7-expressing TC-1 tumor model (Kang et al., 2014). Like the conventional macrophages, MDSCs also polarize toward M1 or M2 phenotypes. As in the case of macrophages, where selenium skews them toward an M2 phenotype to produce proresolution mediators (Fig. 2), it remains to be seen if selenium has a similar impact on MDSC polarization to affect the proliferation of tumor cells by locally blunting inflammation to resolve the tumor lesion.

NF-κB, a redox-sensitive transcription factor and master regulator of inflammatory gene expression, is highly dysregulated in TAMs

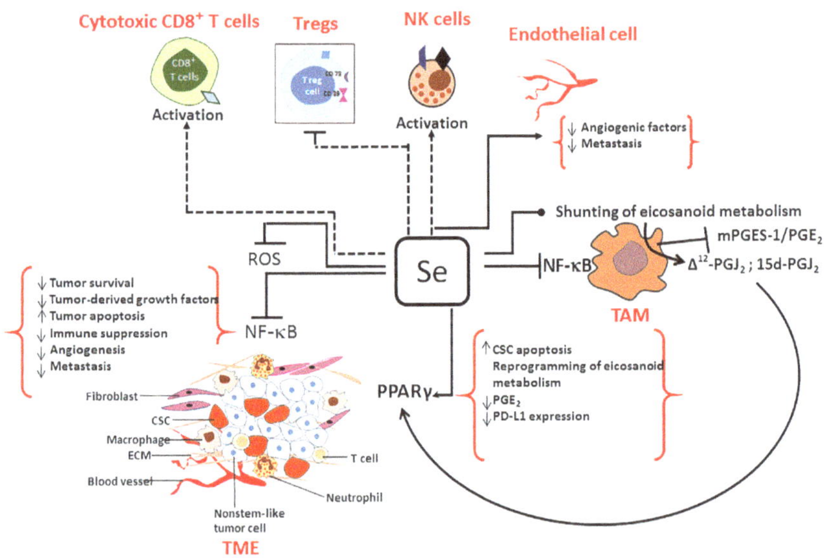

Fig. 2 Selenium reconditions TME by affecting the protumorigenic phenotype of immune infiltrates within the stromal compartment effecting apoptosis of stem cells and decreased survival of bulk tumor cells. TAMs mediate immune suppression, promote growth, angiogenesis, and proliferation of tumor cells through dysregulated activation of NF-κB. Selenium-mediated redox modulation in macrophages inhibits IKKβ leading to decreased activation of NF-κB and its downstream effectors of immune suppression, tumor cell survival, and angiogenesis. Selenium may also sensitize tumor cells to TRAIL-induced apoptosis through NF-κB suppression by increasing the half-life of IKKα in cytosol. The intratumoral angiogenesis is abrogated by selenium through decreased expression of angiopoietin-2, PDGF, and VEGF, which might occur through ROS-mediated DNA damage in endothelial cells. Selenium also enhances the cytotoxic potential of NK cells and CD8^{+} T cells, while negatively regulating Treg cells, which together promote inhibition of tumor growth. Selenium reprograms arachidonic acid metabolism in macrophages leading to proresolving CyPGs, while decreasing inflammatory PGs. CyPGs activate PPARγ in CSCs to block quiescence, while decreased PGE_2 production by TAMs promotes antitumor effects by decreasing PD-L1 expression to affect tumor aggression, angiogenesis, and immune suppression.

(Hagemann et al., 2008; Mantovani et al., 2008; Saccani et al., 2006). Several factors (TNF-α, IL-1, hypoxia, HMGB1, TLR ligands) important for protumor phenotype of TAMs are linked to activation of NF-κB. Interestingly, studies have shown that sodium selenite affects activation of NF-κB by inhibiting IKKβ (Vunta et al., 2007) that is known to be activated in TAMs (Saccani et al., 2006). Thus, it is plausible that perturbation of NF-κB in TAMs by selenium may result in decreased local immune suppression as well as affecting the expression of cell survival factors (IL-6, TNF-α, COX-2,

PGE_2) (Mantovani et al., 2008). Similar effects of selenium were observed in murine macrophages (Vunta et al., 2007). Selenium also impacted angiogenesis through downregulation of VEGF, IL-8, and COX-2, which affects invasion and metastasis (Mantovani et al., 2008). COX-2 appears to be one of the key factors regulated by selenium. The bioactive lipid prostaglandin E_2 (PGE_2) formed from arachidonic acid via COX-2 and microsomal PGE_2 synthase-1 (mPGES-1) facilitates both cancer inflammation and immune suppression. In a recent study, the COX-2/mPGES-1/PGE_2 pathway was demonstrated to regulate the expression of programmed death-ligand 1 (PD-L1), which is associated with increased tumor aggressiveness and immune suppression in TME (Prima, Kaliberova, Kaliberov, Curiel, & Kusmartsev, 2017). Interestingly, selenium through selenoprotein expression reprograms arachidonic acid metabolism by uncoupling COX-2 from mPEGS-1 during inflammation in macrophages to enhance the production of H-PGDS-derived cyclopentenone PGs (CyPGs), Δ^{12}-PGJ_2 and 15d-PGJ_2 (Gandhi et al., 2011; Vunta et al., 2007). A consequence of such an "eicosanoid class switching" mechanism is the significant decrease in PGE_2 by selenium supplementation. It remains to be seen if selenium decreases tumor aggressiveness through the modulation of expression of PD-L1 by altering the levels of PGE_2 in the TME. However, antiproliferatory CyPGs, Δ^{12}-PGJ_2 and 15d-PGJ_2 that are produced by bone marrow-derived macrophages (BMDMs) because of the selenium-dependant eicosanoid class switching were capable of efficiently mediating apoptosis of leukemia stem cells (LSCs) in models of chronic myeloid leukemia (CML), suggesting that similar metabolic reprogramming in TAMs could impact the proliferation of tumor cells. However, this needs to be addressed in the future.

2. SELENIUM-DEPENDENT EFFECTS ON CANCER STEM CELLS IN LEUKEMIA

A causal inverse relationship between selenium and hematological malignancies exists (Azarm, Fazilati, Azarm, & Azarm, 2013; Cai et al., 2016). An early clinical study reported a decrease in total leukocyte counts, immature leukocytes, and spleen size upon oral administration of selenium cystine (diseleno-dialanine) in patients with both CML and AML (Weisberger & Suhrland, 1956). Since then several in vitro studies have demonstrated the antileukemic properties of sodium selenite in leukemia cell lines (Jiang, Macey, Lin, & Newland, 1992; Liao, Bian, Xie, &

Peng, 2017; Misra et al., 2016). Sodium selenite supplementation was effective in decreasing circulating leukemia cells in murine Friend erythroleukemia as well as murine CML models (Gandhi et al., 2014). Lu et al. (2004) demonstrated that the beneficial effects of selenium were conferred, in part, by improved immune surveillance of leukemia cells by NK cells. In addition, selenite treatment-induced cytotoxicity in ex vivo cultures of leukemia cells isolated from peripheral blood of ALL and AML patients (Olm et al., 2009). Intriguingly, no detrimental effects of selenium were observed in cells derived from bone marrow or peripheral blood of normal individuals treated with selenite in this study. The specificity of selenite for LSCs was observed in murine models of CML, which utilized primary hematopoietic stem cells (HSCs) transduced with a retrovirus expressing the fusion oncoprotein BCR–ABL and GFP. Transplantation of these cells leads to disease, suggesting that they act as leukemia stem cells (CML-LSCs). Intriguingly, treatment of these cells with selenite (50–500 nM) led to apoptosis that was only restricted to LSCs, while the HSCs expressing GFP were not affected (Finch et al., 2017; Gandhi et al., 2014, 2011). More importantly, these in vitro studies corroborated well with in vivo studies showing similar efficacy toward selenium supplementation on the ablation of LSCs and remission of the disease. While the basis of the selectivity of selenium toward LSCs is unclear, our studies show that selenium induces a number of effects in CML-LSCs that could account, in part, for its efficacy and selectivity. These effects include antiandrogen activity, DNA damage-induced response involving ATM kinase, and its downstream effectors such as p53, caspases, p21, and inhibition of antiapoptotic pathways via the decrease in Mcl1-dependent control of Bcl2 (Finch et al., 2017; Gandhi et al., 2014; Sanmartin, Plano, & Palop, 2008). Furthermore, p53 activation could also amplify a positive regulatory feedback loop involving long noncoding RNA that enhances tumor suppressor activity of p53 (Sánchez et al., 2014).

CML originates in the bone marrow and is primarily due to a chromosomal translocation [t(9:22); Philadelphia chromosome (Ph+)] in progenitor cells resulting in the chimeric fusion oncoprotein, BCR–ABL1, which exhibits dysregulated tyrosine kinase activity (Koretzky, 2007). BCR–ABL drives the development of a two-stage disease with the overproduction of mature myeloid cells in the chronic phase that rapidly progresses to a blast crisis stage, which is an aggressive acute leukemia (Calabretta & Perrotti, 2004). The standard of care for CML patients is tyrosine kinase inhibitor (TKI) therapy. These maintenance drugs inhibit the BCR–ABL tyrosine kinase activity, which limits the overproduction of myeloid cells. Although these drugs have been very effective in treating CML, they are often associated with side effects. In addition, patients develop resistance to TKI therapies, which has necessitated the development of newer generation of inhibitors. However, majority of patients will relapse when the TKI therapy is discontinued as these drugs primarily target mature cycling cells (that represent the bulk leukemia cells) and remain ineffective against CML LSCs (Druker et al., 2006, 2001; Gorre et al., 2001). Moreover, because TKIs do not target LSCs, there is a significant gap in our ability to specifically target LSCs to treat CML (Bhatia et al., 2003; Copland et al., 2006; Graham et al., 2002; Jorgensen, Allan, Jordanides, Mountford, & Holyoake, 2007). Much like the HSCs that are maintained in specific niches in the bone marrow, cancer stem cells (CSCs) also maintain quiescence with self-renewal capabilities (Bakker, Qattan, Mutti, Demonacos, & Krstic-Demonacos, 2016). LSCs are perhaps the best-studied CSCs. They give rise to bulk leukemia cells, which are responsible for the clinical presentation of leukemias and the pathologies associated with the disease. In addition, LSCs also self-renew, which maintain a population of quiescent LSCs (Huntly & Gilliland, 2005; Riether, Schürch, & Ochsenbein, 2015; Tabe & Konopleva, 2014; Tannishtha, Morrison, Clarke, & Weissman, 2001). Thus, novel therapeutics that eradicate LSCs could provide a durable remission from the disease.

3. INTERACTION OF LSCs WITH IMMUNE CELLS MEDIATED BY CyPGs

LSCs are sensitive to intrinsic and extrinsic factors that include components generated by immune cells within the bone marrow or other tissue microenvironment (Bakker et al., 2016; Konopleva & Jordan, 2011; Tabe & Konopleva, 2014). As introduced in Section 1, the extrinsic pathway of LSC control involves monocytes and macrophages that can be reprogrammed with selenium to promote eicosanoid class switching to produce bioactive CyPGs, which are endowed with antiproliferative activity (Finch et al., 2017; Gandhi et al., 2014, 2011). Utilizing conditioned media BMDMs, studies have shown that expression of the selenoproteome is critical for eicosanoid class switching to produce CyPGs (Gandhi et al., 2014; Vunta et al., 2007).

Coincubation of macrophage-conditioned media with LSCs activated pathways of apoptosis, which were inhibited when the macrophages were treated with NSAIDs prior to addition of selenite (Gandhi et al., 2014). These CyPGs downregulated the antiapoptotic factor, NF-κB

(Gandhi et al., 2011; Vunta et al., 2007). In fact, osteoblasts that are part of osteoblastic niche of bone marrow play an important role in quiescence of LSCs. Selenium (in the form of methylseleninic acid) modified the osteoblast inflammatory stress response to bone metastatic breast cancer cells by downregulating the activation of NF-κB leading to decreased expression of COX-2 and INOS and proinflammatory mediators such as IL-6, MCP-1 (Chen et al., 2009). This observation demonstrates the efficacy of selenium to modify extrinsic factors of TME and the importance of the microenvironment where macrophage-derived factors, including CyPGs, possibly act as key mediators of LSC apoptosis. While the mechanism of action of CyPGs in LSC apoptosis continues to be investigated, studies suggest that CyPGs exacerbate the production of ROS through a mechanism involving NADPH oxidase (Nox1), which in turn leads to the activation of ATM kinase-p53-dependent downstream pathways of apoptosis (Gandhi et al., 2014; Hegde et al., 2011). While the mechanism of activation of Nox1 by CyPGs is intriguing, it appears that certain membrane-bound receptors for CyPGs could be involved in the cross talk, which needs to be substantiated further.

On the other hand, constitutive activation of the NF-κB pathway allows LSCs to evade mechanisms of apoptosis (Bosman, Schuringa, & Vellenga, 2016; Konopleva & Jordan, 2011; Zhou, Ching, & Chng, 2015). As part of the intrinsic pathway, selenium is shown to suppress activation of NF-κB in murine model of esophageal carcinogenesis (Yang, Jia, Chen, Yang, & Li, 2012), human prostate cancer cells (Christensen, Nartey, Hada, Legg, & Barzee, 2007), during hepatocarcinogenesis (Alwahaibi, Mohamed, & Alhamadani, 2010) and in colorectal cancer (Méplan & Hesketh, 2012; Ravasco et al., 2010). While it is known that LSCs can also endogenously produce CyPGs in response to supplementation with selenium, albeit at much lower concentration than macrophages, these CyPGs can diffuse into cells to bind to intracellular receptors such as PPARγ. Interestingly, PPARγ agonists target CML LSCs in vitro by decreasing transcription of STAT5a, which is required for LSC self-renewal (Prost et al., 2015). Furthermore, PPARγ agonists can potentiate the action of TKIs in CML (Glodkowska-Mrowka et al., 2016). In a more recent ACTIM phase 2 clinical trial, PPARγ agonist pioglitazone in combination with imatinib (first-generation TKI) showed positive outcome in treatment of CML. However, pioglitazone is associated with side effects that limits it use (Gavrilova et al., 2003; Nesto et al., 2003). Along these lines, a recent study demonstrated the activation PPARγ-dependent signaling in LSCs by selenium led to decreased expression of STAT5a and its two targets, CITED2 and HIF-2α, which are both known to maintain LSC quiescence (Finch et al., 2017). Treatment of CML mice or isolated LSCs with selenium in the presence GW9662, a PPARγ antagonist, blocked the ability of selenium to downregulate the expression of STAT5a and its targets leading to leukemia. These studies suggest that changes in downstream effectors of PPARγ, which includes metabolic enzymes, could impart some selectivity to ablation of LSCs by selenium.

4. TARGETING LSC METABOLISM WITH SELENIUM

Generally, cancer cells prefer aerobic glycolysis to shunt glucose toward cytoplasmic lactate instead of pyruvate-derived acetyl-CoA that would enter the tricarboxylic acid (TCA) cycle coupled to oxidative phosphorylation (OXPHOS). This unusual capability of tumor cells to reprogram glucose metabolism is an emerging avenue for cancer therapeutics that was originally demonstrated by Warburg (1956). Cancer cells prefer the energetically unfavorable glycolysis that produces 18 times less ATP per mole of glucose than OXPHOS. However, this process is more efficient in generating precursors for nucleotide and lipid biosynthesis for rapid cell division and is several orders of magnitude faster, which aids in metabolic coupling within the tumor niche leading to enhanced proliferation (Martinez-Outschoorn, Peiris-Pagés, Pestell, Sotgia, & Lisanti, 2017). Moreover, acquired oncogenic mutations alter the metabolic network, where many of tumor-promoting genes are intricately associated with metabolic regulation. It is also evident that reprogramming of energy metabolism and evading immune destruction is a hallmark of cancer cells that navigate through the TME (Hanahan & Weinberg, 2011; Hsu & Sabatini, 2008). The size of the tumor niche population in which CSCs exist determines the availability of nutrients and oxygen. The regulation of nutrient availability, in turn, affects the metabolic phenotype of CSCs, which is poorly understood in the context of tumorigenesis. It is established that not all malignant cells are metabolically and functionally equivalent, which has been demonstrated in leukemia, breast cancer, and various other cancers (Sancho, Barneda, & Heeschen, 2016). This functional disparity is often attributed to genetic and epigenetic heterogeneity within multiple subclonal populations, which contribute to various metabolic adaptations specific to each tumor type (Sancho et al., 2016). Consequently, CSCs exhibit diverse metabolic phenotypes. Based on the existing literature, though CSCs are broadly classified to be primarily glycolytic, they are also known to selectively rely on OXPHOS to maintain their stemness. CSCs selectively adopt to nutrient deprivation and tumor niche metabolic phenotype. Mechanistically, the glycolytic program is also associated with stemness by keeping high antioxidative capacity and increased pentose phosphate pathway (Sancho et al., 2016). High aerobic glycolytic rates and glutaminolysis promote fatty acid oxidation for ATP and reducing equivalents such as NADPH. Notably leukemic and pancreatic CSCs utilize fatty acid oxidation regulated by PPARδ pathway (Ito et al., 2012; Samudio et al., 2010; Sancho et al., 2016). In these cells, NADPH generation was key for self-renewal, maintaining stemness as well as the CSC phenotype. To this end, selenoproteins within cells could play a critical role in the modulation of glucose and fatty acid metabolism.

It is well established that deficiency of GPX1 and other selenoproteins can increase the expression of p53, a tumor suppressor gene and a master controller of cellular metabolism (Cheng, Zheng, Quimby, Roneker, & Lei, 2003). Activation of p53 induces the expression of genes associated with DNA repair, apoptosis, cell cycle, and senescence, and more importantly, glucose and fatty acid metabolism (Bensaad et al., 2006). Similarly, nucleolar oxidoreductase selenoprotein H (SelenoH) overexpression in neuronal cells protected against UV-induced DNA damage by suppressing p53 and caspase-mediated apoptosis (Mendelev, Witherspoon, & Li, 2009). Recent studies by Cox et al. demonstrated that SelenoH deficiency exacerbated oxidative stress-dependent activation of p53, where inflammation was a key factor, leading to gastrointestinal tumorigenesis in a zebrafish model (Cox et al., 2016). Given that selenium supplementation can downregulate inflammation, it remains to be seen if SelenoH regulates expression of inflammatory genes such as TNF-α and IL-1β (Fig. 3).

While the studies mentioned earlier suggest an inverse causal relationship between selenium status and p53 activation, these mechanisms may operate differently when the tumor has already been initiated, as in the case of expression of the oncoprotein BCR–ABL in HSCs in the murine CML model. Comparison of HSCs (which represent normal stem cells) and LSCs prior to treatment with selenium showed HSCs to have higher ROS compared to LSCs (Gandhi et al., 2014). Contrary to the prevailing idea with other cell types is the ability of LSCs to tightly regulate ROS and p53 expression, which could enable the LSCs to evade autophagy and/or apoptosis. However, treatment of both these cell types, HSCs and LSCs, with supraphysiological levels of selenium (as selenite; 250–500 nM) selectively

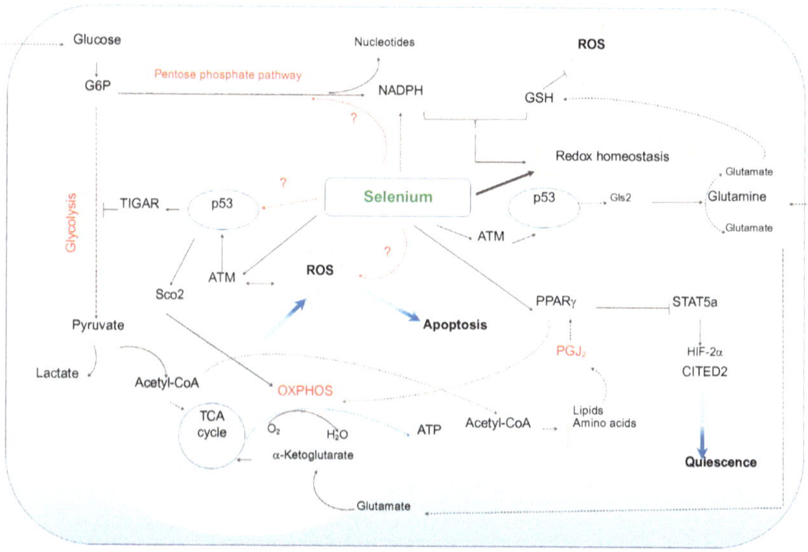

Fig. 3 Significance of selenium in cancer stem cell metabolism. CSCs generally display glycolytic metabolic phenotype selectively coupled with TCA cycle and OXPHOS to generate energy and cellular components (amino acids and lipids) to support their stemness. Metabolically active CSCs generate ROS, which is compensated by production of antioxidants, including GSH, NADPH derived from the pentose phosphate pathway. Selenoproteins through their redox gatekeeper functions can effectively modulate redox homeostasis. Selenium treatment of CSCs activates ATM kinase, which in turn activates p53 leading to metabolic reprogramming. p53 targets pathways via TIGAR-mediated glycolytic arrest, Sco2-mediated potentiation of OXPHOS, and Gls2-mediated regulation of glutamine metabolism to ultimately exacerbate ROS production to activate pathways of apoptosis. Selenium-mediated eicosanoid class switching leads to the production of PGJ_2 as an adaptive response for protection against ROS, which is mediated through PPARγ that transcriptionally represses STAT5a to effectively inhibit downstream target genes, HIF-2α and CITED2, known for maintenance of CSC quiescence, as in CML LSCs.

increased ROS by several orders of magnitude only in LSCs. This increase was also accompanied by increased ATM kinase-p53-caspase 3-dependent pathway of apoptosis (Gandhi et al., 2014). Intriguingly, such an increase in ROS was accompanied by increased expression of many antioxidant genes, including Gpx4 that were unable to compensate and protect the LSCs from ROS. The ability of selenium supplementation at levels where most of the selenoproteins are saturated to increase ROS selectively in one cell type (LSCs) suggests a cross talk between selenium and cysteine thiols in other proteins, particularly phosphatases, in the form of selenopersulfides, which modulates their activities leading to increased cellular oxidative stress.

Of interest to us is the ability of p53 to activate pathways of metabolism in the LSCs that may also contribute to increased cellular oxidative stress. For instance, Gandhi et al. (2014) demonstrated that selenite treatment increased Sco2 (synthesis of cytochrome oxidase-2) and Rm2b (P53r2; ribonucleotide-diphosphate reductase subunit RM2B), which are likely to increase the ability to LSCs to favor OXPHOS that further contributes to increase in ROS. In addition, TP53-inducible glycolysis and apoptosis regulator (TIGAR), a fructose bisphosphatase that negatively regulates glycolysis, was increased in selenite-treated LSCs, which corroborates data from other cancers where TIGAR is overexpressed (Ko et al., 2016; Wanka, Steinbach, & Rieger, 2012). TIGAR overexpression is implicated in inhibition of glycolysis and protective mechanisms against ROS in glioma cells (Wanka et al., 2012).

Metabolic adaptations that govern quiescence in CSCs and processes involving self-renewal, tumor initiation, and chemo-resistance call for thorough studies involving flux of key metabolites through specific pathways in conjunction with the use of genetic methods to elucidate the underpinnings of the resiliency of these cells. A better understanding of these pathways and how they are controlled by altered oncogenes and tumor suppressors needs to be integrated with studies of the alterations in cellular redox tone of cells and how these changes impact metabolomic shifts that ultimately dictate cell fate decisions. It is here that the redox gatekeeper functions of selenoproteins are likely to play a pivotal role leading to mechanistic insights involving metabolomic changes within CSCs in collaboration with immune cells in the tumor niche.

5. SUMMARY

In conclusion, this review highlights various mechanisms by which selenium and selenoproteins affect various stages of tumorigenesis that involves a cross talk with the immune component. In particular, based on the recent work in murine models of CML, the remarkable ability of selenite supplementation to target LSCs for apoptosis is dependent on the enhanced production of CyPGs that drastically decreases the disease burden. The ablation of LSCs in murine models highlights a new area within selenium biology where selenium-mediated changes in metabolic pathways within LSCs as well as immune cells that comprise the microenvironment in the bone marrow and/or spleen underlie the protective effects of selenium supplementation. Even though the underlying mechanism of antileukemic activity of selenium is still not completely understood, the ability of selenium to affect quiescence of CML-LSCs is mediated through the enhanced production of CyPGs that are not only generated within the LSCs but also by immune cells (macrophages) via reprogramming of cellular metabolism is a promising observation. However, one of the key questions that needs to be addressed is the identity of selenoprotein(s) that is pivotal for the eicosanoid class switching, which is currently being pursued. Regardless, these findings suggest new options for the development of new treatments for CML, where dietary selenium supplementation could be substituted for synthetic PPARγ agonists.

ACKNOWLEDGMENTS

This work was supported, in part, by grants from the National Institutes of Health R01CA162665, R01DK077152, Office of Dietary Supplements (ODS-NIH), USDA-NIFA Hatch project numbers 4605 (K.S.P.) and 4736 (R.F.P.). We thank current and former members of the Prabhu laboratory for their invaluable contributions.

REFERENCES

Alwahaibi, N., Mohamed, J., & Alhamadani, A. (2010). Supplementation of selenium reduces chemical hepatocarcinogenesis in male Sprague-Dawley rats. *Journal of Trace Elements in Medicine and Biology*, *24*, 119–123.

Azarm, T., Fazilati, M., Azarm, H., & Azarm, A. (2013). Serum selenium levels in chronic lymphocytic leukemia. *Advanced Biomedical Research*, *2*, 44.

Bakker, E., Qattan, M., Mutti, L., Demonacos, C., & Krstic-Demonacos, M. (2016). The role of microenvironment and immunity in drug response in leukemia. *Biochimica et Biophysica Acta*, *1863*, 414–426.

Bartolini, D., Commodi, J., Piroddi, M., Incipini, L., Sancineto, L., Santi, C., et al. (2015). Glutathione S-transferase pi expression regulates the Nrf2-dependent response to hormetic diselenides. *Free Radical Biology & Medicine*, *88*, 466–480.

Bartolini, D., Piroddi, M., Tidei, C., Giovagnoli, S., Pietrella, D., Manevich, Y., et al. (2015). Reaction kinetics and targeting to cellular glutathione S-transferase of the glutathione peroxidase mimetic PhSeZnCl and its D,L-polylactide microparticle formulation. *Free Radical Biology & Medicine*, *78*, 56–65.

Bates, G. J., Fox, S. B., Han, C., Leek, R. D., Garcia, J. F., Harris, A. L., et al. (2006). Quantification of regulatory T cells enables the identification of high-risk breast cancer patients and those at risk of late relapse. *Journal of Clinical Oncology*, *24*, 5373–5380.

Bensaad, K., Tsuruta, A., Selak, M. A., Vidal, M. N. C., Nakano, K., Bartrons, R., et al. (2006). TIGAR, a p53-inducible regulator of glycolysis and apoptosis. *Cell*, *126*, 107–120.

Bhatia, R., Holtz, M., Niu, N., Gray, R., Snyder, D. S., Sawyers, C. L., et al. (2003). Persistence of malignant hematopoietic progenitors in chronic myelogenous leukemia patients in complete cytogenetic remission following imatinib mesylate treatment. *Blood*, *101*, 4701–4707.

Bosman, M. C., Schuringa, J. J., & Vellenga, E. (2016). Constitutive NF-kappaB activation in AML: Causes and treatment strategies. *Critical Reviews in Oncology/Hematology*, *98*, 35–44.

Bunt, S. K., Yang, L., Sinha, P., Clements, V. K., Leips, J., & Ostrand-Rosenberg, S. (2007). Reduced inflammation in the tumor microenvironment delays the accumulation of myeloid-derived suppressor cells and limits tumor progression. *Cancer Research*, *67*, 10019–10026.

Cai, X., Wang, C., Yu, W., Fan, W., Wang, S., Shen, N., et al. (2016). Selenium exposure and cancer risk: An updated meta-analysis and meta-regression. *Scientific Reports*, *6*, 1–17. Article number 19213. http://dx.doi.org/10.1038/srep19213.

Calabretta, B., & Perrotti, D. (2004). Review in translational hematology the biology of CML blast crisis. *Blood*, *103*, 4010–4022.

Chen, Y. C., Sosnoski, D. M., Gandhi, U. H., Novinger, L. J., Prabhu, K. S., & Mastro, A. M. (2009). Selenium modifies the osteoblast inflammatory stress response to bone metastatic breast cancer. *Carcinogenesis*, *30*, 1941–1948.

Cheng, W.-H., Zheng, X., Quimby, F. R., Roneker, C. A., & Lei, X. G. (2003). Low levels of glutathione peroxidase 1 activity in selenium-deficient mouse liver affect c-Jun N-terminal kinase activation and p53 phosphorylation on Ser-15 in pro-oxidant-induced aponecrosis. *The Biochemical Journal*, *370*, 927–934.

Christensen, M. J., Nartey, E. T., Hada, A. L., Legg, R. L., & Barzee, B. R. (2007). High selenium reduces NF-kappaB-regulated gene expression in uninduced human prostate cancer cells. *Nutrition and Cancer*, *58*, 197–204.

Copland, M., Hamilton, A., Elrick, L. J., Baird, J. W., Allan, E. K., Jordanides, N., et al. (2006). Dasatinib (BMS-354825) targets an earlier progenitor population than imatinib in primary CML but does not eliminate the quiescent fraction. *Blood*, *107*, 4532–4539.

Coussens, L. M., Tinkle, C. L., Hanahan, D., & Werb, Z. (2000). MMP-9 supplied by bone marrow-derived cells contributes to skin carcinogenesis. *Cell*, *103*, 481–490.

Cox, A. G., Tsomides, A., Kim, A. J., Saunders, D., Hwang, K. L., Evason, K. J., et al. (2016). Selenoprotein H is an essential regulator of redox homeostasis that cooperates with p53 in development and tumorigenesis. *Proceedings of the National Academy of Sciences of the United States of America*, *113*, E5562–E5571.

Curiel, T. J., Coukos, G., Zou, L., Alvarez, X., Cheng, P., Mottram, P., et al. (2004). Specific recruitment of regulatory T cells in ovarian carcinoma fosters immune privilege and predicts reduced survival. *Nature Medicine*, *10*, 942–949.

Draghiciu, O., Lubbers, J., Nijman, H. W., & Daemen, T. (2015). Myeloid derived suppressor cells-an overview of combat strategies to increase immunotherapy efficacy. *Oncoimmunology*, *4*, e954829.

Druker, B. J., Guilhot, F., O'Brien, S. G., Gathmann, I., Kantarjian, H., Gattermann, N., et al. (2006). Five-year follow-up of patients receiving imatinib for chronic myeloid leukemia. *The New England Journal of Medicine*, *355*, 2408–2417.

Druker, B. J., Sawyers, C. L., Kantarjian, H., Resta, D. J., Reese, S. F., Ford, J. M., et al. (2001). Activity of a specific inhibitor of the BCR-ABL tyrosine kinase in the blast crisis of chronic myeloid leukemia and acute lymphoblastic leukemia with the Philadelphia chromosome. *The New England Journal of Medicine*, *344*, 1038–1042.

Finch, E. R., Tukaramrao, D. B., Goodfield, L. L., Quickel, M. D., Paulson, R. F., & Prabhu, K. S. (2017). Activation of PPARγ by endogenous prostaglandin J 2 mediates the antileukemic effect of selenium in murine leukemia. *Blood*, *129*, 1802–1810.

Frenkel, G. D., Falvey, D., & MacVicar, C. (1991). Products of the reaction of selenite with intracellular sulfhydryl compounds. *Biological Trace Element Research*, *30*, 9–18.

Fridman, W. H., Pagès, F., Sautès-Fridman, C., & Galon, J. (2012). The immune contexture in human tumours: Impact on clinical outcome. *Nature Reviews Cancer*, *12*, 298–306.

Gandhi, U. H., Kaushal, N., Hegde, S., Finch, E. R., Kudva, A. K., Kennett, M. J., et al. (2014). Selenium suppresses leukemia through the action of endogenous eicosanoids. *Cancer Research*, *74*, 3890–3901.

Gandhi, U. H., Kaushal, N., Ravindra, K. C., Hegde, S., Nelson, S. M., Narayan, V., et al. (2011). Selenoprotein-dependent up-regulation of hematopoietic prostaglandin D2

synthase in macrophages is mediated through the activation of peroxisome proliferator-activated receptor (PPAR) gamma. *Journal of Biological Chemistry*, *286*, 27471–27482.
Gavrilova, O., Haluzik, M., Matsusue, K., Cutson, J. J., Johnson, L., Dietz, K. R., et al. (2003). Liver peroxisome proliferator-activated receptor gamma contributes to hepatic steatosis, triglyceride clearance, and regulation of body fat mass. *Journal of Biological Chemistry*, *278*, 34268–34276.
Glodkowska-Mrowka, E., Manda-Handzlik, A., Stelmaszczyk-Emmel, A., Seferynska, I., Stoklosa, T., Przybylski, J., et al. (2016). PPARγ ligands increase antileukemic activity of second- and third-generation tyrosine kinase inhibitors in chronic myeloid leukemia cells. *Blood Cancer Journal*, *6*, e377.
Gorre, M. E., Mohammed, M., Ellwood, K., Hsu, N., Paquette, R., Rao, P. N., et al. (2001). Clinical resistance to STI-571 cancer therapy caused by BCR-ABL gene mutation or amplification. *Science*, *293*, 876–880.
Graham, S. M., Jorgensen, H. G., Allan, E., Pearson, C., Alcorn, M. J., Richmond, L., et al. (2002). Primitive, quiescent, Philadelphia-positive stem cells from patients with chronic myeloid leukemia are insensitive to STI571 in vitro. *Blood*, *99*, 319–325.
Guillerey, C., Huntington, N. D., & Smyth, M. J. (2016). Targeting natural killer cells in cancer immunotherapy. *Nature Immunology*, *17*, 1025–1036.
Hagemann, T., Lawrence, T., McNeish, I., Charles, K. A., Kulbe, H., Thompson, R. G., et al. (2008). "Re-educating" tumor-associated macrophages by targeting NF-kappaB. *The Journal of Experimental Medicine*, *205*, 1261–1268.
Hanahan, D., & Weinberg, R. A. (2011). Hallmarks of cancer: The next generation. *Cell*, *144*, 646–674.
Hegde, S., Kaushal, N., Ravindra, K. C., Chiaro, C., Hafer, K. T., Gandhi, U. H., et al. (2011). Δ12-Prostaglandin J3, an omega-3 fatty acid-derived metabolite, selectively ablates leukemia stem cells in mice. *Blood*, *118*, 6909–6919.
Hiraoka, N., Onozato, K., Kosuge, T., & Hirohashi, S. (2006). Prevalence of FOXP3+ regulatory T cells increases during the progression of pancreatic ductal adenocarcinoma and its premalignant lesions. *Clinical Cancer Research*, *12*, 5423–5434.
Hsu, P. P., & Sabatini, D. M. (2008). Cancer cell metabolism: Warburg and beyond. *Cell*, *134*, 703–707.
Huntly, B. J. P., & Gilliland, D. G. (2005). Leukaemia stem cells and the evolution of cancer-stem-cell research. *Nature Reviews. Cancer*, *5*, 311–321.
Husbeck, B., Nonn, L., Peehl, D. M., & Knox, S. J. (2006). Tumor-selective killing by selenite in patient-matched pairs of normal and malignant prostate cells. *Prostate*, *66*, 218–225.
Ito, K., Carracedo, A., Weiss, D., Arai, F., Ala, U., Avigan, D. E., et al. (2012). A PML-PPAR-delta pathway for fatty acid oxidation regulates hematopoietic stem cell maintenance. *Nature Medicine*, *18*, 1350–1358.
Jiang, X. R., Macey, M. G., Lin, H. X., & Newland, A. C. (1992). The anti-leukaemic effects and the mechanism of sodium selenite. *Leukemia Research*, *16*, 347–352.
Jorgensen, H. G., Allan, E. K., Jordanides, N. E., Mountford, J. C., & Holyoake, T. L. (2007). Nilotinib exerts equipotent antiproliferative effects to imatinib and does not induce apoptosis in CD34+ CML cells. *Blood*, *109*, 4016–4019.
Kang, T. H., Knoff, J., Yeh, W. H., Yang, B., Wang, C., Kim, Y. S., et al. (2014). Treatment of tumors with vitamin E suppresses myeloid derived suppressor cells and enhances CD8+T cell-mediated antitumor effects. *PLoS One*, *9*, 1–10.
Kiremidjian-Schumacher, L., Roy, M., Wishe, H. I., Cohen, M. W., & Stotzky, G. (1996). Supplementation with selenium augments the functions of natural killer and lymphokine-activated killer cells. *Biological Trace Element Research*, *52*, 227–239.
Ko, Y. H., Domingo-Vidal, M., Roche, M., Lin, Z., Whitaker-Menezes, D., Seifert, E., et al. (2016). TP53-inducible glycolysis and apoptosis regulator (TIGAR) metabolically reprograms carcinoma and stromal cells in breast cancer. *Journal of Biological Chemistry*, *291*, 26291–26303.

Konopleva, M. Y., & Jordan, C. T. (2011). Leukemia stem cells and microenvironment: Biology and therapeutic targeting. *Journal of Clinical Oncology, 29*, 591–599.

Koretzky, G. A. (2007). Review series introduction the legacy of the Philadelphia chromosome. *The Journal of Clinical Investigation, 117*, 2030–2032.

Krasnova, Y., Putz, E. M., Smyth, M. J., & Souza-Fonseca-Guimaraes, F. (2017). Bench to bedside: NK cells and control of metastasis. *Clinical Immunology, 177*, 50–59.

Liao, K., Bian, Z., Xie, D., & Peng, Q. (2017). A selenium-modified ginseng polysaccharide promotes the apoptosis in human promyelocytic leukemia (HL-60) cells via a mitochondrial-mediated pathway. *Biological Trace Element Research, 177*, 64–71.

Lipinski, B. (2005). Rationale for the treatment of cancer with sodium selenite. *Medical Hypotheses, 64*, 806–810.

Lipinski, B. (2016). Sodium selenite as an anticancer agent. *Anti-Cancer Agents in Medicinal Chemistry, 16*, 1–4.

Lu, D. Y., Chi, J., Lin, L. P., Huang, M., Xu, B., & Ding, J. (2004). Effect of anti-cancer drugs on the binding of 125I-fibrinogen to two leukaemia cell lines in vitro. *Journal of International Medical Research, 32*, 488–491.

Mantovani, A., Allavena, P., Sica, A., & Balkwill, F. (2008). Cancer-related inflammation. *Nature, 454*, 436–444.

Mantovani, A., Bottazzi, B., Colotta, F., Sozzani, S., & Ruco, L. (1992). The origin and function of tumor-associated macrophages. *Immunology Today, 13*, 265–270.

Mantovani, A., Sozzani, S., Locati, M., Allavena, P., & Sica, A. (2002). Macrophage polarization: Tumor-associated macrophages as a paradigm for polarized M2 mononuclear phagocytes. *Trends in Immunology, 23*, 549–555.

Martinez, F. O., Sica, A., Mantovani, A., & Locati, M. (2008). Macrophage activation and polarization. *Frontiers in Bioscience, 13*, 453–461.

Martinez-Outschoorn, U. E., Peiris-Pagés, M., Pestell, R. G., Sotgia, F., & Lisanti, M. P. (2017). Cancer metabolism: A therapeutic perspective. *Nature Reviews. Clinical Oncology, 14*, 11–31.

Mendelev, N., Witherspoon, S., & Li, P. A. (2009). Overexpression of human selenoprotein H in neuronal cells ameliorates ultraviolet irradiation-induced damage by modulating cell signaling pathways. *Experimental Neurology, 220*, 328–334.

Méplan, C., & Hesketh, J. (2012). The influence of selenium and selenoprotein gene variants on colorectal cancer risk. *Mutagenesis, 27*, 177–186.

Misra, S., Selvam, A. K., Wallenberg, M., Ambati, A., Matolcsy, A., Magalhaes, I., et al. (2016). Selenite promotes all-trans retinoic acid-induced maturation of acute promyelocytic leukemia cells. *Oncotarget, 7*, 74686–74700.

Nesto, R. W., Bell, D., Bonow, R. O., Fonseca, V., Grundy, S. M., Horton, E. S., et al. (2003). Thiazolidinedione use, fluid retention, and congestive heart failure: A consensus statement from the American Heart Association and American Diabetes Association. *Circulation, 108*, 2941–2948.

Nilsonne, G., Sun, X., Nyström, C., Rundlöf, A. K., Potamitou, Fernandes A., Björnstedt, M., et al. (2006). Selenite induces apoptosis in sarcomatoid malignant mesothelioma cells through oxidative stress. *Free Radical Biology and Medicine, 41*, 874–885.

Ochoa, A. C., Zea, A. H., Hernandez, C., & Rodriguez, P. C. (2007). Arginase, prostaglandins, and myeloid-derived suppressor cells in renal cell carcinoma. *Clinical Cancer Research, 13*, 721–726.

Ojalvo, L. S., Whittaker, C. A., Condeelis, J. S., & Pollard, J. W. (2010). Gene expression analysis of macrophages that facilitate tumor invasion supports a role for Wnt-signaling in mediating their activity in primary mammary tumors. *Journal of Immunology, 184*, 702–712.

Olm, E., Jonsson-Videsater, K., Ribera-Cortada, I., Fernandes, A. P., Eriksson, L. C., Lehmann, S., et al. (2009). Selenite is a potent cytotoxic agent for human primary AML cells. *Cancer Letters, 282*, 116–123.

Prima, V., Kaliberova, L. N., Kaliberov, S., Curiel, D. T., & Kusmartsev, S. (2017). COX2/mPGES1/PGE2 pathway regulates PD-L1 expression in tumor-associated macrophages and myeloid-derived suppressor cells. *Proceedings of the National Academy of Sciences of the United States of America*, *114*, 1117–1122.

Prost, S., Relouzat, F., Spentchian, M., Ouzegdouh, Y., Saliba, J., Massonnet, G., et al. (2015). Erosion of the chronic myeloid leukaemia stem cell pool by PPARγ agonists. *Nature*, *525*, 380–383.

Quail, D. F., & Joyce, J. A. (2013). Microenvironmental regulation of tumor progression and metastasis. *Nature Medicine*, *19*, 1423–1437.

Ravasco, P., Aranha, M. M., Borralho, P. M., Moreira da Silva, I. B., Correia, L., Fernandes, A., et al. (2010). Colorectal cancer: Can nutrients modulate NF-κB and apoptosis? *Clinical Nutrition*, *29*, 42–46.

Riether, C., Schürch, C. M., & Ochsenbein, A. F. (2015). Regulation of hematopoietic and leukemic stem cells by the immune system. *Cell Death and Differentiation*, *22*, 187–198.

Saccani, A., Schioppa, T., Porta, C., Biswas, S. K., Nebuloni, M., Vago, L., et al. (2006). p50 nuclear factor-κB overexpression in tumor-associated macrophages inhibits M1 inflammatory responses and antitumor resistance. *Cancer Research*, *66*, 11432–11440.

Samudio, I., Harmancey, R., Fiegl, M., Kantarjian, H., Konopleva, M., Korchin, B., et al. (2010). Pharmacologic inhibition of fatty acid oxidation sensitizes human leukemia cells to apoptosis induction. *Journal of Clinical Investigation*, *120*, 142–156.

Sánchez, Y., Segura, V., Marín-Béjar, O., Athie, A., Marchese, F. P., González, J., et al. (2014). Genome-wide analysis of the human p53 transcriptional network unveils a lncRNA tumour suppressor signature. *Nature Communications*, *5*, 5812. http://dx.doi.org/10.1038/ncomms6812.

Sancho, P., Barneda, D., & Heeschen, C. (2016). Hallmarks of cancer stem cell metabolism. *British Journal of Cancer*, *114*, 1305–1312.

Sanmartin, C., Plano, D., & Palop, J. A. (2008). Selenium compounds and apoptotic modulation: A new perspective in cancer therapy. *Mini-Reviews in Medicinal Chemistry*, *8*, 1020–1031.

Shen, H. M., Yang, C. F., & Ong, C. N. (1999). Sodium selenite-induced oxidative stress and apoptosis in human hepatoma HepG2 cells. *International Journal of Cancer*, *81*, 820–828.

Tabe, Y., & Konopleva, M. (2014). Advances in understanding the leukaemia microenvironment. *British Journal of Haematology*, *164*, 767–778.

Tannishtha, R., Morrison, S. J., Clarke, M. F., & Weissman, I. L. (2001). Stem cells, cancer, and cancer stem cells. *Nature*, *414*, 105–111.

Vitale, M., Cantoni, C., Pietra, G., Mingari, M. C., & Moretta, L. (2014). Effect of tumor cells and tumor microenvironment on NK-cell function. *European Journal of Immunology*, *44*, 1582–1592.

Vunta, H., Davis, F., Palempalli, U. D., Bhat, D., Arner, R. J., Thompson, J. T., et al. (2007). The anti-inflammatory effects of selenium are mediated through 15-deoxy-Delta12,14-prostaglandin J2 in macrophages. *The Journal of Biological Chemistry*, *282*, 17964–17973.

Wang, H., Chan, Y. L., Li, T. L., Bauer, B. A., Hsia, S., Wang, C. H., et al. (2013). Reduction of splenic immunosuppressive cells and enhancement of anti-tumor immunity by synergy of fish oil and selenium yeast. *PLoS One*, *8*, 1–11.

Wanka, C., Steinbach, J. P., & Rieger, J. (2012). Tp53-induced glycolysis and apoptosis regulator (TIGAR) protects glioma cells from starvation-induced cell death by up-regulating respiration and improving cellular redox homeostasis. *Journal of Biological Chemistry*, *287*, 33436–33446.

Warburg, O. (1956). On respiratory impairment in cancer cells. *Science*, *124*, 269–270.

Weisberger, A. S., & Suhrland, L. G. (1956). Studies on analogues of L-cysteine and L-cystine. III. The effect of selenium cystine on leukemia. *Blood*, *11*, 19–30.

Whiteside, T. L. (2008). The tumor microenvironment and its role in promoting tumor growth. *Oncogene*, *27*, 5904–5912.

Xue, H., Wang, W., Li, Y., Shan, Z., Li, Y., Teng, X., et al. (2010). Selenium upregulates CD4(+)CD25(+) regulatory T cells in iodine-induced autoimmune thyroiditis model of NOD.H-2(h4) mice. *Endocrine Journal*, *57*, 595–601.

Yang, H., Jia, X., Chen, X., Yang, C. S., & Li, N. (2012). Time-selective chemoprevention of vitamin E and selenium on esophageal carcinogenesis in rats: The possible role of nuclear factor kappaB signaling pathway. *International Journal of Cancer*, *131*, 1517–1527.

Zhong, W., & Oberley, T. D. (2001). Redox-mediated effects of selenium on apoptosis and cell cycle in the LNCaP human prostate cancer cell line. *Cancer Research*, *61*, 7071–7078.

Zhou, J., Ching, Y. Q., & Chng, W.-J. (2015). Aberrant nuclear factor-kappa B activity in acute myeloid leukemia: From molecular pathogenesis to therapeutic target. *Oncotarget*, *6*, 5490–5500.

CHAPTER SEVEN

Selenium and Breast Cancer Risk: Focus on Cellular and Molecular Mechanisms

Camile C. Fontelles*, Thomas P. Ong*,†,1

*Faculty of Pharmaceutical Sciences, University of São Paulo, São Paulo-SP, Brazil
†Food Research Center (FoRC), São Paulo-SP, Brazil
[1]Corresponding author: e-mail address: tongusp@gmail.com

Contents

Abstract

Selenium (Se) is a micronutrient with promising breast cancer prevention and treatment potential. There is extensive preclinical evidence of Se mammary carcinogenesis inhibition. Evidence from epidemiological studies is, however, unclear and intervention studies are rare. Here, we examine Se chemoprotection, chemoprevention, and chemotherapy effects in breast cancer, focusing on associated cellular and molecular mechanisms. Se exerts its protective actions through multiple mechanisms that involve antioxidant activities, induction of apoptosis, and inhibition of DNA damage, cell proliferation, angiogenesis, and invasion. New aspects of Se actions in breast cancer have emerged such as the impact of genetic polymorphisms on Se metabolism and response, new functions of selenoproteins, epigenetic modulation of gene expression, and long-term influence of early-life exposure on disease risk. Opportunity exists to design interventional studies with Se for breast cancer prevention and treatment taking into consideration these key aspects.

Advances in Cancer Research, Volume 136
ISSN 0065-230X
http://dx.doi.org/10.1016/bs.acr.2017.08.001

1. BREAST CANCER: GENERAL ASPECTS

Breast cancer is a global health concern. Disease incidence data show that it is the most frequently diagnosed cancer among women worldwide. Breast cancer is the first and second cause of death by cancer in women from less developed and more developed regions, respectively (Ferlay et al., 2015). It has a major social and economic impact, and therefore, preventive strategies should be a main focus for breast cancer control.

An important contributor to breast cancer etiology is family history, with mutations in the DNA repair *BRCA1* and *BRCA2* genes being associated with 5% of all breast cancer cases (Grover & Martin, 2002). Although the genetic background plays a relevant role in breast cancer risk, other factors such as reproductive patterns and lifestyle greatly influence breast cancer incidence (Barnard, Boeke, & Tamimi, 2015; Rudolph, Chang-Claude, & Schmidt, 2016). Importantly, according to recent estimates 25%–30% of all breast cancer cases could be prevented through lifestyle changes (Harvie, Howell, & Evans, 2015).

A healthy dietary pattern has also been associated with a decrease in breast cancer risk (Baglietto et al., 2011), and the protective role of dietary antioxidants has been highlighted (Pan, Zhou, Gibbons, Morrison, & Wen, 2011). One such antioxidant is selenium (Se), an essential micronutrient that has been considered as a promising breast cancer preventive and treatment agent (Cai et al., 2016; Chen, Prabhu, Das, & Mastro, 2013; Combs & Gray, 1998; El-Bayoumy & Sinha, 2004; Ip, Hayes, Budnick, & Ganther, 1991; Yazdi, Mahdavi, Kheradmand, & Shahverdi, 2012). Hence, the aim of the present review is to examine Se chemoprotection, chemoprevention, and chemotherapy effects in breast cancer, focusing on associated cellular and molecular mechanisms.

2. Se: NUTRITIONAL ASPECTS

Se is an essential micronutrient that exerts most of its biological effects through several selenoproteins in which it is incorporated as selenocysteine, considered the 21st amino acid. The human selenoproteome is encoded by 25 genes that are involved in diverse physiological processes (Labunskyy, Hatfield, & Gladyshev, 2014). Glutathione peroxidase (GPx), thioredoxin reductase (TR), and selenoprotein P (SEPP) 1 are the most well-known selenoproteins that are involved with antioxidant defense, immune function, embryogenesis, and fertility (Moghadaszadeh & Beggs, 2006). Among the diverse health benefits ascribed to Se, particular interest has been directed toward its role as a cancer preventive agent (Davis, Tsuji, & Milner, 2012; Rayman, 2012; Vinceti et al., 2014; Wrobel, Power, & Toborek, 2016).

Se food sources include cereals (wheat, maize, and rye), nuts (Brazil nuts and walnuts), broccoli, cabbage, onion, garlic, mushrooms, meats, and fish (Rayman, 2012). Dietary Se is found in different chemical forms, with distinct bioavailability, that include Sec, selenomethionine, selenite, and selenate, that are absorbed through the intestinal lumen and converted to selenide (Ohta & Suzuki, 2008). In the liver Se can be incorporated into SEPP and then distributed to organs (Schweizer, Schomburg, & Kohrle, 2016).

Current recommended Se daily allowances are generally based on levels needed to maximize GPx activity. However, the definition of optimal Se intake levels depends on the ingested Se chemical form and individual's genotype, and the need to consider other selenoproteins as biomarkers of Se nutritional status has been emphasized (Ferguson, Karunasinghe, Zhu, & Wang, 2012). For example, the German, Austrian, and Swiss nutrition societies have updated the reference values for Se intake to 60 and 70 μg/day for women and men, respectively, based on saturation of SEPP in plasma (Kipp et al., 2015). Of notice in several world regions, the recommended dietary Se levels are not met. Examples include New Zealand and some European and Middle Eastern countries (Stoffaneller & Morse, 2015).

3. Se AND BREAST CANCER

3.1 Evidence From Human Studies

Recent meta-analyses were conducted to examine the available epidemiological evidence linking Se to breast cancer. While two meta-analysis found that Se plasma but not toenail levels (Babaknejad et al., 2014) or Se exposure (Cai et al., 2016) was inversely associated with breast cancer risk, other did not confirm such results (Vinceti et al., 2014). Available data concerning Se consumption or biomarkers and breast cancer risk are still inconsistent. The epidemiological aspects of Se and cancer are the subject of the chapter "The epidemiology of selenium and human cancer" by Vinceti.

3.2 Evidence From Animal Models

3.2.1 Se Chemopreventive Effects

While human evidence of Se protective effects on breast cancer is not clear, considerable number of studies conducted since the 1980s in rats and mice subjected mainly to 7,12-dimethylbenz[*a*]anthracene- (DMBA) or *N*-nitroso-*N*-methylurea (MNU)-induced mammary carcinogenesis protocols indicate that Se has prominent breast cancer chemopreventive effects (Hamdy, Latif, Drees, & Soliman, 2012; Ip et al., 1991; Yazdi et al., 2012). This is exemplified by estimations made already in the mid-1980s that more than 100 experimental studies, including those focusing on the mammary gland, had been conducted with Se with 2/3 of them showing cancer preventive effects by then (Combs & Gray, 1998). Overall, these studies focused on Se supplementation of rodent's diets mainly before tumor induction, resulting in protection against breast cancer based on several parameters including longer tumor latency, lower number of tumors, and lower tumor growth (Combs, 2001). The majority of animal studies show that these protective effects by Se occurred at supranutritional doses reaching 20–50 times the rodent nutritional requirements (Chen, Prabhu, Das, et al., 2013; Lü & Jiang, 2005). In addition, the most prevalent evaluated Se forms were sodium selenite and selenomethionine and in general dose–response chemoprevention effects were attained without signs of toxicity (Ip, 1998).

The data obtained in diverse experimental studies indicate that at least in mammary carcinogenesis the production of a methylated Se metabolite is necessary to attain preventive effects and led Ip and Ganther (1990) to propose the Se monomethyl active metabolite pool hypothesis that has been the predominant paradigm in breast cancer research in the last three decades (Lü et al., 2016). *Se*-methylselenocysteine has been identified as one such compound that would exert its protective effects through the critical metabolite methylselenol (Ip et al., 1991). This is the main Se form in high-selenium broccoli that has shown superior mammary carcinogenesis chemoprevention compared to other Se compounds (El-Bayoumy & Sinha, 2004).

Because of the association of increased Se intake and type 2 diabetes risk (Rayman, 2012), it would be important to further investigate Se in lower doses closer to the nutritionally recommended levels of intake, also in transgenic models of breast cancer where key pathways have been disrupted (i.e., *BRCA1* and *RAS*).

3.2.2 Se Chemotherapy Effects

Several preclinical studies have evaluated Se effects using breast cancer cell lines that reflect later steps of mammary carcinogenesis. In addition, in some animal studies Se was also tested in the progression stage of breast cancer models (Chen, Prabhu, Das, et al., 2013; Song, Ren, & Liu, 2015). Inhibition of proliferation and induction of apoptosis in breast cancer cells was reported. In addition, Se promoted cell–cell attachment and inhibited angiogenesis and cell migration in these cells suggesting potential metastasis inhibition actions (Chen, Prabhu, & Mastro, 2013). Combination of Se with established chemotherapy agents such as docetaxel and cisplatin has yielded potentiating or synergistic effects in different breast cancer cell lines (Park, Yoo, Kim, Baek, et al., 2015; Sakallı Çetin, Nazıroğlu, Çiğ, Övey, & Aslan Koşar, 2017). Collectively these studies reinforce Se potential role also in breast cancer chemotherapy.

4. CELLULAR AND MOLECULAR MECHANISMS OF Se IN BREAST CANCER PREVENTION AND TREATMENT

Data mostly from in vitro and animal model systems show that Se exerts protective effects throughout breast carcinogenesis. Several cellular and molecular mechanisms have been implicated in Se chemoprotection, chemoprevention, and chemotherapy effects in breast cancer (Fig. 1). These include antioxidant activities, protection against DNA damage, induction of apoptosis, and inhibition of cell proliferation, angiogenesis, and invasion. New aspects of Se actions in breast cancer have emerged such as the impact of genetic polymorphisms on Se metabolism and response, new functions of selenoproteins, epigenetic modulation of gene expression, and long-term influence of early-life exposure on disease risk.

4.1 Chemoprotection

Se inhibitory effects at the initial stages of breast carcinogenesis involve chemoprotection against DNA instability. Inhibition of the formation of DNA adducts, DNA or chromosome breakage, chromosome gain or loss, and protection of telomere length and mitochondrial DNA has been reported at moderate levels of Se intake in different model systems (reviewed in Ferguson et al., 2012). In breast cancer animal models different forms of Se have been shown to act at the initiation phase of rat mammary carcinogenesis by inhibiting DMBA-induced adducts (El-Bayoumy, 2001). These actions

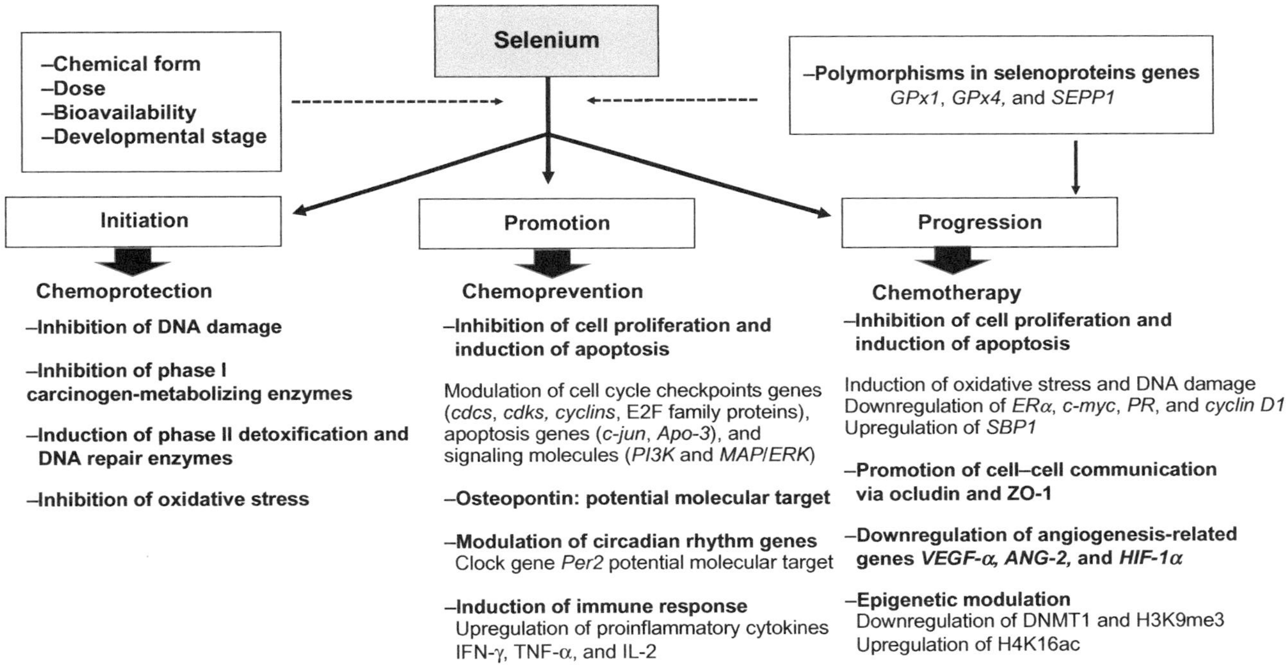

Fig. 1 Cellular and molecular mechanisms associated with Se chemoprotection, chemoprevention, and chemotherapy effects in breast cancer. Se exerts protective effects throughout breast carcinogenesis initiation, promotion, and progression phases. These involve antioxidant activities, induction of apoptosis, and inhibition of DNA damage, cell proliferation, angiogenesis, and invasion. On the molecular levels, these protective actions have been ascribed to Se modulation of different molecular targets, including epigenetic marks. Se efficacy is dependent on developmental and carcinogenesis stage of intervention, nutritional (levels of intake, chemical form of ingested Se, bioavailability), as well as genetic (polymorphisms in genes affecting Se response) factors. Genetic variation in genes coding selenoproteins has been shown to further influence breast cancer risk.

are possibly due to Se regulation of genes related to phase I/II carcinogen metabolism/detoxification enzymes that led to inhibition of DNA adduct formation (El-Bayoumy & Sinha, 2005). The cancer chemopreventive action of sodium selenite in rats initiated with DMBA was also associated with inhibition of lipid peroxidation in both the mammary gland and liver tissues (Takada, Hirooka, Hatano, Hamada, & Yamamoto, 1992) and reduction of the levels of nitric oxide that can induce cytotoxic and mutagenic effects (Hamdy et al., 2012). In humans, supplementation of *BRCA1* mutations carriers decreased chromosome breaks (Kowalska et al., 2005) and reduced the levels of oxidative stress/DNA damage marker 8-oxo-7,8-dihydro-2′-deoxyguanosine (8-oxodG) that was associated with increased base excision repair enzymes activities (Dziaman et al., 2009).

4.2 Chemoprevention

High-throughput screening techniques such as DNA microarray technology were used in different breast cancer model systems to evaluate global gene expression alterations induced by treatment with different Se chemopreventive compounds (reviewed in El-Bayoumy & Sinha, 2005). One such study (Dong, Lisk, Block, & Ip, 2001) with γ-glutamyl-*Se*-methylselenocysteine and *Se*-methylselenocysteine identified in isolated epithelial mammary cells different genes with altered expression that included *superoxide dismutase*, *c-jun*, and *adipocyte fatty acid-binding protein* (Dong et al., 2001). In addition, microarray analysis of the whole rat mammary gland showed that methylseleninic acid altered in both adipocytes and stromal cells the expression of *cyclin A*, *cyclin D1*, *p21*, *p27*, *bcl-2*, and *bax* that was associated with the growth and differentiation of the adjacent epithelial cells (Dong, Ganther, Stewart, & Ip, 2002). Treatment of MCF10AT1 premalignant human breast cells with methylseleninic acid induced alterations in the expression of 30 genes that were associated with regulation of cell cycle checkpoints (cdcs, cdks, cyclins, E2F family proteins), apoptosis (c-jun, Apo-3), and signaling molecules (PI3K and MAP/ERK), leading to inhibition of growth, blockage of cell cycle progression at the G_0–G_1 phase, and induction of apoptosis (Dong, Ip, & Ganther, 2002).

Diet supplementation with *Se*-methylselenocysteine reduced in the mammary tumors of female Balb/c mice the expression of osteopontin that is involved in promoting cell attachment, regulating Ca^{2+} levels and influencing gene expression (Unni, Kittrell, Singh, & Sinha, 2004). A link between circadian rhythm and breast cancer prevention by Se was further shown in a rat study (Zhang & Zarbl, 2008). Authors found that clock genes were modulated by methylselenocysteine in the mammary tissue and that the circadian gene *Per2*, considered a tumor suppressor, was a potential target for Se chemoprevention in early mammary carcinogenesis (Zhang & Zarbl, 2008). In mice Se supplementation further modulated the immune response through the regulation of proinflammatory cytokines, such as IFN-γ, TNF-α, and IL-2, as well as through regulation of NK cell activity, resulting in lower breast tumor volumes and increased survival rate (Yazdi et al., 2012).

4.3 Chemotherapy

GPx1 may be involved in the Se-mediated DNA damage inhibition effects, given that overexpression of GPx1 in (estrogen receptor) ER^+ MCF-7 cells increased the phosphorylation of $p70^{S6K}$, Akt site-specific phosphorylation, and the levels of Gadd45 that is a protein involved in DNA damage response (Nasr, Fedele, Esser, & Diamond, 2004). Moreover, MCF-7 cells treated with sodium selenite presented protection against ultraviolet-induced chromosome damage that was mainly dependent on functional *BRCA1* activity, possibly via GPx1 overexpression (Baliga, Wang, Zhuo, Schwartz, & Diamond, 2007). However, Se also presents ambivalent actions on DNA damage modulation, with sodium selenite inducing DNA damage and apoptosis in breast cancer cells, and selenomethionine increasing DNA repair capacity (Letavayová, Vlčková, & Brozmanová, 2006). These observations reinforce that Se effects in breast cancer are dependent on its chemical form and that this should be considered when designing cancer treatment interventions.

In the ER^- human breast cancer cell line MDA-MB-231, Se could revert an increase in apoptosis and oxidative stress caused by electromagnetic radiation and these protective effects were associated with restoration of the mitochondrial membrane potential (Kahya, Naziroğlu, & Çiğ, 2014). Of notice, depending on the cellular and metabolic context, Se can also behave as a prooxidant agent (Brozmanová, Mániková, Vlčková, & Chovanec, 2010). In human breast cancer cell lines MCF-7 and MDA-MB-231 Se-yeast exposure led to increased Se metabolites that induced the generation of reactive oxygen species, accompanied by loss of mitochondrial membrane potential and increased apoptosis (Guo, Hsia, Shih, Hsieh, & Chen, 2015).

Methylseleninic acid control of MCF-7 cell proliferation was shown to involve inhibition of ERα expression, ER *trans*-activating activity, and

ER-mediated expression of genes such as *C-MYC* (Lee et al., 2005). In a MCF-7 cell-based breast cancer xenograft model in nude mice, combination of methylselenocysteine and tamoxifen led to synergistic inhibition of tumor growth that was associated with inhibition of ERα expression and its targets genes *PR* and *cyclin D1*, as well as inhibition of cell proliferation and induction of apoptosis (Li et al., 2009). Methylseleninic acid was also shown to inhibit in MCF-7 cells the expression of aromatase that is implicated in peripheral aromatization of androgens and increased risk of breast cancer in obese postmenopausal women (Gao et al., 2012).

Se treatment of canine mammary gland tumor cell line (CTM1211) was associated with downregulation of protein and mRNA levels of VEGF-α, ANG-2, and HIF-1α and upregulation of PTEN, which collectively play an essential role in inhibiting angiogenesis, inducing apoptosis and modulating the tumor microenvironment (Liu, Li, Guo, Li, & Qiu, 2016). Furthermore, Se treatment of MCF-7 and MDA-MB-231 cells increased *trans*-epithelial resistance and reduced paracellular permeability in both cell lines, as well as increased in MCF-7 cells relocation of ocludin and ZO-1 to cell–cell junctions, highlighting Se modulation of tight junctions (Martin, Das, Mansel, & Jiang, 2007).

4.4 Emerging Mechanisms

4.4.1 Polymorphisms in Genes Coding Selenoproteins

In addition to Se compounds, selenoproteins have also been implicated in breast cancer development (Chen, Prabhu, Das, et al., 2013). Variations in selenoproteins genes are associated with intraindividual variations in Se status, response to Se supplementation, and oxidative stress modulation (Cominetti et al., 2011). In addition, these SNPs may alter individual Se requirements and the interaction of these polymorphisms with suboptimal Se intake could represent an important factor influencing disease risk (Hesketh, 2008). Because selenoproteins have been implied in Se anticancer actions, particular interest has been directed toward the impact of SNPs in *GPx1*, *GPx4*, and *SEPP1* on breast cancer risk (reviewed in Méplan, 2015).

The majority of studies on polymorphisms in selenoproteins genes and breast cancer have focused on *GPx1* Pro198Leu (rs1050450). An initial study (Hu & Diamond, 2003) showed that the variant Leu allele was associated with increased risk for breast cancer risk. Interestingly, functional analysis revealed that this *GPx1* allele was less responsive to Se supplementation and it was suggested that carriers of the variant genotype would need higher levels of Se to retain basal level of cellular protection (Hu & Diamond, 2003).

In a prospective study with Danish women, *GPx1* Pro198Leu was associated with lower erythrocyte GPx activity and this was suggested to be associated with increased breast cancer risk in this female population (Ravn-Haren et al., 2006).

Contrary to these findings, in a recent study conducted among Polish women (Jablonska et al., 2015), a breast cancer risk protective effect by *GPx1*Pro198Leu variant allele was observed. In addition, other studies did not find an association between *GPx1* Pro198Leu SNP and breast cancer risk (Ahn et al., 2005; Knight et al., 2004). A recent meta-analysis was unable to find an overall significant association between the leucine allele of *GPx1* polymorphism and breast cancer risk (Hu, Zhou, Wang, & Wang, 2010).

Because SNPs in antioxidant genes have been proposed to have low penetrance, it would seem important to analyze their potential interactions. Few studies have adopted such an approach. In a case–control nested within the Nurses' Health Study it was observed that while the Leu allele in *GPx1* Pro198Leu and the Ala variant allele in *manganese superoxide dismutase* (Val16Ala; rs1799752) were not associated to any change in breast cancer risk when considered alone, increased risk was observed in women carrying both variant alleles (Cox, Tamimi, & Hunter, 2006). However, these results were not confirmed in another large case–control study in the United Kingdom (Oestergaard et al., 2006).

SNPs in *GPx4* (rs713041 and rs757229), one of the most active intracellular forms of this enzyme that can regulate leukotriene biosynthesis and inflammatory response, were associated with breast cancer mortality (Udler et al., 2007). In a recent case–control study in a Danish population (Méplan et al., 2013) influence of SNPs in *SEPP1* and *GPx1* on breast cancer risk was observed. Carriers of the variant allele for *SEPP1* Ala234Thr (rs3877899) had lower risk for the disease, while Leu carriers for *GPx1* Pro198Leu had increased risk of developing nonductal breast cancer. In addition, lower erythrocyte GPx activity was only found in breast cancer patients that were homozygous carriers of *GPx1* Pro198Leu variant allele and current hormone replacement therapy users. Based on these results authors (Méplan et al., 2013) proposed a model where breast tissue susceptibility to oxidative damage, and ultimately to carcinogenesis, is determined by the capacity of Se delivery to breast tissue and modulation of GPx1 activity. Interestingly, in a recent study (Ekoue et al., 2017) Se quantification in human breast carcinoma revealed an association between micronutrient tissue levels and *SEPP1* genotype. Homozygous carriers of the variant allele for

SEPP1 Ala234Thr had the highest tissue levels of Se levels as compared to the other genotypes (Ekoue et al., 2017).

Moreover, researchers observed that urinary Se was associated with a decrease in breast cancer risk mainly in female breast cancer patients that displayed the CC genotype of *MTHFR* rs1801133, with MTHFR coding a key enzyme involved in the one'carbon metabolism (Lin et al., 2015). These data indicate that genetic studies of Se and breast cancer risk should also focus on pathways other than the antioxidant defense system.

Thus, in future Se supplementation trials, it would be important to use different biomarkers of the micronutrient status as well as to consider genetic variation that could impact their levels as well as response to Se intake (Jablonska et al., 2009). It has been proposed that in populations with low Se intake selenoprotein genotyping together with erythrocyte GPx activity could be used as biomarkers of breast cancer risk (Méplan et al., 2013).

4.4.2 New Functions of Selenoproteins

In addition to variations in genes coding selenoproteins, the impact of their expression or activity has also been evaluated in breast cancer studies. Higher erythrocyte GPx activity was found in breast cancer patients, possibly due to an increased free radical production associated with tumor promotion (Moradi, Hassan, Talei, & Rajaei, 2009). Immunohistochemical analysis revealed high GPx expression in mammary neoplastic tissue of women with a worse clinical outcome and reduced overall survival (Jardim et al., 2013). On the other hand, downregulation of GPx4 expression was observed in mammary neoplastic tissue of women with breast tumors with poor prognosis (Cejas et al., 2007). Studies in transgenic mice have contributed to the elucidation of selenoproteins role in breast cancer. Knockout of Sec tRNA$^{[Ser]Sec}$ gene (*Trsp*) altered the expression not only of selenoproteins genes (*SEP15*, *GPx1*, and *GPx4*) but also of tumor suppressor genes (*BRCA1* and *p53*) that play an important role in breast carcinogenesis (Kumaraswamy et al., 2003). In addition, transgenic genetic deletion of *Trsp* increased mice susceptibility to DMBA-induced breast carcinogenesis (Hudson et al., 2012). Selenotranscriptome analysis showed decreased expression of *GPx1*, *GPx4*, *GPx5*, and *GPx7* and increased expression of *DIO2*, type II, and *GPx2* and *GPx3* in both MCF-7 and MDA-MB-231 cells (Rusolo et al., 2017). Network analysis revealed that these differentially expressed genes are implicated in common pathways associated to p53, ER1, and catenin β1 (Rusolo et al., 2017). Data from these studies show an interconnection between selenoproteins with genes frequently deregulated in breast cancer development. Elucidation of these interactions may contribute in identifying novel molecular targets of Se.

Selenium-binding protein 1 (SELENBP1 or SBP1), which is involved in cell growth regulation and can be regulated by estrogens, has been associated with Se antibreast cancer actions (Zhang et al., 2013). SBP1 expression is reduced in human breast tumor tissue and in ER^- but not ER^+ breast cancer cell lines. Its expression is progressively decreased throughout breast cancer progression, and this is related to poor survival outcomes (Zhang et al., 2013). Methylseleninic acid inhibitory effects on cell proliferation were dependent on high SBP1 expression, suggesting that this molecular target could represent a potential predictor of the effectiveness of Se intervention for breast cancer prevention (Zhang et al., 2013). The inhibitory effects of selenomethionine on 4T1 mouse mammary carcinoma progression were associated with apoptosis induction, as well as increased and decreased expression of SBP1 and GPx1, respectively (Song et al., 2015). However, the cytotoxic effects of supranutritional doses of selenite in MCF-7 cells were mediated by SBP but not GPx1, with low levels of SBP1 sensitizing these cells to death via elevation of extracellular glutathione (Wang et al., 2015). It was suggested that patients with low levels of SBP1 in mammary tumor tissues would respond better to Se interventions in cancer therapy (Wang et al., 2015).

4.4.3 Epigenetic Mechanisms

Epigenetic mechanisms, which comprises heritable changes in gene expression without any modification of the DNA sequence, are frequently deregulated in breast tumors. Because epigenetic deregulation occurs early in carcinogenesis and is potentially reversible, DNA methylation, histone modifications, and microRNA expression represent promising molecular targets for cancer prevention with Se (Ong, Moreno, & Ross, 2012).

Despite the increasing number of studies directed toward the investigation of Se epigenetic modulation on cancer prevention (Speckmann & Grune, 2015), to the best of our knowledge only one study conducted in our laboratory (de Miranda et al., 2014) considered the association between Se and epigenetic mechanisms in breast cancer. We showed that in MCF-7 cells methylseleninic acid and sodium selenite inhibited DNMT1 expression, while methylseleninic acid decreased H3K9me3 and increased H4K16ac and selenite decreased this latter histone mark. In addition, selenite, but not methylseleninic acid, induced the expression of the tumor suppressor gene *RASSF1A*, although this effect did not involve demethylation of the *RASSF1A* promoter. These data reinforce the importance of the Se chemical form in its breast cancer preventive effects and highlight epigenetic mechanisms as potential molecular targets for Se. Interestingly, GPx1 transcriptional silencing due to promoter hypermethylation at AP-1 site has been observed in 20% of human breast cancers (Kulak et al., 2013), suggesting that epigenetic deregulation of selenoprotein genes could represent an important feature of breast carcinogenesis.

4.4.4 Long-Term Influence of Early-Life Se Exposure on Later Breast Cancer Risk

The fetal period is of particular importance to the etiology of breast cancer, given that it is one of the developmental periods (together with puberty and pregnancy) in which the mammary gland is undergoing extensive modeling (Hilakivi-Clarke, 2007). Factors such as maternal (de Oliveira Andrade et al., 2014) and more recently paternal (Fontelles, Carney, et al., 2016; Fontelles, Guido, et al., 2016) poor nutrition have been shown to program in the female offspring increased risk of breast cancer later in life through altered development of the mammary gland.

Because of Se relevance in early life, increased interest has been directed toward its role in the developmental origins of health and disease. The female offspring of Se-deficient mothers presented increased mortality at birth, decreased viability and survival rates, stunted growth, and altered liver development (Nogales, Ojeda, Fenutría, Murillo, & Carreras, 2013). In a study conducted by Popova with CBA female mice that received sodium selenite in tap water ad libitum, during the last week of pregnancy until 10 days after delivery, their adult male progeny had reduced incidence of spontaneous hepatomas (Popova, 2002). To the best of our knowledge there are no studies in the literature regarding maternal Se consumption and female offspring breast cancer risk. Data from our laboratory (M.P. Rosim et al., unpublished) show that female offspring of Sprague–Dawley female rats fed a Se deficient or a supplemented diet during gestation presented opposing effects in the female offspring, with the former increasing mammary tumor incidence and the latter decreasing the average number of mammary tumors.

Paternal Se deficiency can impact male gonad development through several generations (Behne, Weiler, & Kyriakopoulos, 1996). To the best of our knowledge only one study conducted in our laboratory (Guido et al., 2016) considered the association between paternal Se and breast cancer risk in the female offspring. We showed that a paternal Se-deficient diet could influence the female offspring susceptibility to breast carcinogenesis. More specifically, female offspring of Sprague–Dawley male rats fed a Se-deficient diet (0.05 p.p.m.) presented altered mammary gland developmental patterns, which was associated with a higher risk of developing breast cancer (Guido et al., 2016). Because spermatogenesis involves dynamic chromatin remodeling, it would be important to evaluate whether this paternal breast cancer programming would involve Se-induced epigenetic disturbances in paternal germ cells and daughter mammary gland tissue.

5. Se TOXIC AND PROCARCINOGENIC EFFECTS IN BREAST CANCER

Data from a randomized clinical trial conducted by Lubinski et al. (2011) with 1135 Polish women showed increased number of breast cancer cases in Polish women carrying *BRCA1* mutation and supplemented with sodium selenite (250 μg/day) compared to the placebo group (HR 1.4; 95% CI: 0.9–2.0). This raises concern on the safety of Se for breast cancer, at least in this population of women with this particular genotype. Although selenite at high levels can exert genotoxic effects (Brozmanová et al., 2010), inhibition of DNA damage and increased activity of DNA repair enzymes was observed in women carrying mutation in this same gene and also treated with sodium selenite, although in different doses and length of period (Dziaman et al., 2009; Kowalska et al., 2005). While the potential mechanisms associated with this increased breast cancer risk are not clear, it was recently observed that treatment of mice inoculated with 4T1.2 mammary carcinoma cells with selenite, but not methylseleninic acid or selenomethionine, exacerbated incidence of bone and kidney metastasis (Chen, Prabhu, Das, et al., 2013). In any event, high doses of Se including selenite should be considered with care for breast cancer prevention and its potential procarcinogenic actions better elucidated.

6. CONCLUDING REMARKS AND FUTURE PERSPECTIVES

Despite great initial enthusiasm raised by the Nutritional Prevention of Cancer Trial showing that Se-yeast had prominent prostate cancer prevention potential, failure to confirm these positive effects in the Selenium and Vitamin E Cancer Prevention Trial raised major doubts on the value of this micronutrient for cancer prevention in general. However, more optimistic views have also been vocalized (Lü et al., 2016), highlighting possible reasons for Se "failure" (i.e., wrong target population that was not Se deficient and inadequate choice of selenomethionine as Se chemical form for intervention) and reinforcing the need to progress with Se research in the cancer field by testing other forms of Se such as methylselenol precursors in smaller mechanism-driven phase I/II trials. In the last three decades, a wealth of information has accumulated regarding Se breast cancer chemoprevention efficacy in preclinical model systems. Se potential for breast cancer chemotherapy is also emerging. Understanding of potential cellular and molecular mechanisms underlying theses effects has advanced, although major gaps still exist in such knowledge. In recent years, new aspects of Se actions have emerged such as the impact of genetic polymorphisms on Se metabolism and response, new functions of selenoproteins, epigenetic modulation of gene expression, and long-term influence of early-life exposure on breast cancer risk. More than a decade ago, El-Bayoumy and Sinha (2004) stressed the need of clinical studies to unequivocally evaluate Se effects on breast cancer. Currently this is still a major limitation. Thus, opportunity exists to design new interventional studies with Se for breast cancer prevention and treatment taking into consideration all these points and building up from what has been learnt from human studies with other cancers where Se was shown to be ineffective. In the meantime, as Se deficiency is a frequent nutritional condition that can be a risk factor for breast cancer, efforts should be made for women to attain the recommended daily levels of intake, ideally since early life.

ACKNOWLEDGMENTS

C.C.F. was a recipient of a PhD scholarship CNPq (CNPq; process 153478/2012-8). T.P.O. is the recipient of a researcher fellowship from CNPq (Proc. 307910/2016-4). The Laboratory of Nutrigenomics and Programming is supported by grants from CNPq (Proc. 448501/2014-7), the Food Research Center (FoRC) and the São Paulo State Research Funding Agency (Proc. 2013/07914-8).

REFERENCES

Ahn, J., Gammon, M. D., Santella, R. M., Gaudet, M. M., Britton, J. A., Teitelbaum, S. L., et al. (2005). No association between glutathione peroxidase Pro198Leu polymorphism and breast cancer risk. *Cancer Epidemiology Biomarkers and Prevention, 14*, 2459–2461.

Babaknejad, N., Sayehmiri, F., Sayehmiri, K., Rahimifar, P., Bahrami, S., et al. (2014). The relationship between selenium levels and breast cancer: A systematic review and meta-analysis. *Biological Trace Element Research, 159*, 1–7.

Baglietto, L., Krishnan, K., Severi, G., Hodge, A., Brinkman, M., English, D. R., et al. (2011). Dietary patterns and risk of breast cancer. *British Journal of Cancer, 104*, 524–531.

Baliga, M. S., Wang, H., Zhuo, P., Schwartz, J. L., & Diamond, A. M. (2007). Selenium and GPx-1 overexpression protect mammalian cells against UV-induced DNA damage. *Biological Trace Element Research*, *115*, 227–241.

Barnard, M. E., Boeke, C. E., & Tamimi, R. M. (2015). Established breast cancer risk factors and risk of intrinsic tumor subtypes. *Biochimica et Biophysica Acta—Reviews on Cancer*, *1856*, 73–85.

Behne, D., Weiler, H., & Kyriakopoulos, A. (1996). Effects of selenium deficiency on testicular morphology and function in rats. *Journal of Reproduction and Fertility*, *106*, 291–297.

Brozmanová, J., Mániková, D., Vlčková, V., & Chovanec, M. (2010). Selenium: A double-edged sword for defense and offence in cancer. *Archives of Toxicology*, *84*, 919–938.

Cai, X., Wang, C., Yu, W., Fan, W., Wang, S., Shen, N., et al. (2016). Selenium exposure and cancer risk: An updated meta-analysis and meta-regression. *Scientific Reports*, *6*. 19213.

Cejas, P., García-Cabezas, M. A., Casado, E., Belda-Iniesta, C., De Castro, J., Fresno, J., et al. (2007). Phospholipid hydroperoxide glutathione peroxidase (PHGPx) expression is downregulated in poorly differentiated breast invasive ductal carcinoma. *Free Radical Research*, *41*, 681–687.

Chen, Y. C., Prabhu, K. S., Das, A., & Mastro, A. M. (2013). Dietary selenium supplementation modifies breast tumor growth and metastasis. *International Journal of Cancer*, *133*, 2054–2064.

Chen, Y. C., Prabhu, K. S., & Mastro, A. M. (2013). Is selenium a potential treatment for cancer metastasis? *Nutrients*, *5*, 1149–1168.

Combs, G. F. (2001). Impact of selenium and cancer-prevention findings on the nutrition-health paradigm. *Nutrition and Cancer*, *40*, 6–11.

Combs, G. F., & Gray, W. P. (1998). Chemopreventive agents: Selenium. *Pharmacology and Therapeutics*, *79*, 179–192.

Cominetti, C., de Bortoli, M. C., Purgatto, E., Ong, T. P., Moreno, F. S., Garrido, A. B., et al. (2011). Associations between glutathione peroxidase-1 Pro198Leu polymorphism, selenium status, and DNA damage levels in obese women after consumption of Brazil nuts. *Nutrition*, *27*, 891–896.

Cox, D. G., Tamimi, R. M., & Hunter, D. J. (2006). Gene x Gene interaction between MnSOD and GPX-1 and breast cancer risk: A nested case-control study. *BMC Cancer*, *6*, 217.

Davis, C. D., Tsuji, P. A., & Milner, J. A. (2012). Selenoproteins and cancer prevention. *Annual Review of Nutrition*, *32*, 73–95.

de Miranda, J. X., Andrade, F. D. O., Conti, A., Dagli, M. L. Z., Moreno, F. S., & Ong, T. P. (2014). Effects of selenium compounds on proliferation and epigenetic marks of breast cancer cells. *Journal of Trace Elements in Medicine and Biology*, *28*, 486–491.

de Oliveira Andrade, F., Fontelles, C. C., Rosim, M. P., de Oliveira, T. F., de Melo Loureiro, A. P., Mancini-Filho, J., et al. (2014). Exposure to lard-based high-fat diet during fetal and lactation periods modifies breast cancer susceptibility in adulthood in rats. *Journal of Nutritional Biochemistry*, *25*, 613–622.

Dong, Y., Ganther, H. E., Stewart, C., & Ip, C. (2002). Identification of molecular targets associated with selenium-induced growth inhibition in human breast cells using cDNA microarrays. *Cancer Research*, *62*, 708–714.

Dong, Y., Ip, C., & Ganther, H. (2002). Evidence of a field effect associated with mammary cancer chemoprevention by methylseleninic acid. *Anticancer Research*, *22*, 27–32.

Dong, Y., Lisk, D., Block, E., & Ip, C. (2001). Characterization of the biological activity of gamma-glutamyl-Se-methylselenocysteine: A novel, naturally occurring anticancer agent from garlic. *Cancer Research*, *61*, 2923–2928.

Dziaman, T., Huzarski, T., Gackowski, D., Rozalski, R., Siomek, A., Szpila, A., et al. (2009). Selenium supplementation reduced oxidative DNA damage in adnexectomized BRCA1 mutations carriers. *Cancer Epidemiology Biomarkers and Prevention, 18*, 2923–2928.

Ekoue, D. N., Zaichick, S., Valyi-Nagy, K., Picklo, M., Lacher, C., Hoskins, K., et al. (2017). Selenium levels in human breast carcinoma tissue are associated with a common polymorphism in the gene for SELENOP (Selenoprotein P). *Journal of Trace Elements in Medicine and Biology, 39*, 227–233.

El-Bayoumy, K. (2001). The protective role of selenium on genetic damage and on cancer. *Mutation Research, 475*, 123–139.

El-Bayoumy, K., & Sinha, R. (2004). Mechanisms of mammary cancer chemoprevention by organoselenium compounds. *Mutation Research, 551*, 181–197.

El-Bayoumy, K., & Sinha, R. (2005). Molecular chemoprevention by selenium: A genomic approach. *Mutation Research, 591*, 224–236.

Ferguson, L. R., Karunasinghe, N., Zhu, S., & Wang, A. H. (2012). Selenium and its' role in the maintenance of genomic stability. *Mutation Research, 733*, 100–110.

Ferlay, J., Soerjomataram, I., Dikshit, R., Eser, S., Mathers, C., & Rebelo, M. (2015). Cancer incidence and mortality worldwide: Sources, methods and major patterns in GLOBOCAN 2012. *International Journal of Cancer, 136*, E359–E386.

Fontelles, C. C., Carney, E., Clarke, J., Nguyen, N. M., Yin, C., & Jin, L. (2016). Paternal overweight is associated with increased breast cancer risk in daughters in a mouse model. *Scientific Reports, 6*. 28602.

Fontelles, C. C., Guido, L. N., Rosim, M. P., Andrade, F. D. O., Jin, L., & Inchauspe, J. (2016). Paternal programming of breast cancer risk in daughters in a rat model: Opposing effects of animal- and plant-based high-fat diets. *Breast Cancer Research, 1*, 71.

Gao, R., Zhao, L., Liu, X., Rowan, B. G., Wabitsch, M., & Edwards, D. P. (2012). Methylseleninic acid is a novel suppressor of aromatase expression. *Journal of Endocrinology, 212*, 199–205.

Grover, P. L., & Martin, F. L. (2002). The initiation of breast and prostate cancer. *Carcinogenesis, 23*, 1095–1102.

Guido, L. N., Fontelles, C. C., Rosim, M. P., Pires, V. C., Cozzolino, S. M. F., Castro, I. A., et al. (2016). Paternal selenium deficiency but not supplementation during preconception alters mammary gland development and 7,12dimethylbenz[a]anthracene-induced mammary carcinogenesis in female rat offspring. *International Journal of Cancer, 139*, 1873–1882.

Guo, C., Hsia, S., Shih, M. Y., Hsieh, F. C., & Chen, P. C. (2015). Effects of selenium yeast on oxidative stress, growth inhibition, and apoptosis in human breast cancer cells. *International Journal of Medical Sciences, 12*, 748–758.

Hamdy, S. M., Latif, A. K. M. A., Drees, E. A., & Soliman, S. M. (2012). Prevention of rat breast cancer by genistin and selenium. *Toxicology and Industrial Health, 28*, 746–757.

Harvie, M., Howell, A., & Evans, D. G. (2015). Can diet and lifestyle prevent breast cancer: What is the evidence? *American Society of Clinical Oncology Educational Book, 35*, e66–e73.

Hesketh, J. (2008). Nutrigenomics and selenium: Gene expression patterns, physiological targets, and genetics. *Annual Review of Nutrition, 28*, 157–177.

Hilakivi-Clarke, L. (2007). Nutritional modulation of terminal end buds: Its relevance to breast cancer prevention. *Current Cancer Drug Targets, 7*, 465–474.

Hu, Y. J., & Diamond, A. M. (2003). Role of glutathione peroxidase 1 in breast cancer: Loss of heterozygosity and allelic differences in the response to selenium. *Cancer Research, 63*, 3347–3351.

Hu, J., Zhou, G. W., Wang, N., & Wang, Y. J. (2010). GPX1 Pro198Leu polymorphism and breast cancer risk: A meta-analysis. *Breast Cancer Research and Treatment, 124*, 425–431.

Hudson, T. S., Carlson, B. A., Hoeneroff, M. J., Young, H. A., Sordillo, L., & Muller, W. J. (2012). Selenoproteins reduce susceptibility to DMBA-induced mammary carcinogenesis. *Carcinogenesis*, *33*, 1225–1230.

Ip, C. (1998). Lessons from basic research in selenium and cancer prevention. *The Journal of Nutrition*, *128*, 1845–1854.

Ip, C., & Ganther, H. (1990). Activity of methylated forms of selenium in cancer prevention. *Cancer Research*, *50*, 1206–1211.

Ip, C., Hayes, C., Budnick, R. M., & Ganther, H. E. (1991). Chemical form of selenium, critical metabolites, and cancer prevention. *Cancer Research*, *51*, 595–600.

Jablonska, E., Gromadzinska, J., Peplonska, B., Fendler, W., Reszka, E., Krol, M. B., et al. (2015). Lipid peroxidation and glutathione peroxidase activity relationship in breast cancer depends on functional polymorphism of GPX1. *BMC Cancer*, *15*, 657.

Jablonska, E., Gromadzinska, J., Reszka, E., Wasowicz, W., Sobala, W., Szeszenia-Dabrowska, N., et al. (2009). Association between GPx1 Pro198Leu polymorphism, GPx1 activity and plasma selenium concentration in humans. *European Journal of Nutrition*, *48*, 383–386.

Jardim, B., Moschetta, M. G., Leonel, C., Gelaleti, G. B., Regiani, V. R., Ferreira, L. C., et al. (2013). Glutathione and glutathione peroxidase expression in breast cancer: An immunohistochemical and molecular study. *Oncology Reports*, *30*, 1119–1128.

Kahya, M. C., Nazıroğlu, M., & Çiğ, B. (2014). Selenium reduces mobile phone (900 MHz)-induced oxidative stress, mitochondrial function, and apoptosis in breast cancer cells. *Biological Trace Element Research*, *160*, 285–293.

Kipp, A. P., Strohm, D., Brigelius-Flohé, R., Schomburg, L., Bechthold, A., Leschik-Bonnet, E., et al. (2015). Revised reference values for selenium intake. *Journal of Trace Elements in Medicine and Biology*, *32*, 195–199.

Knight, J., Onay, U. V., Wells, S., Li, H., Shi, E. J. Q., & Andrulis, I. L. (2004). Genetic variants of GPX1 and SOD2 and breast cancer risk at the Ontario site of the breast cancer family registry. *Cancer Epidemiology, Biomarkers and Prevention*, *13*, 146–149.

Kowalska, E., Narod, S. A., Huzarski, T., Zajaczek, S., Huzarska, J., Gorski, B., et al. (2005). Increased rates of chromosome breakage in BRCA1 carriers are normalized by oral selenium supplementation. *Cancer Epidemiology, Biomarkers and Prevention*, *14*, 1302–1307.

Kulak, M., Cyr, A. R., Woodfield, G. W., Bogachek, M., Spanheimer, P. M., Li, T., et al. (2013). Transcriptional regulation of the GPX1 gene by TFAP2C and aberrant CpG methylation in human breast cancer. *Oncogene*, *32*, 4043–4051.

Kumaraswamy, E., Carlson, B., Morgan, F., Miyoshi, K., Robinson, G. W., Su, D., et al. (2003). Selective removal of the selenocysteine tRNA [Ser]Sec gene (Trsp) in mouse mammary epithelium. *Molecular and Cellular Biology*, *23*, 1477–1488.

Labunskyy, V. M., Hatfield, D. L., & Gladyshev, V. N. (2014). Selenoproteins: Molecular pathways and physiological roles. *Physiological Reviews*, *94*, 739–777.

Lee, S. O., Nadiminty, N., Wu, X. X., Lou, W., Dong, Y., Ip, C., et al. (2005). Selenium disrupts estrogen signaling by altering estrogen receptor expression and ligand binding in human breast cancer cells. *Cancer Research*, *65*, 3487–3492.

Letavayová, L., Vlčková, V., & Brozmanová, J. (2006). Selenium: From cancer prevention to DNA damage. *Toxicology*, *227*, 1–14.

Li, Z., Carrier, L., Belame, A., Thiyagarajah, A., Salvo, V. A., Burow, M. E., et al. (2009). Combination of methylselenocysteine with tamoxifen inhibits MCF-7 breast cancer xenografts in nude mice through elevated apoptosis and reduced angiogenesis. *Breast Cancer Research and Treatment*, *118*, 33–43.

Lin, W., Cen, Y. L., Lin, Y., Su, F. X., Wu, B. H., Tang, L. Y., et al. (2015). Joint effects between urinary selenium and polymorphisms in methylation related genes on breast cancer risk. *Neoplasma*, *62*, 491–499.

Liu, Y., Li, W., Guo, M., Li, C., & Qiu, C. (2016). Protective role of selenium compounds on the proliferation, apoptosis, and angiogenesis of a canine breast cancer cell line. *Biological Trace Element Research, 169*, 86–93.

Lü, J., & Jiang, C. (2005). Selenium and cancer chemoprevention: Hypotheses integrating the actions of selenoproteins and selenium metabolites in epithelial and non-epithelial target cells. *Antioxidants & Redox Signaling, 7*, 1715–1727.

Lü, J., Zhang, J., Jiang, C., Deng, Y., Özten, N., & Bosland, M. C. (2016). Cancer chemoprevention research with selenium in the post-SELECT era: Promises and challenges. *Nutrition and Cancer, 68*, 1–17.

Lubinski, J., Jaworska, K., Durda, K., Jakubowska, A., Huzarski, T., Byrski, T., et al. (2011). Selenium and the risk of cancer in BRCA1 carriers. *Hereditary Cancer in Clinical Practice, 9*, A5.

Martin, T. A., Das, T., Mansel, R. E., & Jiang, W. G. (2007). Enhanced tight junction function in human breast cancer cells by antioxidant, selenium and polyunsaturated lipid. *Journal of Cellular Biochemistry, 101*, 155–166.

Méplan, C. (2015). Selenium and chronic diseases: A nutritional genomics perspective. *Nutrients, 7*, 3621–3651.

Méplan, C., Dragsted, L. O., Ravn-Haren, G., Tjonneland, A., Vogel, U., & Hesketh, J. (2013). Association between polymorphisms in glutathione peroxidase and selenoprotein P genes, glutathione peroxidase activity, HRT use and breast cancer risk. *PLoS One, 8*. e73316.

Moghadaszadeh, B., & Beggs, A. H. (2006). Selenoproteins and their impact on human health through diverse physiological pathways. *Physiology, 21*, 307–315.

Moradi, M., Hassan, M. E., Talei, A., & Rajaei, A. F. (2009). A comparative study of selenium concentration and glutathione peroxidase activity in normal and breast cancer patients. *Public Health Nutrition, 12*, 59–63.

Nasr, M. A., Fedele, M. J., Esser, K., & Diamond, A. M. (2004). GPx-1 modulates AKT and P70S6K phosphorylation and Gadd45 levels in MCF-7 cells. *Free Radical Biology and Medicine, 37*, 187–195.

Nogales, F., Ojeda, M. L., Fenutría, M., Murillo, M. L., & Carreras, O. (2013). Role of selenium and glutathione peroxidase on development, growth, and oxidative balance in rat offspring. *Reproduction, 146*, 659–667.

Oestergaard, M. Z., Tyrer, J., Cebrian, A., Shah, M., Dunning, A. M., Ponder, B. J., et al. (2006). Interactions between genes involved in the antioxidant defence system and breast cancer risk. *British Journal of Cancer, 95*, 525–531.

Ohta, Y., & Suzuki, K. T. (2008). Methylation and demethylation of intermediates selenide and methylselenol in the metabolism of selenium. *Toxicology and Applied Pharmacology, 226*, 169–177.

Ong, T. P., Moreno, F. S., & Ross, S. A. (2012). Targeting the epigenome with bioactive food components for cancer prevention. *Journal of Nutrigenetics and Nutrigenomics, 4*, 275–292.

Pan, S. Y., Zhou, J., Gibbons, L., Morrison, H., & Wen, S. W. (2011). Antioxidants and breast cancer risk—A population-based case-control study in Canada. *BMC Cancer, 11*, 372.

Park, S. O., Yoo, Y. B., Kim, Y. H., Baek, K. J., et al. (2015). Effects of combination therapy of docetaxel with selenium on the human breast cancer cell lines MDA-MB-231 and MCF-7. *Annals of Surgical Treatment and Research, 88*, 55–62.

Popova, N. V. (2002). Perinatal selenium exposure decreases spontaneous liver tumorogenesis in CBA mice. *Cancer Letters, 179*, 39–42.

Ravn-Haren, G., Olsen, A., Tjønneland, A., Dragsted, L. O., Nexø, B. A., Wallin, H., et al. (2006). Associations between GPX1 Pro198Leu polymorphism, erythrocyte GPX activity, alcohol consumption and breast cancer risk in a prospective cohort study. *Carcinogenesis*, *27*, 820–825.

Rayman, M. P. (2012). Selenium and human health. *The Lancet*, *379*, 1256–1268.

Rudolph, A., Chang-Claude, J., & Schmidt, M. K. (2016). Gene-environment interaction and risk of breast cancer. *British Journal of Cancer*, *114*, 125–133.

Rusolo, F., Capone, F., Pasquale, R., Angiolillo, A., Colonna, G., et al. (2017). Comparison of the seleno-transcriptome expression between human non-cancerous mammary epithelial cells and two human breast cancer cell lines. *Oncology Letters*, *13*, 2411–2417.

Sakallı Çetin, E., Nazıroğlu, M., Çiğ, B., Övey, İ. S., & Aslan Koşar, P. (2017). Selenium potentiates the anticancer effect of cisplatin against oxidative stress and calcium ion signaling-induced intracellular toxicity in MCF-7 breast cancer cells: Involvement of the TRPV1 channel. *Journal of Receptor and Signal Transduction Research*, *37*, 84–93.

Schweizer, U., Schomburg, L., & Kohrle, J. (2016). Selenoprotein P and selenium distribution in mammals. In D. L. Hatfield, U. Schweizer, P. A. Tsuji, & V. N. Gladyshev (Eds.), *Selenium: Its molecular biology and role in human health* (pp. 261–274). Heidelberg: Springer International Publishing.

Song, H., Ren, X., & Liu, P. (2015). Distribution and inhibition effect of seleno-l-methionine on 4T1 mouse mammary carcinoma. *International Journal of Physiology, Pathophysiology and Pharmacology*, 7, 76–86.

Speckmann, B., & Grune, T. (2015). Epigenetic effects of selenium and their implications for health. *Epigenetics*, *3*, 179–190.

Stoffaneller, R., & Morse, N. L. (2015). A review of dietary selenium intake and selenium status in Europe and the Middle East. *Nutrients*, 7, 1494–1537.

Takada, H., Hirooka, T., Hatano, T., Hamada, Y., & Yamamoto, M. (1992). Inhibition of 7,12-dimethylbenz[a]anthracene-induced lipid peroxidation and mammary tumor development in rats by vitamin E in conjunction with selenium. *Nutrition and Cancer*, *17*, 115–122.

Udler, M., Maia, A. T., Cebrian, A., Brown, C., Greenberg, D., Shah, M., et al. (2007). Common germline genetic variation in antioxidant defense genes and survival after diagnosis of breast cancer. *Journal of Clinical Oncology*, *25*, 3015–3023.

Unni, E., Kittrell, F. S., Singh, U., & Sinha, R. (2004). Osteopontin is a potential target gene in mouse mammary cancer chemoprevention by Se-methylselenocysteine. *Breast Cancer Research*, *6*, R586–92.

Vinceti, M., Dennert, G., Crespi, C. M., Zwahlen, M., Zeegers, M. P. A., Horneber, M., et al. (2014). Selenium for preventing cancer. *Cochrane Database of Systematic Reviews*, *3*, CD005195.

Wang, Y., Fang, W., Huang, Y., Hu, F., Ying, Q., Yang, W., et al. (2015). Reduction of selenium-binding protein 1 sensitizes cancer cells to selenite via elevating extracellular glutathione: A novel mechanism of cancer-specific cytotoxicity of selenite. *Free Radical Biology and Medicine*, *79*, 186–196.

Wrobel, J. K., Power, R., & Toborek, M. (2016). Biological activity of selenium: Revisited. *IUBMB Life*, *68*, 97–105.

Yazdi, M. H., Mahdavi, M., Kheradmand, E., & Shahverdi, A. R. (2012). The preventive oral supplementation of a selenium nanoparticle-enriched probiotic increases the immune response and lifespan of 4T1 breast cancer bearing mice. *Arzneimittel-Forschung*, *62*, 525–531.

Zhang, S., Li, F., Younes, M., Liu, H., Chen, C., & Yao, Q. (2013). Reduced selenium-binding protein 1 in breast cancer correlates with poor survival and resistance to the anti-proliferative effects of selenium. *PLoS One*, *8*, 1–11.

Zhang, X., & Zarbl, H. (2008). Chemopreventive doses of methylselenocysteine alter circadian rhythm in rat mammary tissue. *Cancer Prevention Research*, *1*, 119–127.

CHAPTER EIGHT

Selenium and Epigenetics in Cancer: Focus on DNA Methylation

Ewa Jabłońska[1], Edyta Reszka
Nofer Institute of Occupational Medicine, Lodz, Poland
[1]Corresponding author: e-mail address: ewa.jablonska@imp.lodz.pl

Contents

Abstract

Chemopreventive activity of selenium (Se) may influence epigenome. In this review, we have discussed two aspects of Se and epigenetics in cancer, related to (1) the association between Se and epigenetic regulation in cancer development and prevention; (2) epigenetic modification of selenoprotein-encoding genes in different cancers. In both issues, we focused on DNA methylation as the most investigated epigenetic mechanism. The existing evidence from experimental data in human cancer cell lines, rodents, and human studies in cancer-free subjects indicates that: high Se exposure leads to the inhibition of DNA methyltransferase expression/activity; the association between Se and global methylation remains unclear and requires further investigation with respect to the underlying mechanisms and possible nonlinear character of this relationship; Se affects methylation of specific tumor suppressor genes, possibly in a sex-dependent manner; and cancer phenotype is often characterized by altered methylation of selenoprotein-encoding genes, mainly glutathione peroxidase 3.

Advances in Cancer Research, Volume 136
ISSN 0065-230X
http://dx.doi.org/10.1016/bs.acr.2017.07.002

1. INTRODUCTION

The term "epigenetics" refers to alterations in gene function that do not involve a change in DNA sequence, but modify the structure and organization of chromatin. Such modifications have an impact on the availability of transcription factors toward DNA; thus they directly regulate gene expression (Biswas & Rao, 2017). Mechanisms responsible for epigenetic alterations (Fig. 1) include DNA methylation and histone acetylation, and the key epigenetic modulators are enzymes involved in these biochemical processes: DNA methyltransferases (DNMTs), DNA demethylases (TETs—ten-eleven translocation enzymes), histone acetylases (HATs), and histone deacetylases (HDACs). The other way of gene expression regulation that has been also linked to epigenetic machinery is associated with the action of small noncoding RNAs, called microRNAs (miRNAs) (Biswas & Rao, 2017).

Epigenetic alterations have been recognized to play an important role in human pathologies, including cancer. It is well established that they occur along with genetic mutations in cancer development and progression. Importantly, epigenetic modifications are reversible (contrary to genetic mutations), which makes them promising target for anticancer drugs and

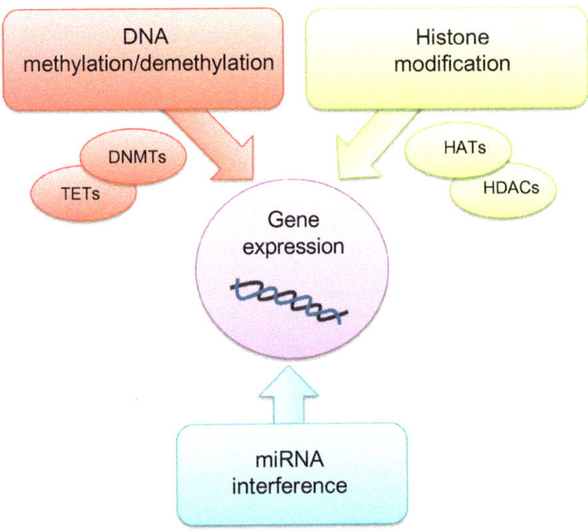

Fig. 1 Epigenetic mechanisms and enzymes involved in gene expression regulation. *DNMTs*, DNA methyltransferases; *HATs*, histone acetylases; *HDACs*, histone deacetylases; *TETs*, ten-eleven translocation enzymes (demethylases).

chemopreventive agents. A number of chemicals have been shown to affect epigenome and there are many ongoing clinical trials testing their efficacy, especially in cancer treatment (Biswas & Rao, 2017). Moreover, several nutrients with potentially anticancer activity have been also shown to target epigenetic alterations (Remely et al., 2015). It is not surprising that the essential trace element, selenium (Se), which has been investigated in terms of its chemoprotective activity over the last five decades, has been also shown to influence gene expression via different epigenetic pathways. In this review, we summarize studies on the link between Se, selenoproteins, and epigenome, particularly in relation to cancer. This topic is identified as one of the research gaps in the Se field (Jablonska & Vinceti, 2015). Since most of epigenetic studies concerned DNA methylation, we focused on this type of epigenetic mechanism. The scope of the topic on Se and DNA methylation, reviewed recently (Speckmann & Grune, 2015), has been expanded here by challenging the inconsistent observations on the epigenetic potential of Se to affect DNA methylation and DNMT activity/expression, proposing that the underlying mechanisms depend on the Se level and, finally, discussing additionally the association between methylation of selenoprotein-encoding genes and cancer.

2. SELENIUM AND DNA METHYLATION

2.1 Selenium, DNA Methyltransferase, and Global DNA Methylation

DNA methylation is a biochemical process of DNA modification, that is transfer of a methyl group to the fifth carbon of cytosine in the 5′-CpG-3′ dinucleotides, resulting in the formation of 5-methylcytosine (5-mC). CpG dinucleotides are typically enriched in gene promoters, where they form CpG islands. In the transcriptionally active genes, CpG islands of promoter regions stay unmethylated, so the transcription factors may bind to DNA. Conversely, methylated CpG islands silence gene expression by preventing binding of transcription factors (Pechalrieu, Etievant, & Arimondo, 2017; Veeck & Esteller, 2010). Enzymes catalyzing DNA methylation in humans belong to two families of DNMTs: (1) DNMT1 (methylates newly synthetized DNA, being responsible for methylation maintenance), and (2) DNMT3a and DNMT3b (responsible for de novo DNA methylation crucial for cellular differentiation and development) (Pechalrieu et al., 2017). The essential substrate and methyl group donor required for catalytic activity of DNMTs is S-adenosylmethionine (SAM), which is formed in one-carbon metabolism pathway, through adenosylation of methionine (Met) (Fig. 2). During methylation process, SAM is transformed to S-adenosylhomocysteine (SAH), which is further transformed to homocysteine (HCys). The pool of SAM is being restored in cells due to remethylation of HCys to Met. DNA methylation status is mainly dependent on dietary intake of vitamins involved in one-carbon metabolism, such as folate and vitamin B12. Both nutrients cooperate with each other in the remethylation of HCys to Met. Folate in the form of 5-methyltetrahydrofolate (5-MTHF) serves as a methyl group donor and B12 vitamin is a coenzyme of methyltransferase catalyzing methylation reaction (Crider, Yang, Berry, & Bailey, 2012; Stover, 2009). Studies on DNA methylation in different human cancers

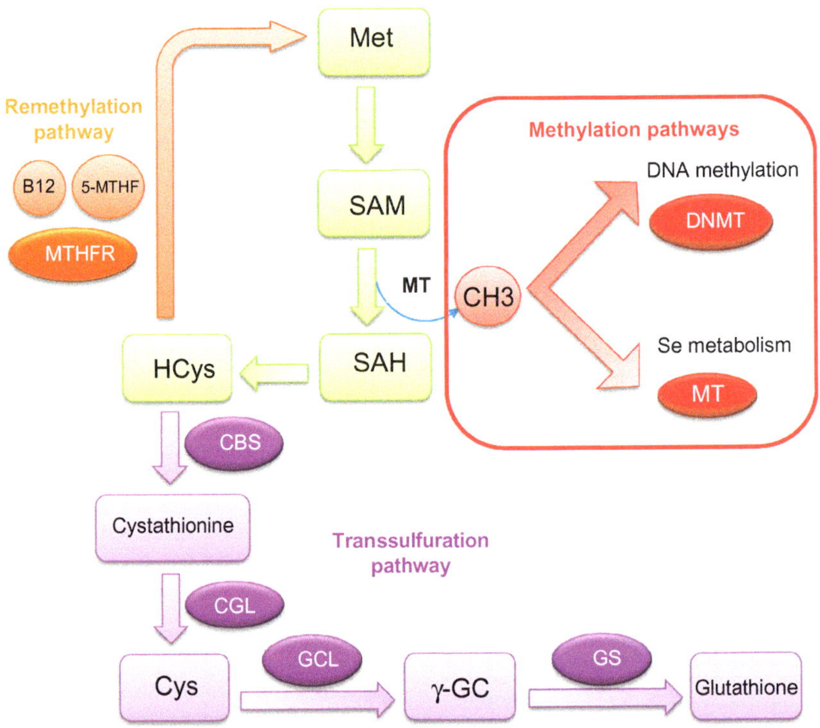

Fig. 2 The association between Se- and one-carbon metabolism-related pathways in the liver, including methylation, remethylation, and transsulfuration. *5-MTHF*, 5-methyltertrahydrofolate; *B12*, vitamin B12; *CBS*, cystathionine β-synthase; *CGL*, cystathionine γ-lyase; *Cys*, cysteine; *DNMT*, DNA methyltransferase; *γ-GC*, γ-glutamylcysteine; *GCL*, glutamate cysteine ligase; *GS*, glutathione synthase; *HCys*, homocysteine; *Met*, methionine; *MT*, methyltransferase; *MTHFR*, methylenetetrahydrofolate reductase; *SAH*, S-adenosylhomocysteine; *SAM*, S-adenosylmethionine.

have shown that tumor cells are characterized with global loss of DNA methylation (which is also called global DNA hypomethylation), but at the same time hypermethylation of promoter regions of specific genes has been observed (Biswas & Rao, 2017; Ehrlich, 2009). Whereas it is still not fully understood what is the exact link between global hypomethylation and cancer development, the role of hypermethylation at specific genes seems clear as it occurs mainly in tumor suppressor genes. Hypermethylation of tumor suppressor genes leading to their silencing results in decreased protection against cell proliferation and increases the risk of tumor development (Fig. 3) (Biswas & Rao, 2017; Ehrlich, 2009).

Markers of DNA methylation refer mostly to the level of global methylation (reflected, for example, by the level of 5-mC or methylation of repetitive elements) or the activity/expression of DNMTs. Summary of studies on the link between Se and DNMT or DNA methylation in different models (cancer and noncancer related) is shown in Table 1. First studies on the association between Se and epigenetic markers concentrated mainly on the effects of Se exposure on DNMTs. Early report published by Cox in 1985 showed that sodium selenite at the concentration of 10 μM inhibited the activity of DNMT extracted from rat liver by 50% and confirmed the inhibitory effect of selenite on liver DNMT in rats supplemented with

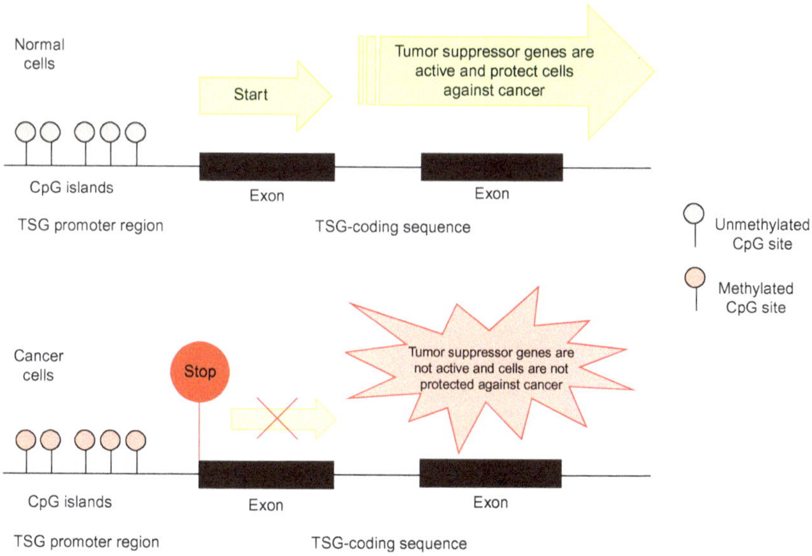

Fig. 3 Role of DNA methylation in cancer development and progression via regulation of TSGs (tumor suppressor genes) expression.

Table 1 The Association Between Se and Global DNA Methylation, DNMT Activity/Expression, and Specific Genes Methylation—Experimental Studies in Cells and Animals, Including Cancer and Noncancer Models

Model	Se Treatment	Effect on DNMT (Activity or Expression)	Effect on Global DNA Methylation	Effect on Specific Genes	References
Colon cancer					
Ex vivo: Human colon carcinomas (nuclear extracts)	Selenite: 0–10 μM (range) BSC: 0–12 μM (range) p-XSC: 0–12 μM (range)	All Se compounds inhibited activity in a dose-dependent manner (50% inhibition of enzyme activity by concentrations: 3.8 μM for selenite, 8.4 for BSC, and 5.2 μM for p-XSC)	N/A	N/A	Fiala, Staretz, Pandya, El-Bayoumy, and Hamilton (1998)
In vitro: HCT116 (human colon carcinoma cells)	p-XSC: 0–40 μM (range)	Inhibition of activity in a dose-dependent manner (50% inhibition of enzyme activity by concentration of 20 μM)	N/A	N/A	Fiala et al. (1998)
In vitro: Caco-2 (human colon adenocarcinoma cells)	Selenite: 0 μM 1 μM 2 μM (basal Se concentration in medium was 0.53 μM)	N/A	Increased DNA methylation	Increased promoter methylation of *TP53*	Davis, Uthus, and Finley (2000)

In vitro: HT-29 (human colon adenocarcinoma cells)	Selenite: 0 μM 1 μM 2 μM (basal Se concentration in medium was 0.53 μM)	Decreased protein expression level at the concentration of 2 μM	Increased DNA methylation in a dose-dependent manner	N/A	Davis and Uthus (2002)
In vitro: Caco-2 cells (human colon adenocarcinoma cells)	SeMeSC 0.0 μM 0.25 μM	N/A	N/A	Decreased methylation of *VHL* promoter region, corresponding to the increased expression of *VHL* mRNA	Uthus, Begaye, Ross, and Zeng (2011)
In vitro: Caco-2, HCT116 (human colon adenocarcinoma cells)	Selenite or SeMeSc 0.2–5 μM (range)	No effect on DNMT mRNA expression	No effect on LINE-1	N/A	Barrera, Johnson, Bao, Cassidy, and Belshaw (2013)
In vivo: Rat model of chemically induced colon carcinogenesis (colon)	Se-enriched milk protein: 1.0 mg Se/kg, either alone, or in combination with GTE	No effect in Se alone group; decreased protein expression in Se + GTE group (the effect was correlated to the decreased number of aberrant crypt foci formation)	N/A	N/A	Hu et al. (2013)

Continued

Table 1 The Association Between Se and Global DNA Methylation, DNMT Activity/Expression, and Specific Genes Methylation—Experimental Studies in Cells and Animals, Including Cancer and Noncancer Models—cont'd

Model	Se Treatment	Effect on DNMT (Activity or Expression)	Effect on Global DNA Methylation	Effect on Specific Genes	References
Prostate cancer					
In vitro: LNCaP (human prostate cancer cells)	SeMet: 150 μM	N/A	N/A	No effect on promoter gene methylation of *GSTP1* or protein level of GSPT1	Ramachandran, Navarro, Gordian, Das, and Singal (2007)
In vitro: LNCaP (human prostate cancer cells)	Selenite: 0.5 μM 1.5 μM	Decreased mRNA and protein expression in a dose-dependent manner	Decreased global DNA methylation (only 1.5 μM concentration was tested)	Demethylation of promoter regions in *GSTP1*, *APC*, and *CSR1*, corresponding to: increased mRNA and protein expression for GSTP1 in a time- and dose-dependent manner (both doses tested), and also increased mRNA for *APC* and *CSR1* (only 1.5 μM tested)	Xiang, Zhao, Song, and Zhong (2008)
Breast cancer					
In vitro: MCF-7 (human breast adenocarcinoma cells)	Selenite: 8 μM 10 μM MSA: 1.6 μM 2.0 μM	Decreased protein expression level in a dose-dependent manner	N/A	N/A	de Miranda et al. (2014)

Erythroleukemia					
In vitro: FELC (Friend erythroleukemia cells, transformed murine cells) nuclear extracts	Selenite: 0–100 μM (range)	Inhibition of activity in a dose-dependent manner (50% inhibition of enzyme activity by concentration of 5 μM)	Decreased DNA methylation in a dose-dependent manner (concentrations analyzed: 0, 10, 20, and 40 μM)	N/A	Cox and Goorha (1986)
Noncancer-related models					
Ex vivo: Rat liver (nuclear extracts)	Selenite: 1–100 μM (range)	Inhibition of activity in a dose-dependent manner (50% inhibition of enzyme activity by concentration of 10 μM)	N/A	N/A	Cox (1985)
In vivo: Rats (liver, nuclear extracts)	Selenite: 3 mg/kg diet	Inhibition of DNA methylase in the liver by 57%	N/A	N/A	Cox (1985)
In vivo: Rats (colon, liver)	Selenite or SeMet 0.0 mg/kg diet 0.1 mg/kg diet 2.0 mg/kg diet for 6 weeks (basal diet contained <0.0016 mg/kg Se)	N/A	Increased DNA methylation in both tissues. No differences between 0.1 and 2.0 mg/kg doses. Chemical form influenced the effect in colon but not in the liver	N/A	Davis et al. (2000)

Continued

Table 1 The Association Between Se and Global DNA Methylation, DNMT Activity/Expression, and Specific Genes Methylation—Experimental Studies in Cells and Animals, Including Cancer and Noncancer Models—cont'd

Model	Se Treatment	Effect on DNMT (Activity or Expression)	Effect on Global DNA Methylation	Effect on Specific Genes	References
In vivo: Rats (colon, liver)	Selenite: 0.0 mg/kg diet 2.0 mg/kg diet	Increased DNMT activity in liver and colon at dose 2.0 mg/kg	Increased DNA methylation in colon and liver	N/A	Davis and Uthus (2003)
In vivo: Rats (colon, liver)	Selenite: 0.0 mg/kg diet 0.2 mg/kg diet	Nonsignificantly increased DNMT activity in the colon and liver at dose 0.2 mg/kg	Increased global DNA methylation in colon but not in the liver	N/A	Uthus, Ross, and Davis (2006)
In vivo: Mice (colon, liver)	0.08 mg Se/kg 0.88 mg Se/kg (high fat/carbohydrate diet)	N/A	Decreased global DNA methylation in the liver but not in the colon at dose 0.88 mg/kg	N/A	Armstrong et al. (2011)
In vivo: Rats (colon, liver, blood)	SeMet 0.0 mg/kg diet 0.15 mg/kg diet 4.0 mg/kg diet	N/A	Decreased global DNA methylation in liver and nonsignificant decrease in colon mucosa at dose 4 mg/kg. No changes in blood	Increased exon-specific methylation of *Tp53* in colon and liver in a dose-dependent manner	Zeng, Yan, Cheng, and Uthus (2011)

In vivo: Mice (heart)	Selenite: <0.025 mg/kg 0.15 mg/kg 0.5 mg/kg	Decreased activity at dose 0.5 as compared to 0.15 after 3 weeks but not after 12 weeks	Decreased global DNA methylation in a time-dependent manner	N/A	Metes-Kosik et al. (2012)
In vitro: Primary neonatal rat cardiomyocytes	Selenite: 0.5 μM 1.5 μM	Decreased protein expression	N/A	Se increased methylation of promoter regions in two inflammatory response genes (not discussed in this review)	Yang et al. (2014)
In vivo: Rat model of Keshan disease (rat myocardium)	Selenite: 0.0 mg/kg 0.1 mg/kg 2.0 mg/kg (basal diet contained 0.004 mg of Se/kg diet)	Decreased protein expression	N/A	Se increased methylation of promoter regions in two inflammatory response genes (not discussed in this review)	Yang et al. (2014)

APC, adenomatous polyposis coli; *BSC*, benzyl selenocyanate; *CSR1*, cellular stress response 1; *GSTP1*, glutathione *S*-transferase P; *GTE*, green tea extract; *LINE-1*, long interspersed nucleotide element 1; *MSA*, methylseleninic acid; *N/A*, not analyzed; *p-XSC*, 1,4-phenylenebis (methylene)selenocyanate; *SeMeSC*, *Se*-methylselenocysteine; *SeMet*, selenomethionine; *TP53*, tumor protein 53 (human); *Tp53*, tumor protein 53 (rodent); *VHL*, von Hippel–Lindau gene.

inorganic Se at the dose of 3 mg/kg diet (Cox, 1985). Similarly, sodium selenite at the concentration of 5 μM inhibited the activity of DNMT by 50% in the transformed murine erythroleukemia cells (FELC), and furthermore, this effect was accompanied by decreased global methylation (Cox & Goorha, 1986). Majority of further studies on cancer cell lines, mainly derived from colon cancer, confirmed the inhibitory effect of Se on DNMT. Fiala et al. showed that the enzyme activity was inhibited in human colon carcinoma cells (HCT116) by the synthetic Se compound p-XSC (1,4-phenylenebis(methylene)selenocyanate) with an IC_{50} of 20 μM (Fiala et al., 1998). Similarly, DNMT activity was inhibited in nuclear extracts from human colon carcinomas by p-XSC (IC_{50}=5.2 μM); the same was also observed for other synthetic Se compound, such as benzyl selenocyanate (BSC; IC_{50}=8.1 μM) and selenite (IC_{50}=3.8 μM) (Fiala et al., 1998). Se, mainly in the form of selenite, decreased DNMT also at the protein level as shown in human colon adenocarcinoma cells (HT-29 cell line; Se concentration 2 μM) (Davis & Uthus, 2002), human prostate cancer cells (LNCaP cell line; Se concentrations 0.5 and 1.5 μM) (Xiang et al., 2008), and human breast adenocarcinoma cells (MCF-7 cell line; Se concentration: 8 and 10 μM for selenite and 1.6 and 2.0 μM for methylseleninic acid) (de Miranda et al., 2014). In vivo study of chemically induced colon cancer showed that DNMT protein level was decreased by Se-enriched milk (1 mg Se/kg diet) when combined with green tea extract (Hu et al., 2013). On the contrary, Barrera et al. did not observe any effect of Se compounds (selenite and methylselenocysteine: 0.2–5 μM) on DNMT mRNA expression (neither did they affect the level of global DNA methylation) in two human colon cancer cell lines (Caco-2, HCT116) (Barrera et al., 2013). Nevertheless, majority of the existing data give substantial evidence that Se affects DNMT activity/protein expression in a negative manner in cancer cells. Notably, in most studies the inhibitory effect was observed in a dose-dependent manner (Cox, 1985; Cox & Goorha, 1986; de Miranda et al., 2014; Fiala et al., 1998; Hu et al., 2013; Xiang et al., 2008).

Now the question rising from these observations is how this effect corresponded to the potentially anticancer activity of Se? In the mentioned rat study of chemically induced colon cancer (Hu et al., 2013), decreased DNMT observed after supplementation with Se combined with green tea extract was accompanied by decreased formation of precancerous lesions (aberrant crypt foci). Thus, it was suggested that DNMT inhibition underlies the anticancer activity of Se (combined with green tea extract) in the colon (Hu et al., 2013). In human cancer cell lines from breast (MCF-7) and colon (HCT116; both studies discussed earlier), decreased DNMT activity/expression upon Se exposure was also observed along with direct anticancer effects, such as decreased proliferation or growth (de Miranda et al., 2014; Fiala et al., 1998). More indirect anticancer effects of Se (activation of tumor suppressor genes) accompanied by decreased DNMT were observed in prostate cancer cells (PLNCaP) (Xiang et al., 2008) that will be discussed later in detail. Overall, these studies suggest that DNMT inhibition may be linked to anticancer activity of Se at the postinitiation stage of cancer development, such as during promotion and progression. It should be noted that none of the studies focused on the anticancer effect typically at the initiation stage (the study models were either cancer cell lines or rodent studies of chemically induced carcinogenesis). As the first stage of cancer development, initiation is relevant to dietary chemoprevention strategy. The only human study, which investigated the effects of Se on DNMT, failed to indicate any changes in *DNMT1* mRNA expression in rectal biopsies of nine subjects supplemented with Se at 48 μg/day (as six Brazil nuts) (Hu et al., 2016). Nevertheless, a significant decrease in *DNMT1* mRNA expression was observed in that study in the second group of subjects ($n = 12$) who consumed Brazil nuts in combination with green tea extract. It is worth to mention that the baseline Se status of all these 21 subjects was relatively high (see Table 2 for details).

It should be expected that decreased DNMT activity upon Se treatment will result in a decreased level of global DNA methylation. However, the authors rarely analyzed both markers together, so a clear conclusion on this link is not feasible. Unexpectedly, selenite administration at 2 μM was associated with the increased level of global DNA methylation in human colon cancer cells (HT-29), which was inconsistent with the observed decreased protein level of DNMT (Davis & Uthus, 2002). Similarly, selenite treatment at 2 μM increased the level of global DNA methylation in the other human colon cancer cell line (Caco-2) (Davis et al., 2000). In fact, only two in vitro studies, conducted in LNCaP and FELC cell lines, showed a direct correlation between the decreased DNMT activity/expression and the decreased level of global DNA methylation under selenite treatment (1.5 μM) (Cox & Goorha, 1986; Xiang et al., 2008).

In vivo studies in healthy animals or noncancer-related studies provided inconsistent data on the ability of Se to inhibit DNMT or affect DNA global methylation. In line with the first report by Cox, showing DNMT activity to be inhibited by selenite in rat liver in vivo (at high dose of

Table 2 Human Studies on the Association Between Se Status and DNA Methylation or DNMT

Study/Group	Se Status	Association With Epigenetic Markers of DNA Methylation	References
Cross-sectional study: 287 subjects, age 37.9 years	Plasma Se: 45.4–148.8 ng/mL	Inverse association between plasma Se concentration and global DNA methylation in blood leukocytes	Pilsner et al. (2011)
Cross-sectional study: 185 subjects, age 50 years	Plasma Se: 48.2–147.7 ng/mL	Positive association between plasma Se and LINE-1 methylation (index of global methylation) in rectal biopsies only in females, negative correlation between plasma Se and *WIF1* methylation in all subjects regardless of sex, positive in males but negative in females correlation between plasma Se and methylation of *APC*, *SFRP1*, *SFRP2*, *ESR1*, *N33*, *PCA1*	Tapp et al. (2013)
Intervention study: supplementation with Se as Brazil nuts for 6 weeks, estimated daily dose 48 μg): 9 subjects, age 63.3 years, received only Se 12 subjects, age 62.4 years, received Se + GTE	Plasma Se: Se group: Before—101.8 ng/mL After—124.8 ng/mL Se + GTE group: Before—106.6 ng/mL After—127.1 ng/mL	No effect on *DNMT1* mRNA expression in rectal biopsies in Se group, significant decrease was observed only in the group Se + GTE	Hu et al. (2016)

APC, adenomatous polyposis coli; *ESR1*, estrogen receptor 1; *GTE*, green tea extract; *LINE-1*, long interspersed nucleotide element 1; *N33*, tumor suppressor candidate 3; *PCA1*, alkB homolog 3, alpha-ketoglutarate-dependent dioxygenase; *SFRP1*, secreted frizzled-related protein 1; *SFRP2*, secreted frizzled-related protein 2; *WIF1*, WNT inhibitory factor 1.

3 mg Se/kg diet) or ex vivo using rat liver extracts (Cox, 1985), DNMT protein or activity was decreased upon Se treatment in primary neonatal rat cardiomyocytes (0.5 and 1.5 μM) (Yang et al., 2014) and in the heart tissue of rodents: rats fed 0.1 or 2.0 mg of Se/kg diet (model of Keshan disease) (Yang et al., 2014) and mice fed 0.5 mg Se/kg diet (Metes-Kosik et al., 2012). Opposite observations were made by Davis, Uthus, and colleagues who conducted several in vivo experiments in rats, focusing mainly on the effect of severe Se deficiency on DNA methylation in rat tissues. They observed that rats supplemented with Se had higher DNMT activity in colon and liver when compared to rats fed Se-deficient diet (Davis & Uthus, 2003; Uthus et al., 2006). Consistently, Se supplementation was also associated with the increased level of global DNA methylation in rat colon (Davis et al., 2000; Davis & Uthus, 2003; Uthus et al., 2006) and liver (Davis et al., 2000; Davis & Uthus, 2003). Se deficiency in these studies was compared to the effects of Se supplementation (either in the form of selenite or in the form of SeMet) at doses: 0.1 or 0.2 mg/kg diet (regarded as Se adequate intakes) and 2 mg/kg diet (regarded as supranutritional intake). Thus, the authors concluded that Se deficiency was associated with global DNA hypomethylation, and since the methylation level was not different between 0.1 and 2.0 mg/kg doses, Se deficiency was more important modifier of methyl metabolism than either adequate or supplemental Se (Davis et al., 2000; Uthus et al., 2006). Davis et al. have suggested also that altered DNA methylation may be a potential mechanism of tumorigenesis associated with Se deficiency (as a result of global hypomethylation), but it does not underlie chemoprotective effects of high Se intake (Davis et al., 2000). However, conclusion regarding high Se intake and DNA methylation has been undermined by further in vivo studies that compared animals fed a high-Se diet, including rats receiving 4 mg of Se/kg diet (Zeng et al., 2011) and mice fed with 0.88 mg of Se/kg diet (Armstrong et al., 2011). Notably, these doses exceeded by several times (from 6 to 26, respectively) the adequate intake for laboratory animals proposed by the NRC (National Research Council (US), 1995) that corresponds to 0.15 mg/kg diet for rats and mice. In both cases, it was clearly shown that high Se intake decreased DNA methylation in liver, but not in the colon, so the overall effect (at least in the liver) was similar as in case of Se deficiency (Zeng et al., 2011). As already mentioned earlier, it is not clear why Davis et al. did not observe changes in DNA methylation between rats fed with Se at 0.1 and 2.0 mg/kg (Davis et al., 2000). Possibly the second (higher) dose was still within the optimal range for rats.

The authors of the study using 4 mg Se/kg diet in rats (who did not expect to observe decreased DNA methylation upon Se exposure but rather the opposite effect) explained that the inconsistency between reports on supranutritional Se intake and global DNA methylation could be attributed to differences in diet composition and the fact that different animal species/strains were used in these studies (Zeng et al., 2011). As suggested by Speckmann and Grune, the additional explanation may be also related to the methodology, since studies differed between each other in terms of techniques used for global DNA methylation assessment (Speckmann & Grune, 2015).

Surprisingly, the possible role of Se dose in the association between Se and DNA methylation as well as different levels of exposure in the presented studies have not been challenged so far. The fact that the decreased level of global methylation was observed under both Se deficiency (as compared to adequate Se intake) and high Se intake/exposure suggests a nonlinear character of the relationship between Se and DNA methylation (Fig. 4). Another

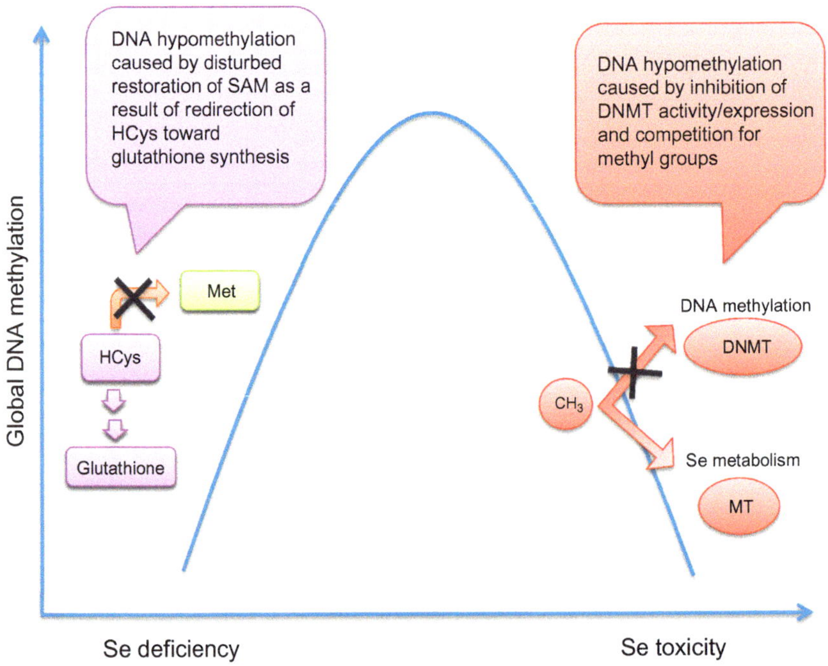

Fig. 4 Proposed nonlinear association between Se intake/exposure and global DNA methylation and the underlying mechanisms beyond the effects of different Se levels on DNA methylation. *DNMT*, DNA methyltransferase; *HCys*, homocysteine; *Met*, methionine; *MT*, methyltransferase; *SAM*, S-adenosylmethionine.

issue that should also be challenged here is tissue specificity of the reported effects; indeed, the effects of high Se were more specific in the liver than in the colon. Certainly, Se is present to much higher concentrations in the liver than in the colon. Actually the liver is among the richest reservoirs of Se of the human organism (Zachara et al., 2001), and as such it may be most affected by any additional Se supply.

Apparently, more studies need to be done in order to investigate the effects of Se deficiency as well as high Se intake on DNMT and DNA methylation, both in healthy and in cancer cells. More insights on the above issues could have been added by human observational studies; however, only few investigators analyzed the association between Se and DNA methylation status in human tissues (Table 2) and, in addition, none of them focused on cancer patients. In a cross-sectional study of 287 (48 males, 239 females) subjects from Bangladesh, plasma Se concentration was inversely correlated with genomic global DNA methylation in blood leukocytes (Pilsner et al., 2011). Conversely, another cross-sectional study of 185 (84 males, 101 females) subjects from the United Kingdom indicated significant and positive correlation between plasma Se and global DNA methylation in rectal mucosa biopsies, but only in females (Tapp et al., 2013). Though both study groups did not differ in terms of Se status (mean or median values: 87.6 and 84.5 ng/mL; see Table 2 for range values), it is difficult to compare the results due to several differences. Those two studies used different methods of DNA methylation assessment: the first one used methyl group acceptance assay, whereas in the other study the authors analyzed methylation of repetitive element, LINE-1 (long interspersed nucleotide element 1). Different tissues were also used to measure the marker (plasma vs colon mucosa). In addition, there were differences in the age and sex ratio between groups, and subjects from Bangladesh were characterized by high level of environmental exposure to arsenic (As), a carcinogenic metal known to affect DNA methylation, and interact with Se at the metabolic level (Pilsner et al., 2011; Reichard & Puga, 2010).

2.2 Selenium and Methylation of Tumor Suppressor Genes

Whereas the decreased level of global DNA methylation has been linked to Se deficiency, the effect on specific genes under low Se intake/status seems less clear (details in Table 1). For example, increased methylation of promoter region at *VHL* (von Hippel–Lindau gene, tumor suppressor gene) was observed under severe Se deficiency in human colon cancer cells

(Caco-2) (Uthus et al., 2011). The authors of that study confirmed the effect of Se deficiency on rodent *Vhl* downregulation in colon mucosa of rats fed a Se-deficient diet (Uthus et al., 2011). This observation was generally in line with previous observations of these authors (already discussed in Section 2.1), according to which Se deficiency was associated with decreased DNMT1 activity and decreased DNA methylation in human colon cancer cells or in rat colon mucosa, and that Se-deficient rats formed more aberrant crypts under exposure to carcinogen (Davis & Uthus, 2002, 2003; Davis et al., 2000; Uthus et al., 2006). Altogether, these results suggest that the altered DNA methylation, including decreased expression of *VHL* tumor suppressor gene as a result of hypermethylated promoter region, may underlie the mechanism of colon cancer development associated with Se deficiency.

Contrary to *VHL*, another tumor suppressor gene, *TP53* (encoding p53; tumor protein p53), was shown to have reduced promoter methylation in Caco-2 human colon cancer cells under Se deficiency (Davis et al., 2000). Consequently, treatment with selenite increased *TP53* promoter methylation in these cells (Davis et al., 2000). Moreover, Se supplementation at the supranutritional level (4 mg/kg diet, in the form of SeMet) increased exon-specific methylation of p53 gene in rat colon and liver (Zeng et al., 2011). Interestingly, in this case increased methylation of gene exon was supposed to have protective effects as it was suggested to preserve the intact *TP53* sequence. It is generally recognized that p53 gene is commonly mutated in different cancers and its mutations have been shown to correlate with hypomethylation as a possible underlying event in genomic instability. *TP53* mutations often change the phenotype of p53 from tumor suppressor protein toward new oncogenic function. Thus hypermethylation of the *TP53* exon sequence in order to prevent these mutations has been regarded as a novel important mechanism of protection against cancer (Herrero, Rojas, Misiewicz-Krzeminska, Krzeminski, & Gutierrez, 2016; Perri, Pisconti, & Della Vittoria Scarpati, 2016).

Apart from colon cancer, the effect of Se on methylation of specific genes was investigated also in prostate cancer. In cells derived from human prostate cancer (LNCaP), exposure to nontoxic dose of Se (1.5 μM) led to demethylation of promoter regions of three genes known to be silenced in LNCaP cells, including *GSTP1* (phase II detoxification enzyme), *APC* (tumor suppressor gene), and *CSR1* (protein with antioxidant and tumor suppressor function) (Xiang et al., 2008). Consequently, demethylation of *GSTP1*, *APC*, and *CSR1* was associated with reexpression of these genes, as shown at the mRNA and/or protein level. As already mentioned, treatment with selenite had a broad effect on the epigenome of LNCaP cells in this study, since it was also shown to significantly decrease the level of global DNA methylation and to reduce *DNMT1* mRNA/protein expression and *DNMT3A* mRNA expression. Moreover, the effect on the DNMT1 protein level was dose dependent. On the other hand, Ramachandran et al. failed to demonstrate the demethylation of *GSTP1* in LNCaP cells treated with the organic compound seleno-DL-methionine, which may point to the importance of chemical form in this process (Ramachandran et al., 2007). Alternatively, different times of exposure to Se could have been responsible for inconsistent results between these studies (Xiang et al., 2008).

According to human observational studies, Se status has been found to be correlated with methylation of several cancer-related genes. Tapp et al. found a negative correlation between plasma Se and promoter methylation of the tumor suppressor gene *WIF1* (WNT inhibitory factor 1) in colon mucosa of 185 subjects shown to be free from colorectal disease. For other six genes, including tumor suppressor gene *APC* (adenomatous polyposis coli; see Table 2 for details), the relationship was apparently sex dependent, showing a negative correlation with plasma Se in females, but positive in males (Tapp et al., 2013).

2.3 Possible Mechanisms Underlying Se-Induced DNA Hypomethylation Under Different Levels of Exposure/Intake

It is generally recognized that mechanisms through which Se affects DNA methylation are associated with the ability of Se to interfere with the one-carbon metabolism (Met–HCys cycle, dependent on folate and vitamin B12 due to which methyl group donor, SAM, is formed) (Speckmann & Grune, 2015). Se is involved in the methyl metabolism since this nutrient is itself metabolized through methylation via an SAM-dependent pathway, so it competes with DNMT for methyl donation from SAM (Fig. 2). Numerous studies, as already discussed, indicated that Se inhibits DNMT activity/expression though it is not known what is the exact mechanism of such inhibition and whether it is directly linked with utilization of SAM. Importantly, most of the reports showed that the inhibitory effect of Se on DNMT is dose dependent. This is somewhat conflicting with the proposed concept of nonlinear association between Se and global DNA methylation because the inhibitory effect observed on DNMT and DNA methylation seems to occur only at high Se exposure/intake levels, whereas it is completely not clear how the decreased level of DNA methylation may correspond with DNMT activity under conditions of Se deficiency. Apparently, Se deprivation decreases global DNA methylation via other mechanisms, not directly related to DNMT activity. The underlying mechanism here may be linked to the metabolism of HCys, metabolite used for restoration of SAM in the remethylation pathway and also for synthesis of glutathione during the transsulfuration pathway (Fig. 2). Notably, Se-deficient condition is known to increase synthesis of glutathione, possibly as a compensatory mechanism counteracting oxidative stress (Davis & Uthus, 2003; Hill & Burk, 1982). Glutathione is a molecule crucial for antioxidant defense and it is synthesized in the liver during transsulfuration pathway. The first step of transsulfuration utilizes HCys which is condensed with serine (Ser) in an irreversible reaction catalyzed by cystathionine β-synthase (CBS), in order to form cystathionine (Fig. 2). Cystathionine is further transformed by cystathionine β-lyase to cysteine (Cys), which is the first sulfur amino acid precursor of glutathione. Synthesis of glutathione involves two enzymatic steps, catalyzed by glutamate cysteine ligase (GCL) and glutathione synthase (GS), respectively (Lu, 2009). Hill and Burk reported in 1982 that Se deficiency was associated with the increased glutathione synthesis in rat hepatocytes and this observation was directly associated with the increased activity of GCL (Hill & Burk, 1982). Furthermore, in vivo study conducted in rats by Davis and Uthus indicated that Se deficiency led to the increased plasma glutathione concentration and this effect was correlated with decreased liver activities of betaine–homocysteine methyltransferase (BHMT) and methionine synthase (MS) (Davis & Uthus, 2003). Both BHMT and MS are enzymes required for remethylation of HCys to SAM; thus decreased activities of these enzymes along with the increased glutathione synthesis observed under Se deprivation may indicate that Se deficiency leads to redirecting of HCys toward transsulfuration pathway (Davis & Uthus, 2003).

Human observational studies have shown Se to be negatively correlated with plasma/serum levels of HCys (Angelova, Atanassova, Chalakova, & Dimitrov, 2008; Bates, Thane, Prentice, & Delves, 2002; Bekaert et al., 2008; Belanger et al., 2006; Gonzalez et al., 2004; Klapcinska, Poprzecki, Danch, Sobczak, & Kempa, 2005). Interestingly, Se supplementation (up to 300 μg/day) did not affect the plasma HCys concentration in humans (Bekaert et al., 2008; Venn, Grant, Thomson, & Green, 2003). Furthermore, rodent studies indicated that the HCys level was significantly altered not by Se supplementation, but by Se deficiency; HCys was reduced in particular by Se deficiency in plasma (Metes-Kosik et al., 2012; Uthus & Ross,

2007, 2009; Uthus, Yokoi, & Davis, 2002) and increased in the liver (Geillinger et al., 2014; Wolf et al., 2010). In rodent studies, Se deficiency, but not Se supplementation, was also reported to affect liver expression of enzymes involved in the one-carbon metabolism pathway (Uthus & Ross, 2007; Uthus et al., 2006, 2002). Collectively, these data suggest that HCys metabolism may be more affected by Se deficiency than by Se supranutritional supply, and that the effect of Se deficiency on the HCys level is tissue specific. Thus, we may hypothesize that alterations in the HCys metabolism mediate the negative effect of Se deficiency on global DNA methylation, the underlying mechanism of which in the different tissues remains to be elucidated.

To conclude, both Se deficiency and supranutritional/toxic Se intake may decrease global DNA methylation, but the underlying mechanisms are different (Fig. 4). The existing data suggest that Se deficiency leads to redirection of HCys toward transsulfuration pathway in order to synthesize glutathione which makes HCys less available for restoration of SAM and methylation processes are inhibited. On the other hand, excess Se intake is associated with competition for SAM in order to utilize methyl groups required for Se metabolism; consequently, SAM is less available for DNMT and methylation process is inhibited as well. It is worth to add that both transsulfuration pathway (glutathione synthesis) and Met cycle are pathways exclusively occurring in the liver (Lu, 2009). Moreover, as the main reservoir of essential contributors to one-carbon cycle such as folate and vitamin B12 (Cui et al., 2016), and at the same the richest source of Se in the organism (Zachara et al., 2001), liver seems to play a central role in the interference between Se and methylation reactions.

3. METHYLATION OF SELENOPROTEIN-ENCODING GENES

As already mentioned, epigenetic alterations are present in all human cancers and are known to be associated with genetic alterations which drive the cancer phenotype. These changes mainly involve DNA methylation and histone modifications. Hypermethylation of promoter regions of various genes, accompanied by downregulation of those genes, is often observed with human malignant tissues (Kulis & Esteller, 2010). Since now, comprehensive studies, mainly on glutathione peroxidase 3 (*GPX3*) promoter methylation, but also single studies on *GPX1*, selenoprotein X or methionine sulfoxide reductase B1 (*SEPX1* or *MSRB1*), and selenium-binding protein 1 (*SBP1* or *SELENBP1*) promoter methylation have been conducted using tumor, adjacent, and normal tissue specimens (Table 3) and also in vitro studies (Table 4).

GPX3 belongs to the selenoprotein-encoding gene family (Hatfield, Tsuji, Carlson, & Gladyshev, 2014), and it has been primarily found to be ubiquitously downregulated by promoter hypermethylation in the wide spectrum of solid tumors, including head and neck (HNC), urinary bladder (BC), prostate (PC), kidney, ovarian (OC), breast (BrC), gastric (GC), liver, and skin tumors that were observed in tissue specimens (Table 3). *GPX3* hypermethylation with the concomitant decrease of gene or protein expression was also observed in the above-mentioned cancer cell lines (Table 4). It is generally recognized that *GPX3* promoter hypermethylation may explain a mechanism of *GPX3* gene silencing in various solid tumors (Cao et al., 2015; Lee et al., 2005). However, recently, *GPX3* hypermethylation was also observed in bone marrow of acute and chronic myeloid leukemia (AML and CML) (Yao et al., 2015; Zhou et al., 2015) that suggest *GPX3* hypermethylation as a common epigenetic feature of neoplasms.

3.1 Methylation Status in Human Cancer Tissues

First finding regarding *GPX3* methylation in solid tumors was reported in esophageal Barrett's adenocarcinoma (BAC) by Lee et al. in 2005. Mono- and biallelic promoter methylation of *GPX3* in Barrett's adenocarcinoma, dysplasia, and metaplasia samples were observed. Normal adjacent samples showed an unmethylated *GPX3* promoter region. Importantly, *GPX3* hypermethylation was correlated with the downregulation of *GPX3* expression (Lee et al., 2005). In another study, 100 BAC samples presented *GPX3* methylation in less than 20% of tissue specimens. Importantly, promoter hypermethylation was not observed for other glutathione peroxidase-encoding genes, namely *GPX1* and *GPX4* (Lee et al., 2005; Peng et al., 2009). *GPX3* gene was aberrant methylated also in esophageal squamous cell carcinoma (ESCC) (He et al., 2011; Li et al., 2014). Methylation of *GPX3* with accompanied downregulation of *GPX3* expression was also observed in papillary thyroid cancer (PTC) in 47% of tumor samples. Moreover, *GPX3* methylation in PTC specimens was associated with tumor size and regional lymph node metastasis (Zhao et al., 2015). Very elegant study of Li et al. (2014), conducted in 2014, indicated that *GPX3* aberrant methylation detected not in tumor tissue, but in plasma, can be used for the early diagnosis of ESCC. In this study the "methylome" status was

Table 3 Methylation of Selenoprotein-Encoding Genes in Human Cancer Tissues

Cancer Type	Gene	Methylation Location	Gene Methylation in Tissue Samples	References
Head and neck cancer				
BAC	*GPX3*	Exon 1 (position 1984–2201 base) with 17 CpGs	88.2%, $N=34$ BAC 81.8%, $N=11$ dysplasia 61.9%, $N=21$ Barrett's esophagus 0%, $N=10$ normal mucosa	Lee et al. (2005)
BAC	*GPX3*	Promoter region	$<20\%$, $N=100$ BAC 60%, $N=11$ dysplasia 40%, $N=11$ Barrett's esophagus $<9\%$, $N=36$ normal mucosa	Peng et al. (2009)
ESCC	*GPX3*	Promoter region	54.8%, $N=42$ ESCC 9.5%, $N=42$ adjacent mucosa	Li et al. (2014)
ESCC	*GPX3*	Promoter region	71.4%, $N=28$ ESCC 10.7%, $N=28$ adjacent mucosa	He et al. (2011)
HNC	*GPX3*	21 CpGs around TSS	61%, $N=18$ HNC 8%, $N=12$ normal mucosa	Chen et al. (2011)
PTC	*GPX3*	Promoter region	46.8%, $N=94$ PTC 0%, $N=15$ normal mucosa	Zhao et al. (2015)
Urogenital cancer				
BC	*GPX3*	21 CpGs around TSS	54%, $N=13$ BC 8%, $N=12$ normal tissue	Chen et al. (2011)

Continued

Table 3 Methylation of Selenoprotein-Encoding Genes in Human Cancer Tissues—cont'd

Cancer Type	Gene	Methylation Location	Gene Methylation in Tissue Samples	References
ccRCC	*GPX3*	Exon 1, 14 CpGs	77.1%, $N=210$ ccRCC 14.6%, $N=48$ adjacent mucosa	Liu et al. (2015)
PC	*GPX3*	Exon1/intron 1 (+53 to +793 base from TSS)	90%, $N=30$ PC	Yu et al. (2007)
PC	*GPX3*	Promoter region	93%, $N=41$ PC	Lodygin, Epanchintsev, Menssen, Diebold, and Hermeking (2005)
Gynecologic cancer				
BrC	*GPX3*	Promoter region	Methylated, $N=40$ BrC Unmethylated, $N=6$ unrelated individuals	Mohamed et al. (2014)
BrC	*GPX1*	35 CpGs	20%, $N=8$ primary BrC	Kulak et al. (2013)
Gastrointestinal cancer				
GC	*GPX1*, *GPX3*	Promoter region	*GPX3* 30.1%, *GPX1* 16.7%, $N=566$ GC	Jee, Kim, Jung, Kim, and Kim (2009)
GC	*GPX1*, *GPX3*	Promoter region	*GPX1* 31%, *GPX3* 50%, $N=16$ GC	Min, Kim, Jung, Jee, and Kim (2012)
GC	*GPX3*	8 CpGs	60%, $N=60$ GC 39%, $N=23$ adjacent tissue	Peng et al. (2012)

GC	*GPX3*	Promoter region	18%, $N=39$ GC 13%, $N=15$ adjacent tissue	de Melo et al. (2014)
CRC	*SBP1*	Promoter region	CRC with higher methylation vs adjacent tissue, $N=3$	Pohl et al. (2009)
HCC	*GPX3*	Promoter region	76.7%, $N=60$ HCC 0%, $N=60$ adjacent tissue	Cao et al. (2015)
Skin cancer				
MEL	*GPX3*	Promoter region	100% (70% partially), $N=10$ primary MEL 0%, $N=3$ normal skin	Chen et al. (2016)
Blood cancer				
CML	*GPX3*	18 CpGs	15%, 0.019 (0.000–11.919), $N=80$ CML BM 2%, 0.005 (0.000–1.000), $N=44$ unrelated controls BM	Yao et al. (2015)
AML	*GPX3*	18 CpGs	26%, 0.012 (0.000–7.493), $N=181$ de novo AML BM 2%, 0.005 (0.000–1.000), $N=44$ unrelated controls BM	Zhou et al. (2015)
MDS	*GPX3*	18 CpGs	15%, $N=110$ MDS BM	Zhou et al. (2017)

AML, acute myeloid leukemia; *BAC*, esophageal Barrett's adenocarcinoma; *BC*, bladder cancer; *BM*, bone marrow; *BrC*, breast cancer; *ccRCC*, clear cell renal cell carcinoma; *CML*, chronic myeloid leukemia; *CRC*, colorectal cancer; *ESCC*, esophageal squamous cell carcinoma; *GC*, gastric cancer; *HCC*, hepatocellular carcinoma; *HNC*, head and neck; *MDS*, myelodysplastic syndrome; *MEL*, melanoma; *PC*, prostate cancer; *PTC*, papillary thyroid cancer; *TSS*, transcription start site.

Table 4 Methylation of Selenoprotein-Encoding Genes Identified in Human Cancer Cell Lines

Cancer Type	Gene	Methylation Location	Gene Methylation in Cell Lines	References
Head and neck cancer				
EC	*GPX3*	Promoter region	3 EC cell lines: **TE-7**: 90%; **SKGT4** < 5%; **OE33**: 30%	Peng et al. (2009)
ESCC	*GPX3*	Promoter region	7 ESCC cell lines: Methylated: **TE-1**, **KYSE-30**, **KYSE-510**, **EC-9706** Partially methylated: KYSE-410 Unmethylated: HEEpiC, Het-1A	He et al. (2011)
HNC	*GPX3*	21 CpGs around TSS	10 HNC cell lines: 70%: methylated: SCC25, SCC25cp, PCI51, O11, 012 Partially methylated: HN38, Fadu Unmethylated: PCI13, HN17B, HN22A	Chen et al. (2011)
Urogenital cancer				
BC	*GPX3*	21 CpGs around TSS	6 BC cell lines: 83%: methylated: HTB-3, CRL-1749 Partially methylated: HTB-4, CRL-1472 Unmethylated: HTB-1, SW-780	Chen et al. (2011)
ccRCC	*GPX3*	Exon 1: 14 CpGs	6 ccRCC cell lines: 83.3%: methylated: **786-0**, Kert-3 Partially methylated: Caki-1, 769P, **Osrc-2** Unmethylated: Caki-2	Liu et al. (2015)

PC	*GPX3*	Exon1/intron 1 (+53 to +793 base from TSS)	3 PC cell lines: Methylated: **PC3**, **Du-145**, **LNCaP**	Yu et al. (2007)
PC	*GPX3*	Promoter region	7 PC cell lines: Methylated: Du-145, **LNCaP**, **PC3**, PPC1, TSU-Pr1, LAPC4, BPH-1	Lodygin et al. (2005)
Gynecologic cancer				
OC	*GPX3*	21 CpGs around TSS	8 OC cell lines: 38%: methylated: IGOVR1, IGOVR1/cp, SKOV3 Unmethylated: OVCAR38, PA-1, OVCAR5, 2008, 2008C	Chen et al. (2011)
BrC	*GPX1*	35 CpGs	4 BrC cell lines: Methylated: **MCF-7**, **SKBR-3** Unmethylated: BT-474, MDA-MB-453	Kulak et al. (2013)
BrC	*SEPX1*	Promoter region	2 BrC cell lines: Low methylation: MCF-7 High methylation: **MDA-MB231**	De Luca, Sacchetta, Nieddu, Di Ilio, and Favaloro (2007)
CC	*GPX3*	Promoter region	4 CC cell lines: 100%: methylated: **HeLa**, **SiHa** Partially methylated: **CaSki**, **Me-180**	Zhang et al. (2014)

Continued

Table 4 Methylation of Selenoprotein-Encoding Genes Identified in Human Cancer Cell Lines—cont'd

Cancer Type	Gene	Methylation Location	Gene Methylation in Cell Lines	References
Gastrointestinal cancer				
GC	*GPX1*, *GPX3*	Promoter region	3 GC cell lines: *GPX1* methylated: SNU-1 *GPX1* unmethylated: SNU601, SNU719 *GPX3* methylated: SNU-1, SNU-601, SNU-719	Jee et al. (2009)
GC	*GPX3*	Exon 1 (1216–1302 base, TSS: 928 base)	9 GC cell lines: Methylated: **MKN-74**, **MKN-45**, **SNU-1**, **SNU-719** Unmethylated: NCI-N87, SNU484, KATOIII, SNU601, MKN1	Zhang et al. (2010)
GC	*GPX1*, *GPX3*	Promoter region	10 GC cell lines: *GPX1* methylated: SNU1, 484 *GPX3* methylated: SNU1, 216, 484 (partially), 601, 620 (partially), 638, 719 *GPX1* or *GPX3* unmethylated: SNU5, 16, 668	Min et al. (2012)
GC	*GPX3*	8 CpGs	9 GC cell lines: Methylated: AGS: 3%; **MKN28**: 91%: MKN45: 92%; MKN75: 91%; KATO3: 88%; **SNU1**: 94%; SNU5: 4%; SNU16: 6%; RF1: 24%	Peng et al. (2012)

CRC	*SBP1*	Promoter region	5 CRC cell lines: Methylated: **HCT116**, SW480 Partially methylated: Caco-2, HT-29 Unmethylated: LST147T	Pohl et al. (2009)
Skin cancer				
MEL	*GPX3*	Promoter region	3 MEL cell lines: Methylated: **SK-MEL-2** Partially methylated: **SK-MEL-24** Unmethylated: HEMs	Chen et al. (2016)
Blood cancer				
CML	*GPX3*	18 CpGs	Methylated: **K562**: 98%	Yao et al. (2015)
AML	*GPX3*	18 CpGs	Methylated: **THP1**: 97%	Zhou et al. (2015)

AML, acute myeloid leukemia; *BC*, bladder cancer; *BrC*, breast cancer; *CC*, cervical cancer; *ccRCC*, clear cell renal cell carcinoma; *CML*, chronic myeloid leukemia; *CRC*, colorectal cancer; *EC*, esophageal adenocarcinoma; *ESCC*, esophageal squamous cell carcinoma; *GC*, gastric cancer; *HEMs*, human epidermal melanocytes; *HNC*, head and neck cancer; *MEL*, melanoma; *OC*, ovarian cancer; *PC*, prostate cancer; *TSS*, transcription start site.

Bold type denotes cell lines presenting restoration of gene/protein expression after 5-aza-dC treatment.

determined in four ESCC tissues paired with four normal surrounding tissues and also normal mucosa from four healthy individuals. One hundred and sixty-eight genes had differentially methylated CpG sites in their promoter regions. Additionally, inverse gene expression based on the Gene Expression Omnibus (GEO) database was noticed. Among all genes, three genes presented the most significant difference between tumor and normal tissue. *EPB41L3*, *GPX3*, and *COL14A1* were chosen for further methylation-specific PCR (MSP) analyses in 42 pairs of samples from ESCC patients. Methylation of *EPB41L3*, *GPX3*, and *COL14A1* was observed in 59.5%, 54.8%, and 45.2% in tumor samples and in 4.8%, 9.5%, and 11.9% in adjacent ESCC samples, respectively. The higher methylation of *GPX3* was correlated with advanced pN tumor stage. Interestingly, the methylation of these three genes was found only in ESCC patients' plasma, but not in normal individuals upon testing 42 ESCC patients and 50 healthy individuals (Li et al., 2014). Hypermethylation of *GPX3* was observed in clinical HNC and also BC tissue samples. Furthermore, *GPX*3 methylation correlated with HNC chemoresistance and may serve as a potential prognostic biomarker for HNC patients treated with cisplatin-based chemotherapy (Chen et al., 2011).

Methylation in the *GPX3* promoter region was also detected in other urogenital cancers. *GPX3* hypermethylation was observed in 93% of investigated primary prostate tumors (Lodygin et al., 2005) and 90% of PC tissue specimens (Yu et al., 2007). Aberrant methylation of *GPX3* was observed in 77% of clear cell renal cell carcinoma (ccRCC) tumor samples and in 15% of adjacent tissues. Concomitantly, *GPX3* expression and gene and protein levels were significantly lower in ccRCC tumors than in adjacent renal tissues. Moreover, *GPX3* methylation status was associated with the higher tumor nuclear grade (Liu et al., 2015). BrC samples also showed aberrant *GPX3* methylation, while it was absent in normal breast specimens from unrelated individuals. Moreover, *GPX3* mRNA and protein expression were downregulated in tissues from inflammatory breast ductal carcinoma (IBC) compared to non-IBC (Mohamed et al., 2014). Methylation of 35 CpG sites of cytosolic glutathione peroxidase—*GPX1*—promoter was identified in approximately 20% among eight primary breast cancers (Kulak et al., 2013).

The other intensively investigated group of cancers comprises gastrointestinal cancer. Reduced *GPX1* and *GPX3* expression was found to be associated with gene methylation in GC. In GC tumors, more samples with promoter hypermethylation were found for *GPX3* (30.1%) than for

GPX1 (16.7%) (Jee et al., 2009). Other study has also confirmed that in the investigated GC samples in 31% for *GPX1* and in 50% for *GPX3* were methylated (Min et al., 2012). *GPX3* hypermethylation detected in GC specimens presented significant association with the downregulation of *GPX3* expression. Interestingly, tumor-adjacent samples in one third also presented promoter hypermethylation (5% of samples), while 60% of GC specimens showed *GPX3* methylation. Moreover, the downregulation of *GPX3* and methylation were significantly associated with the gastric cancer lymph node metastasis (Peng et al., 2012). To add, a nonsignificant difference in *GPX3* methylation of GC tumors compared to that of normal mucosa (18% vs 13%) was also noted (de Melo et al., 2014). One study showed that *GPX3* hypermethylation was observed in hepatocellular carcinoma (HCC) in 76.6% of tumor samples that were concomitant with the significantly lower *GPX3* mRNA level and protein. No methylation was observed in the matched adjacent samples. Moreover, *GPX3* methylation was also associated with portal tumor thrombosis, metastasis, and HCC differentiation (Cao et al., 2015). Similarly, only one investigation conducted skin tumors analysis. Ten investigated primarily melanoma tissue specimens showed *GPX3* methylation, while human epidermal melanocytes and normal skin samples presented unmethylated *GPX3* (Chen et al., 2016).

Recently, it has been also noted that *GPX3* methylation may present an epigenetic feature in other than solid tumors. Bone marrow (BM) mononuclear cells of patients with CML (Yao et al., 2015) and AML (Zhou et al., 2015) presented higher methylation status than BM mononuclear cells of healthy controls. There were 15% of CML patients with methylation, while *GPX3* methylation presented 26% of AML patients. To compare, only 1%–2% of unrelated healthy individuals possessed *GPX3* methylation. A negative correlation was also observed between methylation and gene expression in CML; however, in AML there was no significant link between *GPX3* expression and methylation (Yao et al., 2015). Interestingly, 15% of methylated samples were observed in BM of patients with myelodysplastic syndrome (MDS), which was significantly higher than in controls, and lower than in AML patients (Zhou et al., 2015). Cases with *GPX3* methylation showed a significantly shorter survival time than patients with unmethylated *GPX3* promoter region. *GPX3* methylation significantly increased during the progression from MDS to secondary AML (Zhou et al., 2017).

3.2 Methylation Status in Human Cancer Cell Lines

It is widely recognized that methylation status of *GPX3* gene directly affects its gene expression because treatment of several cell lines with a chemical analogue of cytidine (5-aza-2′-deoxycytidine, 5-aza-dC) to demethylate the gene, resulted in the increase of its mRNA level. Inhibition of DNMT activity, mediated by 5-aza-dC treatment, leads to reversion of CpG site methylation. 5-Aza-dC intercalates with DNA during S phase of cell cycle, and after being recognized by DNMT, then it forms a covalent irreversible complex with the enzyme, resulting in the blocked DNMT function (Biswas & Rao, 2017). Treatment of three ESCC cell lines with 5-aza-dC and trichostatin A, an inhibitor of histone deacetylase, led to reversal of *GPX3* methylation pattern and reexpression of *GPX3* gene at the mRNA and protein levels (Peng et al., 2009). The highest methylation status of *GPX3* promoter region accompanied by lack of gene expression was observed among four of seven investigated ESCC cell lines. After 5-aza-dC treatment, four silenced cell lines expressed *GPX3* (He et al., 2011).

The most comprehensive analysis of *GPX3* methylation in the promoter region covering CpG dinucleotides was performed by Chen et al. (2011). They analyzed *GPX3* expression and promoter methylation in human cell lines ($n=63$) derived from head and neck, lung, breast, bladder, ovary, esophagus, and melanoma cancers. Significant downregulation of *GPX3* expression or complete *GPX3* gene silencing was found in all investigated cell lines. However, gene downregulation accompanied by *GPX3* hypermethylation was observed in HNC, OC, BC, and BrC cell lines with the diverse methylation pattern identified as methylated, partially methylated, or unmethylated (Chen et al., 2011).

The CpG methylation was also associated with the reduced *GPX3* expression in PC cells. Importantly, all of analyzed PC cell lines have been characterized by hypermethylation of *GPX3* promoter. Moreover, treatment of these cell lines with 5-aza-dC resulted in increased mRNA levels (Lodygin et al., 2005; Yu et al., 2007). *GPX3* methylation and downregulation of the gene were detected in five out of six ccRCC cell lines. Restoration of *GPX3* expression in two partially methylated and methylated ccRCC cell lines was also detected after treatment with 5-aza-dC (Liu et al., 2015). *GPX3* methylation of 21 CpG sites was observed in 38% among OC cell lines (Chen et al., 2011), while cervical cancer cell lines presented total DNA hypermethylation (100%) (Zhang et al., 2014). Among investigated 4 BrC cell lines, 2 of them presented *GPX1* promoter region with 35 CpG sites' methylation. Moreover, treatment with the demethylation agent 5-aza-dC resulted in the increase of *GPX1* mRNA expression (Kulak et al., 2013).

Identification of novel methylation-silenced genes in GC was performed using three cell lines. Upregulation of gene expression was determined using oligonucleotide microarrays after 5-aza-dC treatment of cells. Six of 143 upregulated genes were identified, including *GPX1* and *GPX3* (Jee et al., 2009). Downregulation of *GPX3* was observed in eight GC cell lines. It was accompanied by heavy CpG methylation of *GPX3* exon 1 region with close proximity to TSS region (Zhang et al., 2010). In vitro data showed also lower methylation of *GPX1* gene promoter than the *GPX3* promoter region. It was found only 2 of 10 investigated cell lines with *GPX1* hypermethylation and 5 of 10 cell lines with hypermethylation of *GPX3* (Min et al., 2012). Examination of *GPX3* promoter hypermethylation within eight CpG sites showed hypermethylation in six of nine GC from 3% to 94%. Again, 5-aza-dC restored *GPX3* gene expression with significant demethylation of *GPX3* promoter in two cell lines SNU1 and MKN28 that primarily presented the highest methylation (94% and 91%, respectively) (Peng et al., 2012). The Cancer Genome Atlas (TCGA) analysis presented the prevalence of *GPX3* methylation in five CpGs of promoter in 30% of samples of colorectal cancer (CRC) that led to reduced *GPX3* expression (Pelosof et al., 2017). Importantly, analysis of nine CRC cell lines showed association between *GPX3* methylation and gene expression. The methylation pattern was strictly associated to platinum sensitivity, because the cisplatin IC_{50} in *GPX3* methylated cell lines was approximately sixfold lower than that in *GPX3*-unmethylated cell lines. Moreover, *GPX3* methylation was associated with tumor xenograft sensitivity to platinum treatment, leading to reduction of tumor growth. Thus *GPX3* methylation may predict platinum treatment response in CRC (Pelosof et al., 2017).

To add, in melanoma cell lines SK-MEL-2 and SK-MEL-24, differential *GPX3* methylation was observed (Chen et al., 2016). In the erythroleukemia cell line K562 and leukemic cell line THP1, massive hypermethylation (around 100%) of *GPX3* promoter was observed. Additionally, the treatment with 5-aza-dC restored *GPX3* expression and decreased DNA methylation level (Yao et al., 2015; Zhou et al., 2015).

Selenium-binding protein 1 (*SBP1* or *SELENBP1*) was also found to be significantly downregulated in human cancers. *SBP1* promoter hypermethylation was observed in CRC tissues and five human CRC cell lines. Interestingly, modified methylation levels were observed in different cell lines, which was accompanied by changes in the SBP1 protein level of expression. For example, LST174T cell line expressed highest level of protein with the lowest level of *SBP1* promoter methylation. In HCT116 cells, SBP1 protein was undeletable and methylation was increased. Treatment of HCT116 cells with 5-aza-dC restored SBP1 protein level (Pohl et al., 2009).

Selenoprotein X (SEPX1), or methionine sulfoxide reductase B1 (MSRB1), was epigenetically modified in the advanced stage BrC. High levels of *MsrB1* transcript, protein, and promoter activity were detected in low metastatic MCF-7 human BrC cells. On the contrary, very low levels of both *MsrB1* transcript and promoter activity were detected in the highly metastatic MDA-MB231 cell line. Interestingly, MCF-7 cells possess low methylation levels of *MsrB1* promoter compared with the metastatic breast cancer cells. Moreover, the treatment with 5-aza-dC of highly metastatic BrC cell line caused upregulation of the *MsrB*1 expression (De Luca et al., 2007).

4. CONCLUSIONS

In this review, we discussed two aspects of Se and epigenetics in cancer: (1) the association between Se and DNA methylation in cancer, with majority of the data focusing on colon cancer; (2) methylation of selenoprotein-encoding genes in different cancers, pointing mainly to the importance of *GPX3*. With respect to the first issue, the existing evidence indicated mostly that Se inhibited the activity or expression of DNMT in cancer cells or animal models of chemically induced carcinogenesis. Importantly, DNMT inhibition by Se corresponded to such effects as reduced formation of precancerous lesions upon carcinogen treatment (Hu et al., 2013) and decreased progression or growth of cancer cells (de Miranda et al., 2014; Fiala et al., 1998). Se was also shown to affect methylation of selected TSGs, such as *VHL*, *TP53*, *GSTP1*, *APC*, and *CSR1* in cancer cell lines, and as shown in cancer-free subjects, the association between Se and methylation of TSG may be sex dependent. The link with global DNA methylation in terms of cancer prevention by Se remains unclear. Taking into account in vitro and in vivo observations on decreased DNA methylation induced by both Se deficiency (Davis & Uthus, 2002, 2003; Davis et al., 2000; Uthus et al., 2006) and high Se exposure of animals fed a high-Se diet (0.88 mg/kg in mice and 4 mg/kg in rats) (Armstrong et al., 2011; Zeng et al., 2011), a nonlinear character of this relationship can be suggested. Such an interpretation fits with the concept of a U-shaped association between the Se status and cancer in humans (Bleys,

Navas-Acien, & Guallar, 2008). On the basis of experimental data, we proposed also different mechanisms of DNA hypomethylation under different Se levels, including (1) redirection of HCys toward transsulfuration pathway and glutathione synthesis in case of Se deficiency and (2) competition for SAM in order to utilize methyl groups required for Se metabolism in case of excess Se intake.

Overall, limited human data on the general association between Se and DNA methylation are available, which are also inconsistent. These aspects do not allow to draw a clear conclusion on the effect of Se on epigenome (Pilsner et al., 2011; Tapp et al., 2013). Thus, the role of supranutritional Se intake as a cancer prevention strategy targeting epigenetic mechanisms deserves further investigation, with a special emphasis on human studies. Notably, none of the epidemiological studies carried out so far in humans and related to Se and cancer, concerned DNA methylation. Additionally, other epigenetic mechanisms, such as histone modification or miRNA expression, were also not investigated in the context of Se supplementation. Future studies should challenge the possible nonlinear character of the relationship between Se and epigenome as well as the effect of other modifiers of DNA methylation such as dietary intake of folates and other nutrients involved in the methyl metabolism. Diet composition associated with methyl metabolism and methyl status may significantly impact Se activity, as shown, for example, in the rat model of chemically induced colon carcinogenesis, in which Se supplementation unexpectedly increased the number of precancerous lesions when diet was deficient in folate (Davis & Uthus, 2003). Similarly, genetic susceptibility involved in methyl metabolism should not be neglected, as suggested by a recent report showing that genetic polymorphisms of *MTHFR* (methylene tetrahydrofolate reductase) modifies the association between urinary Se and breast cancer risk (Lin et al., 2015).

Se was identified as epigenetic regulator itself, but concomitantly specific selenoproteins, such as GPX1, GPX3, SBP1, and SEPX1, have been found to be regulated epigenetically in various cancer types. Nearly all of studies showed modified methylation status of *GPX3* promoter region and only a few studies reported on *GPX1*, *SBP1*, and *SEPX1* promoter region methylation. The majority of selenoprotein-encoding gene methylation studies in cancer were conducted on head and neck and gastrointestinal cancers followed by urogenital and gynecologic cancers. *GPX3* methylation was observed in all of those cancer types, but with different magnitudes. Therefore, *GPX3* promoter methylation seems to be tissue specific and cancer specific. Moreover, it was also shown that exposure to DNMT inhibitor 5-aza-dC was related to gene and protein upregulation in selected cancer cell lines. Such selective modulation of DNA methylation may also indicate cell-type-specific epigenetic feature. The very high *GPX3* methylation status of some tumors, such as those of head and neck and prostate, encourages to consider it as a tumor biomarker for these cancers. Despite the putative importance of other selenoproteins in carcinogenesis, their promoter methylation has rarely been investigated in human cancer tissues and also cancer cell lines. *GPX1* hypermethylation is frequently observed in gastric and breast cancers, *SBP1* promoter activity appears to be controlled by epigenetic modification in CRC, while *SEPX1* promoter could be epigenetically changed in breast cancer cell lines.

GPX1, *GPX3*, *SEPX1*, and *SBP1* promoter constitutes a target for epigenetic regulation in cancer. Epigenetic modification of selenoproteins observed in specific cancer sites may reveal a complicated issue. Interestingly, CpG sites (CpGs) in *GPX1*, *GPX3*, *SEPX1*, and *SBP1* promoter region vary in their structure (Fig. 5), indicating differentiated epigenetic regulation of selenoprotein-encoding genes. *GPX1* gene possesses two CpG islands within gene body (covering two exons) and first CpG island covers TSS in the promoter region. Interestingly, *GPX3* gene presents three CpG islands within gene body first CpG island localized downstream from transcription start sites, near exon 1 of *GPX3* gene. To compare, *SEPX1* gene also possesses three CpG islands that cover promoter region with TSS and exon 1, while *SBP1* presents only one CpG island upstream from TSS in the promoter region.

It should be mentioned that *GPX1*, *GPX3*, and *SEPX1* gene expression relies on Se status, but SBP1 is a cytosolic protein with function unknown that binds Se. Recently, it has been found that SAH, a precursor of HCys, inhibits the methylation of the tRNA [Ser]Sec which is necessary for efficient translation of GPX1 and other selenoproteins requiring methylated form of tRNA [Ser]Sec to biosynthesis. Thus, the hypomethylation stress induced by accumulation of SAH may affect GPX1 protein expression, but not mRNA expression (Barroso et al., 2014). Interestingly, selenoproteins are known to present different sensitivity to the loss of methylated form of tRNA [Ser]Sec with high (e.g., *GPX1*, *GPX3*, *SEPW1*) and low (e.g., *TRXR1*, *TRXR2*) sensitivity (Hatfield, Carlson, Xu, Mix, & Gladyshev, 2006). Therefore, hypomethylation stress has a potential to differentially modulate selenoprotein expression.

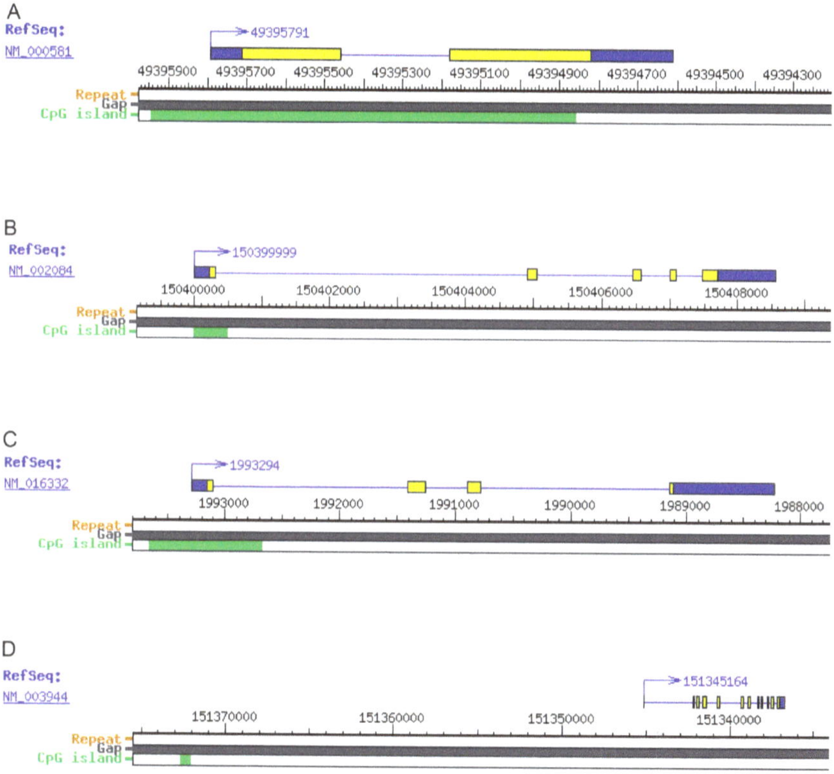

Fig. 5 CpG islands within gene structure of selenoproteins according to DBTSS database of transcription start sites (http://dbtss.hgc.jp) (Yamashita, Wakaguri, Sugano, Suzuki, & Nakai, 2010). Detailed number of CpG islands [from] MethPrimer2.0 online platform (http://www.urogene.org/cgi-bin/methprimer2/MethPrimer.cgi) (Li & Dahiya, 2002). Exons highlighted in *yellow rectangles*, and 5′UTR and 3′UTR regions highlighted in *blue rectangles*, TSS presented as *arrow*. (A) ***GPX1*** (Entrez Gene ID: 2876, gene location: 3p21.3 from 49394492 to 49395909 base, ID of transcript: NM_000581 (921 bases, 2 exons), **2 CpG islands**. (B) ***GPX3*** (Entrez Gene ID: 2878, gene location: 5q33.1 from 150399143 to 150409408 base, ID of transcript: NM_002084 (1779 bases, 5 exons), **3 CpG islands**. (C) ***SEPX1*** (Entrez Gene ID: 51734, gene location: 16p13.3 from 1987728 to 1993806 base, ID of transcript: NM_016332 (1386 bases, 4 exons), **3 CpG islands**. (D) ***SBP1*** (Entrez Gene ID 8991, gene location 1q21.3 from 1513332159 to 151375516 base, ID of transcript: NM_003944 (1721 bases, 13 exons), **1 CpG island**.

To conclude, more studies encompassing different human, animal, and in vitro models are required to show complete insights into Se-associated epigenetic mechanisms in cancer. In order to elucidate that epigenetic regulation, it is very important to focus not only on methyl metabolism and methylation status determination but also on Se status in target tissue.

Acknowledgments

This work was supported by Nofer Institute of Occupational Medicine Internal Grants IMP 1.39 and IMP 14.1 and the National Science Centre Grant 2012/05/B/NZ5/01406.

REFERENCES

Angelova, E. A., Atanassova, P. A., Chalakova, N. T., & Dimitrov, B. D. (2008). Associations between serum selenium and total plasma homocysteine during the acute phase of ischaemic stroke. *European Neurology*, *60*, 298–303.

Armstrong, K. M., Bermingham, E. N., Bassett, S. A., Treloar, B. P., Roy, N. C., & Barnett, M. P. (2011). Global DNA methylation measurement by HPLC using low amounts of DNA. *Biotechnology Journal*, *6*, 113–117.

Barrera, L. N., Johnson, I. T., Bao, Y., Cassidy, A., & Belshaw, N. J. (2013). Colorectal cancer cells Caco-2 and HCT116 resist epigenetic effects of isothiocyanates and selenium in vitro. *European Journal of Nutrition*, *52*, 1327–1341.

Barroso, M., Florindo, C., Kalwa, H., Silva, Z., Turanov, A. A., Carlson, B. A., et al. (2014). Inhibition of cellular methyltransferases promotes endothelial cell activation by suppressing glutathione peroxidase 1 protein expression. *The Journal of Biological Chemistry*, *289*, 15350–15362.

Bates, C. J., Thane, C. W., Prentice, A., & Delves, H. T. (2002). Selenium status and its correlates in a British national diet and nutrition survey: People aged 65 years and over. *Journal of Trace Elements in Medicine and Biology*, *16*, 1–8.

Bekaert, B., Cooper, M. L., Green, F. R., McNulty, H., Pentieva, K., Scott, J. M., et al. (2008). Effect of selenium status and supplementation with high-selenium yeast on plasma homocysteine and B vitamin concentrations in the UK elderly. *Molecular Nutrition & Food Research*, *52*, 1324–1333.

Belanger, M. C., Dewailly, E., Berthiaume, L., Noel, M., Bergeron, J., Mirault, M. E., et al. (2006). Dietary contaminants and oxidative stress in Inuit of Nunavik. *Metabolism*, *55*, 989–995.

Biswas, S., & Rao, C. M. (2017). Epigenetics in cancer: Fundamentals and beyond. *Pharmacology & Therapeutics*, *173*, 118–134.

Bleys, J., Navas-Acien, A., & Guallar, E. (2008). Serum selenium levels and all-cause, cancer, and cardiovascular mortality among US adults. *Archives of Internal Medicine*, *168*, 404–410.

Cao, S., Yan, B., Lu, Y., Zhang, G., Li, J., Zhai, W., et al. (2015). Methylation of promoter and expression silencing of GPX3 gene in hepatocellular carcinoma tissue. *Clinics and Research in Hepatology and Gastroenterology*, *39*, 198–204.

Chen, B., Rao, X., House, M. G., Nephew, K. P., Cullen, K. J., & Guo, Z. (2011). GPx3 promoter hypermethylation is a frequent event in human cancer and is associated with tumorigenesis and chemotherapy response. *Cancer Letters*, *309*, 37–45.

Chen, H., Zheng, Z., Kim, K. Y., Jin, X., Roh, M. R., & Jin, Z. (2016). Hypermethylation and downregulation of glutathione peroxidase 3 are related to pathogenesis of melanoma. *Oncology Reports*, *36*, 2737–2744.

Cox, R. (1985). Selenite: A good inhibitor of rat-liver DNA methylase. *Biochemistry International*, *10*, 63–69.

Cox, R., & Goorha, S. (1986). A study of the mechanism of selenite-induced hypomethylated DNA and differentiation of Friend erythroleukemic cells. *Carcinogenesis*, *7*, 2015–2018.

Crider, K. S., Yang, T. P., Berry, R. J., & Bailey, L. B. (2012). Folate and DNA methylation: A review of molecular mechanisms and the evidence for folate's role. *Advances in Nutrition*, *3*, 21–38.

Cui, L. H., Quan, Z. Y., Piao, J. M., Zhang, T. T., Jiang, M. H., Shin, M. H., et al. (2016). Plasma folate and vitamin B12 levels in patients with hepatocellular carcinoma. *International Journal of Molecular Sciences*, *17*, E1032.

Davis, C. D., & Uthus, E. O. (2002). Dietary selenite and azadeoxycytidine treatments affect dimethylhydrazine-induced aberrant crypt formation in rat colon and DNA methylation in HT-29 cells. *The Journal of Nutrition*, *132*, 292–297.

Davis, C. D., & Uthus, E. O. (2003). Dietary folate and selenium affect dimethylhydrazine-induced aberrant crypt formation, global DNA methylation and one-carbon metabolism in rats. *The Journal of Nutrition*, *133*, 2907–2914.

Davis, C. D., Uthus, E. O., & Finley, J. W. (2000). Dietary selenium and arsenic affect DNA methylation in vitro in Caco-2 cells and in vivo in rat liver and colon. *The Journal of Nutrition*, *130*, 2903–2909.

De Luca, A., Sacchetta, P., Nieddu, M., Di Ilio, C., & Favaloro, B. (2007). Important roles of multiple Sp1 binding sites and epigenetic modifications in the regulation of the methionine sulfoxide reductase B1 (MsrB1) promoter. *BMC Molecular Biology*, *8*, 39.

de Melo, C. F., Gigek, C. O., da Silva, J. N., Cardoso Smith Mde, A., de Araujo, R. M., Burbano, R. R., et al. (2014). Association of COX2 gene hypomethylation with intestinal type gastric cancer in samples of patients from northern Brazil. *Tumour Biology*, *35*, 1107–1111.

de Miranda, J. X., Andrade Fde, O., Conti, A., Dagli, M. L., Moreno, F. S., & Ong, T. P. (2014). Effects of selenium compounds on proliferation and epigenetic marks of breast cancer cells. *Journal of Trace Elements in Medicine and Biology*, *28*, 486–491.

Ehrlich, M. (2009). DNA hypomethylation in cancer cells. *Epigenomics*, *1*, 239–259.

Fiala, E. S., Staretz, M. E., Pandya, G. A., El-Bayoumy, K., & Hamilton, S. R. (1998). Inhibition of DNA cytosine methyltransferase by chemopreventive selenium compounds, determined by an improved assay for DNA cytosine methyltransferase and DNA cytosine methylation. *Carcinogenesis*, *19*, 597–604.

Geillinger, K. E., Rathmann, D., Kohrle, J., Fiamoncini, J., Daniel, H., & Kipp, A. P. (2014). Hepatic metabolite profiles in mice with a suboptimal selenium status. *The Journal of Nutritional Biochemistry*, *25*, 914–922.

Gonzalez, S., Huerta, J. M., Alvarez-Uria, J., Fernandez, S., Patterson, A. M., & Lasheras, C. (2004). Serum selenium is associated with plasma homocysteine concentrations in elderly humans. *The Journal of Nutrition*, *134*, 1736–1740.

Hatfield, D. L., Carlson, B. A., Xu, X. M., Mix, H., & Gladyshev, V. N. (2006). Selenocysteine incorporation machinery and the role of selenoproteins in development and health. *Progress in Nucleic Acid Research and Molecular Biology*, *81*, 97–142.

Hatfield, D. L., Tsuji, P. A., Carlson, B. A., & Gladyshev, V. N. (2014). Selenium and selenocysteine: Roles in cancer, health, and development. *Trends in Biochemical Sciences*, *39*, 112–120.

He, Y., Wang, Y., Li, P., Zhu, S., Wang, J., & Zhang, S. (2011). Identification of GPX3 epigenetically silenced by CpG methylation in human esophageal squamous cell carcinoma. *Digestive Diseases and Sciences*, *56*, 681–688.

Herrero, A. B., Rojas, E. A., Misiewicz-Krzeminska, I., Krzeminski, P., & Gutierrez, N. C. (2016). Molecular mechanisms of p53 deregulation in cancer: An overview in multiple myeloma. *International Journal of Molecular Sciences*, *17*, E2003.

Hill, K. E., & Burk, R. F. (1982). Effect of selenium deficiency and vitamin E deficiency on glutathione metabolism in isolated rat hepatocytes. *The Journal of Biological Chemistry*, *257*, 10668–10672.

Hu, Y., McIntosh, G. H., Le Leu, R. K., Nyskohus, L. S., Woodman, R. J., & Young, G. P. (2013). Combination of selenium and green tea improves the efficacy of chemoprevention in a rat colorectal cancer model by modulating genetic and epigenetic biomarkers. *PLoS One*, *8*. e64362.

Hu, Y., McIntosh, G. H., Le Leu, R. K., Somashekar, R., Meng, X. Q., Gopalsamy, G., et al. (2016). Supplementation with Brazil nuts and green tea extract regulates targeted biomarkers related to colorectal cancer risk in humans. *The British Journal of Nutrition*, *116*, 1901–1911.

Jablonska, E., & Vinceti, M. (2015). Selenium and human health: Witnessing a Copernican revolution? *Journal of Environmental Science and Health. Part C, Environmental Carcinogenesis & Ecotoxicology Reviews*, *33*, 328–368.

Jee, C. D., Kim, M. A., Jung, E. J., Kim, J., & Kim, W. H. (2009). Identification of genes epigenetically silenced by CpG methylation in human gastric carcinoma. *European Journal of Cancer*, *45*, 1282–1293.

Klapcinska, B., Poprzecki, S., Danch, A., Sobczak, A., & Kempa, K. (2005). Selenium levels in blood of upper Silesian population: Evidence of suboptimal selenium status in a significant percentage of the population. *Biological Trace Element Research*, *108*, 1–15.

Kulak, M. V., Cyr, A. R., Woodfield, G. W., Bogachek, M., Spanheimer, P. M., Li, T., et al. (2013). Transcriptional regulation of the GPX1 gene by TFAP2C and aberrant CpG methylation in human breast cancer. *Oncogene*, *32*, 4043–4051.

Kulis, M., & Esteller, M. (2010). DNA methylation and cancer. *Advances in Genetics*, *70*, 27–56.

Lee, O. J., Schneider-Stock, R., McChesney, P. A., Kuester, D., Roessner, A., Vieth, M., et al. (2005). Hypermethylation and loss of expression of glutathione peroxidase-3 in Barrett's tumorigenesis. *Neoplasia*, *7*, 854–861.

Li, L. C., & Dahiya, R. (2002). MethPrimer: Designing primers for methylation PCRs. *Bioinformatics*, *18*, 1427–1431.

Li, X., Zhou, F., Jiang, C., Wang, Y., Lu, Y., Yang, F., et al. (2014). Identification of a DNA methylome profile of esophageal squamous cell carcinoma and potential plasma epigenetic biomarkers for early diagnosis. *PLoS One*, *9*. e103162.

Lin, W., Cen, Y. L., Lin, Y., Su, F. X., Wu, B. H., Tang, L. Y., et al. (2015). Joint effects between urinary selenium and polymorphisms in methylation related genes on breast cancer risk. *Neoplasma*, *62*, 491–499.

Liu, Q., Jin, J., Ying, J., Sun, M., Cui, Y., Zhang, L., et al. (2015). Frequent epigenetic suppression of tumor suppressor gene glutathione peroxidase 3 by promoter hypermethylation and its clinical implication in clear cell renal cell carcinoma. *International Journal of Molecular Sciences*, *16*, 10636–10649.

Lodygin, D., Epanchintsev, A., Menssen, A., Diebold, J., & Hermeking, H. (2005). Functional epigenomics identifies genes frequently silenced in prostate cancer. *Cancer Research*, *65*, 4218–4227.

Lu, S. C. (2009). Regulation of glutathione synthesis. *Molecular Aspects of Medicine*, *30*, 42–59.

Metes-Kosik, N., Luptak, I., Dibello, P. M., Handy, D. E., Tang, S. S., Zhi, H., et al. (2012). Both selenium deficiency and modest selenium supplementation lead to myocardial fibrosis in mice via effects on redox-methylation balance. *56*, 1812–1824.

Min, S. Y., Kim, H. S., Jung, E. J., Jee, C. D., & Kim, W. H. (2012). Prognostic significance of glutathione peroxidase 1 (GPX1) down-regulation and correlation with aberrant promoter methylation in human gastric cancer. *Anticancer Research*, *32*, 3169–3175.

Mohamed, M. M., Sabet, S., Peng, D. F., Nouh, M. A., El-Shinawi, M., & El-Rifai, W. (2014). Promoter hypermethylation and suppression of glutathione peroxidase 3 are associated with inflammatory breast carcinogenesis. *Oxidative Medicine and Cellular Longevity*, *2014*, 787195.

National Research Council (US). (1995). *Nutrient requirements of laboratory animals: Fourth revised edition, 1995*. Washington, DC: The National Academies Press.

Pechalrieu, D., Etievant, C., & Arimondo, P. B. (2017). DNA methyltransferase inhibitors in cancer: From pharmacology to translational studies. *Biochemical Pharmacology*, *129*, 1–13.

Pelosof, L., Yerram, S., Armstrong, T., Chu, N., Danilova, L., Yanagisawa, B., et al. (2017). GPX3 promoter methylation predicts platinum sensitivity in colorectal cancer. *Epigenetics*. (in press). http://dx.doi.org/10.1080/15592294.2016.1265711.

Peng, D. F., Hu, T. L., Schneider, B. G., Chen, Z., Xu, Z. K., & El-Rifai, W. (2012). Silencing of glutathione peroxidase 3 through DNA hypermethylation is associated with lymph node metastasis in gastric carcinomas. *PLoS One, 7*. e46214.

Peng, D. F., Razvi, M., Chen, H., Washington, K., Roessner, A., Schneider-Stock, R., et al. (2009). DNA hypermethylation regulates the expression of members of the Mu-class glutathione S-transferases and glutathione peroxidases in Barrett's adenocarcinoma. *Gut, 58*, 5–15.

Perri, F., Pisconti, S., & Della Vittoria Scarpati, G. (2016). P53 mutations and cancer: A tight linkage. *Annals of Translational Medicine, 4*, 522.

Pilsner, J. R., Hall, M. N., Liu, X., Ahsan, H., Ilievski, V., Slavkovich, V., et al. (2011). Associations of plasma selenium with arsenic and genomic methylation of leukocyte DNA in Bangladesh. *Environmental Health Perspectives, 119*, 113–118.

Pohl, N. M., Tong, C., Fang, W., Bi, X., Li, T., & Yang, W. (2009). Transcriptional regulation and biological functions of selenium-binding protein 1 in colorectal cancer in vitro and in nude mouse xenografts. *PLoS One, 4*. e7774.

Ramachandran, K., Navarro, L., Gordian, E., Das, P. M., & Singal, R. (2007). Methylation-mediated silencing of genes is not altered by selenium treatment of prostate cancer cells. *Anticancer Research, 27*, 921–925.

Reichard, J. F., & Puga, A. (2010). Effects of arsenic exposure on DNA methylation and epigenetic gene regulation. *Epigenomics, 2*, 87–104.

Remely, M., Lovrecic, L., de la Garza, A. L., Migliore, L., Peterlin, B., Milagro, F. I., et al. (2015). Therapeutic perspectives of epigenetically active nutrients. *British Journal of Pharmacology, 172*, 2756–2768.

Speckmann, B., & Grune, T. (2015). Epigenetic effects of selenium and their implications for health. *Epigenetics, 10*, 179–190.

Stover, P. J. (2009). One-carbon metabolism-genome interactions in folate-associated pathologies. *The Journal of Nutrition, 139*, 2402–2405.

Tapp, H. S., Commane, D. M., Bradburn, D. M., Arasaradnam, R., Mathers, J. C., Johnson, I. T., et al. (2013). Nutritional factors and gender influence age-related DNA methylation in the human rectal mucosa. *Aging Cell, 12*, 148–155.

Uthus, E., Begaye, A., Ross, S., & Zeng, H. (2011). The von Hippel-Lindau (VHL) tumor-suppressor gene is down-regulated by selenium deficiency in Caco-2 cells and rat colon mucosa. *Biological Trace Element Research, 142*, 223–231.

Uthus, E. O., & Ross, S. A. (2007). Dietary selenium affects homocysteine metabolism differently in Fisher-344 rats and CD-1 mice. *The Journal of Nutrition, 137*, 1132–1136.

Uthus, E. O., & Ross, S. (2009). Dietary selenium (Se) and copper (Cu) interact to affect homocysteine metabolism in rats. *Biological Trace Element Research, 129*, 213–220.

Uthus, E. O., Ross, S. A., & Davis, C. D. (2006). Differential effects of dietary selenium (se) and folate on methyl metabolism in liver and colon of rats. *Biological Trace Element Research, 109*, 201–214.

Uthus, E. O., Yokoi, K., & Davis, C. D. (2002). Selenium deficiency in Fisher-344 rats decreases plasma and tissue homocysteine concentrations and alters plasma homocysteine and cysteine redox status. *The Journal of Nutrition, 132*, 1122–1128.

Veeck, J., & Esteller, M. (2010). Breast cancer epigenetics: From DNA methylation to microRNAs. *Journal of Mammary Gland Biology and Neoplasia, 15*, 5–17.

Venn, B. J., Grant, A. M., Thomson, C. D., & Green, T. J. (2003). Selenium supplements do not increase plasma total homocysteine concentrations in men and women. *The Journal of Nutrition, 133*, 418–420.

Wolf, N. M., Mueller, K., Hirche, F., Most, E., Pallauf, J., & Mueller, A. S. (2010). Study of molecular targets influencing homocysteine and cholesterol metabolism in growing rats by manipulation of dietary selenium and methionine concentrations. *The British Journal of Nutrition*, *104*, 520–532.

Xiang, N., Zhao, R., Song, G., & Zhong, W. (2008). Selenite reactivates silenced genes by modifying DNA methylation and histones in prostate cancer cells. *Carcinogenesis*, *29*, 2175–2181.

Yamashita, R., Wakaguri, H., Sugano, S., Suzuki, Y., & Nakai, K. (2010). DBTSS provides a tissue specific dynamic view of transcription start sites. *Nucleic Acids Research*, *38*, D98–104.

Yang, G., Zhu, Y., Dong, X., Duan, Z., Niu, X., & Wei, J. (2014). TLR2-ICAM1-Gadd45alpha axis mediates the epigenetic effect of selenium on DNA methylation and gene expression in Keshan disease. *Biological Trace Element Research*, *159*, 69–80.

Yao, D. M., Zhou, J. D., Zhang, Y. Y., Yang, L., Wen, X. M., Yang, J., et al. (2015). GPX3 promoter is methylated in chronic myeloid leukemia. *International Journal of Clinical and Experimental Pathology*, *8*, 6450–6457.

Yu, Y. P., Yu, G., Tseng, G., Cieply, K., Nelson, J., Defrances, M., et al. (2007). Glutathione peroxidase 3, deleted or methylated in prostate cancer, suppresses prostate cancer growth and metastasis. *Cancer Research*, *67*, 8043–8050.

Zachara, B. A., Pawluk, H., Bloch-Boguslawska, E., Sliwka, K. M., Korenkiewicz, J., Skok, Z., et al. (2001). Tissue level, distribution, and total body selenium content in healthy and diseased humans in Poland. *Archives of Environmental Health*, *56*, 461–466.

Zeng, H., Yan, L., Cheng, W. H., & Uthus, E. O. (2011). Dietary selenomethionine increases exon-specific DNA methylation of the p53 gene in rat liver and colon mucosa. *The Journal of Nutrition*, *141*, 1464–1468.

Zhang, X., Yang, J. J., Kim, Y. S., Kim, K. Y., Ahn, W. S., & Yang, S. (2010). An 8-gene signature, including methylated and down-regulated glutathione peroxidase 3, of gastric cancer. *International Journal of Oncology*, *36*, 405–414.

Zhang, X., Zheng, Z., Yingji, S., Kim, H., Jin, R., Renshu, L., et al. (2014). Down-regulation of glutathione peroxidase 3 is associated with lymph node metastasis and prognosis in cervical cancer. *Oncology Reports*, *31*, 2587–2592.

Zhao, H., Li, J., Li, X., Han, C., Zhang, Y., Zheng, L., et al. (2015). Silencing GPX3 expression promotes tumor metastasis in human thyroid cancer. *Current Protein & Peptide Science*, *16*, 316–321.

Zhou, J. D., Lin, J., Zhang, T. J., Ma, J. C., Yang, L., Wen, X. M., et al. (2017). GPX3 methylation in bone marrow predicts adverse prognosis and leukemia transformation in myelodysplastic syndrome. *Cancer Medicine*, *6*, 267–274.

Zhou, J. D., Yao, D. M., Zhang, Y. Y., Ma, J. C., Wen, X. M., Yang, J., et al. (2015). GPX3 hypermethylation serves as an independent prognostic biomarker in non-M3 acute myeloid leukemia. *American Journal of Cancer Research*, *5*, 2047–2055.

CHAPTER NINE

Selenium and Cancer Stem Cells

Giuseppe Murdolo*[,1], Desirée Bartolini[†], Cristina Tortoioli*, Marta Piroddi[†], Pierangelo Torquato[†], Francesco Galli[†]

*Section of Internal Medicine, Endocrine and Metabolic Sciences, University of Perugia, Perugia, Italy
[†]University of Perugia, Perugia, Italy
[1]Corresponding author: e-mail address: gmurdolo@tiscali.it

Contents

Abstract

Selenium (Se) is an essential micronutrient that functions as "redox gatekeeper" and homeostasis factor of normal and cancer cells. Epidemiology and experimental studies, in the last years suggested that both inorganic and organic forms of Se may have favorable health effects. In this regard, a protective action of Se on cellular systems that may help preventing cancer cell differentiation has been demonstrated, while the hypothesis that Se compounds may cure cancer and its metastatic diffusion appears speculative and is still a matter of investigation. Indeed, the overall actions of Se compounds in carcinogenesis are controversial.

The recognition that cancer is a stem cell disease instigated major paradigm shifts in our basic understanding of cancer and attracted a great deal of interest. Although current treatment approaches in cancer are grounded in the need to kill the majority of cancer cells, targeting cancer stem cells (CSCs) may hold great potential in improving cancer treatment. In this respect, Se compounds have been demonstrated modulating numerous signaling pathways involved in CSC biology and these findings are now stimulating further research on optimal Se concentrations, most effective and

Advances in Cancer Research, Volume 136
ISSN 0065-230X
http://dx.doi.org/10.1016/bs.acr.2017.07.006

cancer-specific Se compounds, and inherent pathways involved in redox and metabolic regulation of CSCs.

In this review, we summarize the current knowledge about the effects of Se compounds on CSCs, by focusing on redox-dependent pathways and main gene regulation checkpoints that affect self-renewal, differentiation, and migration responses in this subpopulation of cancer cells.

1. INTRODUCTION

Selenium (Se) is an essential and unique micronutrient in that it is cotranslationally incorporated into polypeptides to form selenoproteins; these are components of many biochemical pathways and control key cellular processes (Chen, Prabhu, & Mastro, 2013). Foods (such as cereals, grain, and vegetables) contain different amount and chemical forms of Se (Bartolini et al., 2014; Navarro-Alarcon & Cabrera-Vique, 2008). An adequate Se intake has been estimated at 50 μg/die, while intake in the order of 350–700 μg/die are associated with toxic effects (Council, 2000), and some adverse effects were observed even at lower doses (260 μg/die) (Vinceti et al., 2017).

Over the past decades, research has provided evidence that administering Se as a supplement to the human diet may have both beneficial and harmful health effects (Clark et al., 1996; Rayman, 2012; Vinceti et al., 2014, 2017). From a functional standpoint, the metabolic pathway of dietary Se in human cells is of outstanding importance. Various Se compounds, by entering the metabolic pathway at different points, generate specific forms of Se that could have individual and specific cellular effects and/or mechanisms of action (Boudreau, Conrad, & Hoskin, 2007; Jiang, Ganther, & Lu, 2000; Park, Kim, Oh, & Chung, 2007). Indeed, organic Se has been hypothesized to have more potential as an anticancer therapeutic than inorganic Se (Chen et al., 2013). These observations emphasize the urgent need to search for the optimal formulation and dose of Se that should be used as supplement to achieve beneficial effects in cancer prevention as well as on other aspects of human health. At the same time, Se compounds have been proposed to hold great potential in anticancer protocols. This option became even more intriguing as such potential was identified to include emerging aspects of tumor biology, such as counteracting the role of cancer stem cells (CSCs), the focus in this chapter.

2. SELENIUM, HUMAN HEALTH, AND CANCER

Selenium supplementation became an issue of public interest since a remarkable decrease in cancer mortality in subjects consuming 200 μg of Se daily (approximately four times the recommended daily requirement) was reported (Clark et al., 1996). The fascinating hypothesis that increased intake of Se might be protective against the development of cancer has thus attracted a great deal of interest. However, the relationships between Se supplementation and cancer development/progression yielded divergent and apparently paradoxical results that are discussed in detail elsewhere in this volume.

Although an inverse association between Se exposure and the risk of some types of cancer was found in several studies (reviewed in Vinceti et al., 2014 and elsewhere in this volume), this cannot be taken as evidence of a cause-and-effect relationship. Conflicting results, including inverse, null, and direct associations, have been reported for some cancer types. On the other hand, some trials suggest harmful effects of Se on cancer development (Vinceti et al., 2017). Altogether, due to inherent study limitations (assessment of exposure times to Se; chemical forms of Se; and heterogeneity of the population), these data must be interpreted with caution.

From a mechanistic standpoint, Se may affect carcinogenesis by regulating a variety of cellular functions/processes, which have been reviewed elsewhere (Whanger, 2004). These include effects on the expression of selenoproteins and the reaction with cellular thiols and reactive species, which cumulatively influence the cellular redox status, DNA stability and oncogene activation, signal transduction through cell cycle and and death checkpoints, the activity of inflammatory pathways and other aspects of the immune response, angiogenesis, and the expression of drug metabolizing enzymes and other detoxifying genes.

It is generally assumed that Se may produce these cellular responses and more in general its anticancer effects at "low" concentrations, whereas at levels higher than those suggested for adequate dietary intake, it can be genotoxic and even carcinogenic (Bronzetti et al., 2001), with toxic levels relatively close to those recommended as health-promoting (Vinceti et al., 2017). Nevertheless, the mechanisms of action of Se compounds, either via prooxidant effects, as ascertained in cytotoxicity and apoptosis, or via the activation of antioxidant responses, as proposed in cancer prevention, are still unclear (Chen et al., 2013; Diehn et al., 2009; Hurst, Elliott, Goldson, & Fairweather-Tait, 2008).

Different factors might help explain such apparently divergent and paradoxical view on the activities of Se in the pathobiology of cancer. First, Se might be more important in preventing progression than in the development of cancer, and hence its stronger effects may be against advanced rather than early stages of the disease (Rayman, 2005). Second, Se might show an effect on cancer risk only over a finite concentration range, as may occur when the Se concentrations in the group under investigation range from below to above the concentration needed to optimize the activity of selenoenzymes, for example glutatione peroxidases (GPx) and selenoprotein P plasma 1 gene (SEPP1). Indeed Se shows a U-shaped relationship between dietary intake or status and cancer protection (Duffield-Lillico et al., 2002; Peters & Takata, 2008; Rayman, 2008). Third, the interaction between selenium intake or status and genetic background might also be of importance. Single-nucleotide polymorphisms (SNPs) in selenoprotein genes can affect the efficiency with which a selenoprotein is synthesized, its activity and concentration in plasma, and risk of disease (Rayman, 2009). Different SNPs in GPx and selenoprotein S (SEPS) have been linked with the risk of prostate, lung, breast, colorectal, bladder, and thyroid cancers and with survival. Moreover, these selenoprotein SNPs (and SNPs in related pathways) can interact with Se status such that individuals with specific gene variants might benefit more than others from additional Se.

A further level of complexity in the biological response to Se molecules could be associated with the heterogeneity of tumor cell populations that characterize the different types of cancers, and with its mutability during the tumor process. In this context, cancer-initiating (or stem) cells are assuming great relevance and deserve further investigation as a therapeutic target of Se compounds.

Overall, the question whether Se intake in humans may actually prevent cancer risk was considered "wide open" at the end of the 1980s (Birt, 1989), and still no conclusive evidence has been obtained to recommend Se supplements to actually prevent cancer in humans (Vinceti et al., 2014). Hopefully, the research on CSCs discussed in the next sections would help shed new light on both the chemoprevention and therapeutic properties of Se.

3. CSCs: A DEBATED THEORY

Despite many years of searching for effective therapies, cancer still remains the most common cause of death worldwide (Ferlay et al., 2015).

One of the main reasons for this is the complexity of cancer development and progression processes, which has daunted most prevention approaches.

Over the past two decades, some important steps forward have been made to understanding the malignant population responsible for tumor maintenance, initiation of relapse, and/or drug resistance have been developed. It has been postulated that many types of cancers can develop from cells with biological properties analogous to organ stem cells, such as ability to self-renew and creation of heterogeneous lineages (Clarke et al., 2006). This population of cells is referred to as cancer-initiating cells or CSCs.

The cellular origin of CSCs and their potential for trans-differentiation and renewal are important and still debated aspects in this context. There are two theories as to the origin of CSCs (extensively discussed in Bomken, Fiser, Heidenreich, & Vormoor, 2010; Cabrera, Hollingsworth, & Hurt, 2015; Visvader & Lindeman, 2012). One is that oncogenic events strike the stem cell of a tissue, thus giving rise to uncontrolled proliferation. The other is that normal cells over time accumulate oncogenic changes that lead to reversion toward "stemness."

Like normal tissue stem cells, CSCs are capable of self-renewal and differentiation into cancer progenitors or mature cancer cells. CSCs can repopulate clonally by cell division (symmetric or asymmetric) or uncontrolled proliferation (Eaves, 2008; Li, Wicha, Schwartz, & Sun, 2011). Thus, it is thought that CSCs may derive either from normal stem cells that undergo genetic or epigenetic alterations, or from cancer cells (not fully differentiated: the cancer progenitor cells) that acquire the potential for unrestrained proliferation (Reya, Morrison, Clarke, & Weissman, 2001). In this respect, other and quite consolidated models anticipated the postulation of the aforementioned theories suggesting that cancer cells within a tumor consist of various clonal subpopulations, thereby exhibiting heterogeneity across many properties, such as genetic variation, marker expression, proliferative and metastatic potential, and drug sensitivity (Heppner & Miller, 1983). Such heterogeneity was used to elaborate two models of carcinogenesis: (1) the "stochastic model" and (2) the "hierarchical model" (Dick, 2009). The former is based on the assembly of diverse cancer clones, while the latter on the generation of multiple subclones from a single clonal population of cells. In the stochastic model, most cancer cells are capable of proliferating extensively and forming new tumors in cooperation with intrinsic and extrinsic factors. According to this model, carcinogenesis occurs randomly from somatic cells undergoing transformation. In the hierarchical model, only a distinct subpopulation of cancer cells, the CSCs, has the ability to extensively proliferate and initiate tumor formation and growth. According to this model, carcinogenesis originates from the CSCs, which can be enriched based on unique cellular features (Dick, 2008). According to this theory, only the CSCs possess the cellular capacity to replenish the tumor population. Notably, it has also been demonstrated that cancer nonstem and cancer stem cells may trans-differentiate and then interconvert each other under particular conditions (Cabrera et al., 2015). However, this does not exclude that the achievement of CSC status, either naturally or by cellular plasticity, is required per se for tumor formation.

Current treatment approaches in cancer are grounded in the need to kill the majority of cancer cells, based on the stochastic model. However, in many instances such efforts have not been successful in the treatment of solid cancers, and this has suggested to refocus our thinking around the hierarchical model in trying to explain resistance to anticancer therapeutics and tumor recurrence (Visvader & Lindeman, 2012).

The hierarchical model for carcinogenesis maintains that CSCs are the origin of tumor formation, metastasis, and relapse. In the past two decades, a wealth of findings supported such model; compelling evidence demonstrated the existence of CSCs more than two decades ago (Bonnet & Dick, 1997) and since that time CSCs from human tumors, such as hematologic, breast, prostate, and brain tumors, were isolated and identified on the basis of their tumorigenic capability and cell surface antigen expression (Oh, Hlatky, Jeong, & Kim, 2016). These pieces of evidence and others recently appeared in the scientific literature, conclusively established that the CSC hypothesis represents a fundamental underpinning of cancer biology and, as such, it is a strategic aspect to be considered in the development of therapeutic strategies (Visvader & Lindeman, 2008).

In terms of phenotypes, CSCs can be recognized by the identification of antigens that however are not always expressed on the cell surface. At the same time, none of these surface markers are exclusively made by CSCs. Furthermore, CSC phenotypes, even from the same tumor, can exhibit different markers owing to the possible presence of multiple CSC pools or to the occurrence of epigenetic alterations (Visvader & Lindeman, 2012). It is thus plausible that a combination of different molecular markers, together with epigenetic profiling, may refine the identification strategy for the CSC phenotype.

The recognition that cancer is a stem cell disease instigated major paradigm shifts in our basic understanding of cancer with profound clinical implications on personalized care. However, studies supporting such a hypothesis were in large part based on xenotransplantation experiments.

Under these circumstances, human cancer cells are grown in immunocompromised mice and only CSCs, often representing less than 1% of the malignancy, generate tumors (Yoo & Hatfield, 2008). It has thus been postulated that, rather than CSC alone, the majority of malignant cells actually sustain tumors, implying that a careful reevaluation of the CSC theory may be needed.

One hypothesis that may partly reconcile such a disputation may come from the evidence that CSC properties dependent on cellular interactions with adjacent stromal cells that appear to organize in a specialized microenvironment or "niche" (Calabrese et al., 2007). Each niche has a complex architecture and is composed of various stromal cells, soluble factors, blood vessels, and extracellular matrix components. Under physiological conditions, tissue stem cells reside in such niches and a complex cellular cross talk with other cells maintains them in a quiescent state. The dependence of CSCs on a similar environment, i.e., the CSC niche, allow them retain their ability to self-renew and give rise to more differentiated progenitor cells (Calabrese et al., 2007). Moreover, the CSC niche appears to exert a protective role and, by sheltering cells, can contribute to resistance against antitumor therapy.

Signaling pathways and genes with prominent roles in stem cell biology have been identified in various tissues, including the Wnt/β-catenin, SOX2, Oct-4, Hedgehog (Hh), Notch, and PI3K/Akt/mTOR, which are involved in the regulation of self-renewal and differentiation, apoptotic cell death and autophagy, as well as in epithelial–mesenchymal transition (EMT) and other CSC-related processes (Gao, Kong, Yang, Gao, & Shi, 2016; Jia et al., 2017; Mao et al., 2014; Reya & Clevers, 2005). Some of these signaling pathways are abnormally activated in CSCs and in other cell populations of a wide array of human cancers, thus making them an attractive target for anticancer therapy, which will be discussed in detail later in this chapter.

4. CELLULAR PROPERTIES OF CSCS: FOCUS ON REDOX REGULATION

Metabolic and homeostatic processes of tissues are under the influence of a series of redox-dependent regulatory interactions that occur both at the intracellular and extracellular level. The flux of superoxide and then of hydrogen peroxide, the most important reactive oxygen species (ROS) of cellular systems (Sies, 2017), as well as of nitric oxide-derived reactive species (RNS), controls redox-sensitive elements in virtually all signal transduction and transcription pathways, some of which are involved in stem cell self-renewal and differentiation (Ding et al., 2015; Ryoo, Lee, & Kwak, 2016). Dysregulations of ROS-dependent cellular processes are often associated with the pathogenesis of several human diseases including many forms of cancer. ROS are generated at high levels in several cancer cell models partially due to their sustained metabolism (Ambrosone, 2000; Szatrowski & Nathan, 1991), which suggests a modified threshold for redox-sensitive processes in these cells and then in the tumor microenvironment. At the same time, increasing the flux of ROS over that threshold is expected to activate death pathways of cancer cells by either apoptotic programs or necrosis (Halliwell, 2007).

The inherent mechanisms by which cancer cells reprogram their metabolism and redox balance in response to such an increased flux of ROS are still matter of investigation. However, a higher detoxification/antioxidant potential to counteract and scavenge ROS enhances the survival and proliferation properties of the transformed cell also impairing cellular responses to anticancer therapies (recently reviewed in Ding et al., 2015; Ryoo et al., 2016). Induction of ROS-mediated damage in cancer cells by pharmacological agents that either promote ROS generation beyond the cellular antioxidative capacity or disable the cellular antioxidant systems, has been considered as a "radical" therapeutic strategy to preferentially kill cancer cells, also including CSCs (Ding et al., 2015; Ryoo et al., 2016; Trachootham, Alexandre, & Huang, 2009).

Although our understanding of CSC redox biology is increasing, aspects of redox homeostasis and signaling relevant to CSC self-renewal and protection from ROS-mediated cell killing remain nascent (Ding et al., 2015; Shi, Zhang, Zheng, & Pan, 2012). Similarly to normal stem cells, are dormant, quiescent, slow-cycling cells with the lower levels of intracellular ROS and upregulation of groups of detoxification/antioxidant genes (Diehn et al., 2009; Moore & Lyle, 2011; Ryoo et al., 2016), which may account for their self-renewal capacities and recurrence or resistance to chemotherapy or radiation therapy. The lower flux of ROS in CSCs could be attributed to a restrained production and/or enhanced scavenging by detoxification and antioxidant genes; the latter include a series of drug metabolism genes regulated through the transcriptional activity of NF-E2-related factor-2 (Nrf2) (Jia, Chen, Zhu, Jia, & Cui, 2015; Ryoo et al., 2016) and other stress response elements, that include for instance some hypoxia-inducible factors investigated in breast cancer (Semenza, 2017).

These aspects demonstrate the existence of a redox reprogramming in CSCs, which is different from, but possibly influenced by, the other cellular components in the tumor microenvironment. In this sense, reprogramming the flux of ROS and the cellular redox of CSCs could be either the consequence or the cause of maintaining the cell cycle at a lower progression rate. The redox couples $NADPH/NADP^+$ and GSH/GSSH show characteristic levels in CSCs and their evaluation is emerging as an important aspect in the characterization of CSC biology as well as in the clinical practice (Carroll et al., 2016; Zaccarin et al., 2017). As recently reported in a CSC model stably overexpressing the TAZ oncogene (Zaccarin et al., 2017), increasing the flux of oxidants in CSCs stimulates effects of redox resilience with a more reduced state of the $NADPH/NADP^+$ couple and of several cellular proteins, while just few were more oxidized; the latter included the redox-sensitive enzyme protein G6PDH, which emerged as both more expressed and activated by oxidation. This enzyme is considered the rate-limiting step in the pentose phosphate pathway, which may account for the observed increase in the $NADPH/NADP^+$ ratio in this CSC model. In turn, these experiments demonstrated the role of this pathway in a series of dynamic redox transitions that affect a complex (and so far poorly understood) series of redox targets and switches in CSC signaling and gene regulation. Therefore, available preliminary data suggest that manipulating ROS metabolism and redox-sensitive proteins in CSCs may hold some potential in interfering with the peculiar redox signaling of these cells, thus improving cancer treatment (Ding et al., 2015). Further insights on this aspects are expected from current lines of research on molecular and functional aspects of CSC redox that include selenoprotein investigation.

5. SELENIUM AND CSCs

Gene expression analysis in tumor cells demonstrated that Se-induced growth inhibition can be associated with modulation of the cell cycle, apoptosis, and signaling (Redman et al., 1998). However, despite some promising investigation, few data are available to understand the exact mechanism and the molecular determinants underlying the antitumor effects of the different Se compounds, including those on CSCs. That lack of knowledge is also the consequence of the difficulty in translating in this population of cancer cells all the available knowledge on anticancer properties of Se compounds with their different chemical forms, molecular mechanisms and degree of toxicity and anticarcinogenic potential shown in the different experimental protocols and clinical trials (Korbut, Ptak-Belowska, & Brzozowski, 2012). However, the aberrant regulation of some cell growth and proliferation pathways in CSCs is of particular interest and may hold potential for developing selective chemotherapy agents (Liu, Dontu, & Wicha, 2005). In this respect, only a few studies have focused on the involvement of Se on CSC biology and on the development of CSC-targeted therapies based on Se-compounds, which are summarized in this section.

5.1 GPx-2 and Other Selenoproteins in CSC Biology

Among the different selenoproteins investigated in cancer cells, the glutathione peroxidase (GPx) family of enzymes has been proposed to influence some cellular properties of CSC, such as differentiation. In this context, GPx-2 has been the most investigated isoform; its expression is regulated through the cellular availability of Se and the activity of oxidant-sensitive and protein kinase-regulated transcription factors, such as Nrf2 or some members of the signal transducers and activators of transcription (STAT) family (Emmink et al., 2014). In human colon cells grown and maintained in culture as colonospheres, silencing of GPx-2 increased the flux of cellular ROS then leading to generation of stem-like cancer cells. This evidence suggests that GPx-2 enzymatic activity influences the earliest steps of tumor development, sustaining stem cell differentiation, which is regulated with a threshold effect by the burden of cellular oxidants, such as H_2O_2, and their signal transduction effects through redox-sensitive pathways. The signaling associated with GPx-2 activity also affects advanced steps in the apoptotic and metastatic capacity of these cells (Emmink et al., 2014) as well as of other types of cancer cells (Suzuki et al., 2013).

These properties of GPx-2 and its Nrf2-dependent regulation, along with that of other genes involved in ROS detoxification and redox regulation processes, such as TrxRd1, highlight the importance of Nrf2 expression and activity in CSCs. As introduced above in Section 4, the transcriptional activity of Nrf2 is one of the main factors that may lower the flux of oxidants in these cells in comparison with differentiated cancer cells. Such redox reprogramming of CSC to maintain lower the intracelluair levels of oxidants suggests specific features of these cells and may represent a weak point of their metabolic framework to selectively intervene with the thiol peroxidase and ROS-generating activity of reactive Se compounds (Bartolini et al., 2015). A decreased expression of GPx-2 has recently been shown to affect differentiation and senescence pathways, possibly acting in a concerted manner with other proteins, such as FOXM1 (Smirnov et al., 2016). Reduced mRNA expression of GPx-2 is associated with the differentiation of embryonic stem cells (Saretzki et al., 2008) and, as discussed later, this may lead to development of stem-like cancer cells in human colorectal cancer spheroid models that lack multilineage differentiation capacity (Speckmann et al., 2011). Conversely, GPx-2 overexpression increased the expression of markers associated with the differentiation of different colon lineages, such as MUC2 (goblet cells), FABP1 (enterocytes), and chromogranin A (enteroendocrine cells) (Emmink et al., 2014). Redox-sensitive nodes of differentiation are candidates for a role in these GPx-2-dependent responses. These include Wnt signaling and Notch, which are important players in the pathophysiology of normal intestinal epithelium and its malignant counterpart (Pérez, Talens-Visconti, Rius-Perez, Finamor, & Sastre, 2017). Other forms of GPx and selenoproteins could share with GPx-2 these redox-dependent effects on differentiation pathways, and particularly on Wnt/β-catenin, such as GPx-3 and Selenop (discussed later in detail). TrxRd is another potential target of Se compounds and as such it may deserve further investigation in CSC. The prototypical Se-organic molecule Ebselen is a potent competitive inhibitor of this selenoprotein investigated in human breast carcinoma (Engman et al., 1997) with proposed role in CSC response to oxidants and anticancer therapies (Rodman et al. 2016; Zaccarin et al., 2017). This molecule was recently investigated in our laboratories and preliminary data demonstrated its time- and dose-dependent activity as cell viability inhibitor of primary cultures of human adipose-derived mesenchymal stem cells prepared as described in (Murdolo et al., 2016) (Fig. 1; Murdolo G. and Galli F., unpublished data). Interesting enough, in these experiments the inhibitory activity of Ebselen on stem cell viability was superior with respect to PhSeZnCl. The latter is a novel air-stable selenolate synthesized in our laboratories with higher thiol peroxidase activity in cancer cells than Ebselen (Bartolini et al., 2015). Therefore, these data support a selective mechanism for the effect of Ebselen on stem cell viability that could not simply depend on the generic redox cycling effect on cellular thiols of Se compounds. If that could be a TrxRd-dependent mechanism is currently under investigation. Importantly, these Se compounds are also irreversible inhibitors of glutathione S-transferase π (GSTP) (Batrolini et al., 2015)

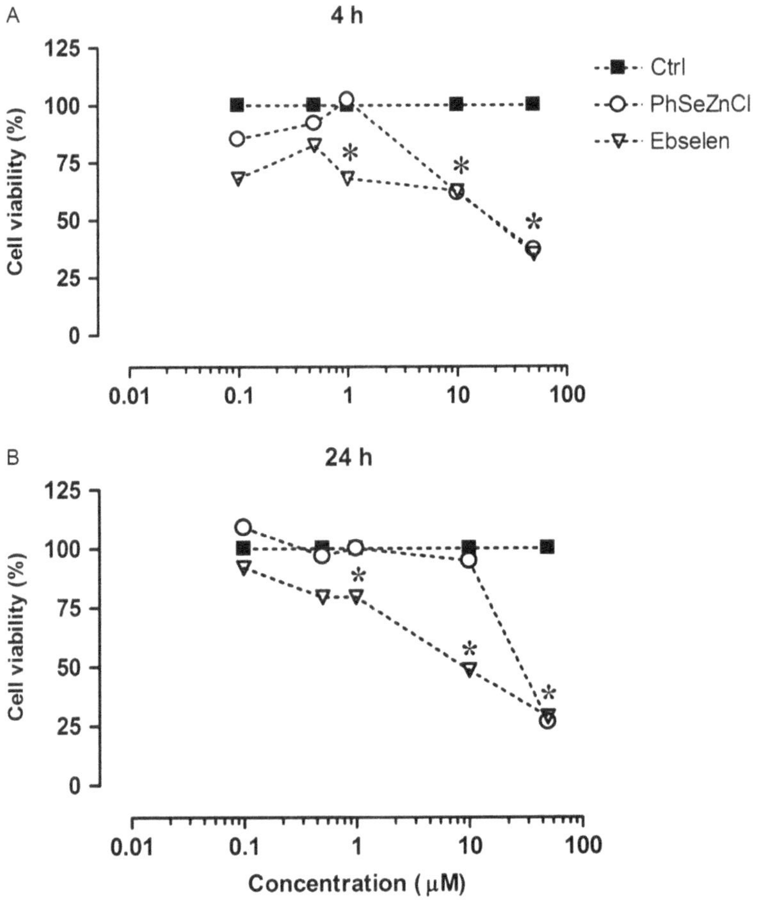

Fig. 1 Cell viability of human adipose mesenchymal stem cells after 4 h (panel A) and 24 h (panel B) of treatment with increasing concentrations (0.01, 0.1, 1, 10 and 50 mM) of the Se-compounds Ebselen and PhSeZnCl as determined by the MTT reduction test. Data were as percentage (%) of control (untreated) cell tests. $^*P < 0.01$ vs the corresponding control cell test.

and as such they interfere with the Nrf2-dependent response to these anticancer agents and to other cellular stressors (Bartolini & Galli, 2016). Recently, the dual inhibition of GSTP and TrxRd has been demonstrate to synergize in decreasing the tumor mass and CD44v9-positive fraction mouse xenografts derived from colorectal carcinoma patients (Tanaka, Inoue, Shimizu, Akimoto, & Kubota, 2016).Disrupting the redox regulation of CSC by the targeting Se-sensitive nodes of both the glutathione and Trx dependent pathways is therefore a very promising anticancer strategy now under investigation by different groups.

5.2 Targeting the Wnt/β-Catenin Pathway With Se Compounds

Some similarities between normal adult stem cells and CSCs suggest that Wnt and other signaling pathways involved in regulating somatic stem cell properties, such as Hedgehog and Notch, are also involved in the regulation of CSC (reviewed elsewhere Takahashi-Yanaga & Kahn, 2010). Wnt/β-catenin pathways are ubiquitously important in signaling and gene regulation, from embryonic development to control of homeostatic self-renewal of normal and even pathological cells (reviewed in Luchetti et al., 2014; Murdolo et al., 2013). Importantly, Wnt/β-catenin is abnormally activated in a wide array of human cancers as well as in CSCs (Cai & Zhu, 2012; Cheng et al., 2013; Mao et al., 2014) presumably functioning in self-renewal and stemness transition networks by the control of groups of genes associated with pluripotency and tumor suppression pathways, drug resistance, and CSC marker expression, the latter involving many of the cell surface markers (LGR5/GPR49, CD44, CD24, and Epcam) used to identify and isolate putative tumor stem cell populations in different tissues (Takahashi-Yanaga & Kahn, 2010). Worth of note, the shut off of Wnt/β-catenin signal transduction after differentiation is important to prevent malignant transformation of human embryonic stem cells (Blum, Bar-Nur, Golan-Lev, & Benvenisty, 2009).

Once activated Wnt signaling inhibits the proteolytic degradation complex of the oncogene protein β-catenin (composed of axin, glycogen synthetase kinase 3β, and adenomatosis polyposis coli protein), which thereby accumulates in the cytoplasm to entering the nucleus and implement its transcriptional response by the interaction with the nuclear proteins T cell factor and lymphoid-enhancing transcription factor (Grumolato et al., 2013). Target genes of this signal transduction pathway are essential players in embryonic development and stem cell maintenance producing effects on cell cycle progression, apoptosis, and proliferation (Luchetti et al., 2014). This array of biological responses explains how inappropriate regulation and activation of Wnt/β-catenin pathways can be associated with pathogenic cues of many disorders including cancer (Clevers, 2006; Moon, Kohn, De Ferrari, & Kaykas, 2004). For example, canonical Wnt signaling is believed to play an important role in the maintenance of hematopoietic progenitors and also in their lineage commitment during hematopoiesis, e.g., erythroid vs megakaryocytic, by the transcriptional control of *survivin*, a Wnt/CBP/β-catenin-regulated gene. Wnt/β-catenin transcriptional activity also influences the progression of CML (chronic myeloid leukemia) from chronic phase to blast crisis and may sustain drug resistance toward the tyrosine kinase inhibitor imatinib (reviewed in Takahashi-Yanaga & Kahn, 2010). Multidrug resistance genes, mediated through MDR-1, ABCG2, ABCA3, *ABCB1*, and BRCP1, are intrinsically expressed in stem and/or progenitor cells from multiple adult tissues and may contribute to the side population phenotype of malignant cells. Importantly, as with other CSC populations, multiple myeloma CSCs have been found to be relatively resistant to existing chemotherapies, displaying high expression of multidrug resistance genes, such as transporters and intracellular detoxification enzymes (Gao et al., 2016; Issa, Cretton, & Cuendet, 2017).

For all these characteristics, the Wnt/β-catenin signaling pathway represents an attractive target for anticancer therapy. Se compounds can affect this molecular target in CSC to promote secondary chemoprevention effects and improve treatment's efficacy and outcome. Besides the roles of GPX-2 described in the previous section, other GPx isoforms and selenoproteins are reported to intervene in the redox-dependent and -independent control of this pathway. GPx3 knockout led to increased levels of β-catenin in intestinal tumors (Barrett et al., 2013), and this isoenzyme is reported to inhibit Wnt signaling in other models of human cancer, such as TPC-1 and FTC133 thyroid cancer cells (Zhao et al., 2015). At the same time, Se deficiency induces Wnt pathway activation and reduced *Selenop* enhanced expression of Wnt targets and stem cell phenotypes in intestinal tumors and intestinal organoid models, respectively (Barrett et al., 2015; Kipp et al., 2009). The molecular mechanisms of this selenoprotein remain elusive.

The antitumor activity of Se compounds in multiple malignancies includes suppression of β-catenin signaling as a underlying mechanism; for instance, this is reported to represent the mechanism of action for the selenium compound 1,4-phenylene bis(methylene)selenocyanate (Narayanan, Narayanan, Desai, Pittman, & Reddy, 2004; Rao et al., 2000). Methylseleninic acid treatment of cancer cells also inhibits β-catenin accumulation in the nucleus by the increased degradation of this oncogene protein in the cytoplasm (Saifo, Rempinski, Rustum, & Azrak, 2010). Moreover, sodium selenite significantly inhibited intestinal tumor formation decreasing cell proliferation and inducing apoptosis by the activation JNK1 stress-activated MAPK and subsequent inhibition of β-catenin signaling and transcriptional activity (Fang, Han, Bi, Xiong, & Yang, 2010).

The Wnt pathway also influences EMT of cancer cells, a process in which epithelial cells lose their cell–cell adhesions and apicobasal polarity, but also acquire more mesenchymal and invasive/metastatic phenotype (reviewed in Yoshida, 2016). EMT is known to generate stromal-like cancer cells which acquire the invasion potential into the extracellular matrix mediated by matrix metalloproteinase (MMP) expression which again is demonstrated to fall under the transcriptional control of the Wnt/β-catenin pathway in gastric cancer and hepatocellular carcinoma (HCC) (Jia et al., 2017; Yoshida, 2016). Seleno-L-methionine was reported to influence the expression of both prometastatic and antimetastatic genes of the matrix metalloproteinase family (pro-MMP-2 and pro-MMP-9 and the tissue inhibitors TIMP-1 and TIMP-2) by the generation of its metabolite methylselenol, which in turn resulted in lowered migration and invasion potential of tumor cells (Zeng, Briske-Anderson, Idso, & Hunt, 2006). Du et al. revealed that brachyury, an essential EMT regulator altering the expression level of E-cadherin and Slug, which is highly conserved through evolution, promotes EMT of HCC via activation of Akt/mTOR signal pathway (Du et al., 2014), which is a main target of several Se compounds to alternatively activate the apoptotic and autophagy signaling of cancer cells (Sanmartin, Plano, Sharma, & Palop, 2012). Combining the effect on Wnt with that reported on the Akt/mTOR pathway, Se compounds may represent promising agents to target EMT and possibly other critical aspects of CSC biology. Accordingly, EMT contributes to the increased population of CSCs, which corresponds to the highly tumorigenic subpopulation of tumor cells, thereby affecting tumor heterogeneity and therapeutic resistance (Yoshida & Saya, 2016). For instance, Zhang et al. recently demonstrated that the resistance to FOLFOX, the conventional anti-HCC chemotherapeutic regimen, correlated with nestin expression that promotes EMT and the increase of CSC number associated with Wnt/β-catenin activation (Zhang et al., 2016). Collectively, accumulating evidence is suggesting that Wnt/β-catenin activity and EMT represent emerging therapeutic targets of difficult-to-cure human malignancies (Yoshida & Saya, 2016) and as such are worthy of further investigation as targets of Se compounds.

5.3 Se Compounds Modulate Arachidonic Acid Metabolism and PPAR-Gamma Activity in CSCs

Consistent evidence in literature shows that Se compounds can interfere with the arachidonic acid metabolism of leukemia CSCs (Gandhi et al., 2014). Such molecular response commits cells to apoptotic death influencing selenoprotein expression then leading to lowered PGE_2 levels and enhanced production of H-PGDS-derived cyclopentenone PGs (CyPGs) (Gandhi et al., 2014; Vunta et al., 2007). This switching effect of Se on arachidonate metabolism appears to occur by the uncoupling of COX2 and mPEGS1 proteins in inflammatory macrophages and its ultimate consequence is the promotion of PD-L1-mediated antiproliferative and apoptotic effects in tumor cells. These aspects were investigated in models of chronic myeloid leukemia (CML) in which leukemia stem cells (LSCs) were also observed to endogenously produce CyPGs in response to selenium supplementation. Therefore, both autocrine and paracrine (through inflammatory cells) effects could be proposed for CyPGs in LSCs. These bioactive lipids act as PPAR-gamma agonists, and as such, affect LSC metabolism and other transcription factors downstream of this nuclear receptor, thus providing a selective mechanism of action in LSC ablation and treatment of CML by Se compounds. In fact, pharmacologic agonists of this receptor, namely glitazones, have been proposed for CML therapy in combination with protein tyrosine kinase inhibitors (Glodkowska-Mrowka et al., 2016) and the selenium-induced activation of PPAR-gamma in LSCs decreased STAT-5a expression, required for LSC self-renewal (Prost et al., 2015), and that of its downstream targets, CITED2 and HIF2α, both known to maintain LSC quiescence (Finch et al., 2017). In this context, the pharmacological inhibition of PPAR-gamma prevented these PPAR-gamma-dependent responses to selenium in CML mice or isolated LSCs, thus promoting leukemia.

6. CONCLUSIONS

Although clinical trials provided conflicting results on Se therapy in human cancers, in vitro results and animal studies strongly suggest that Se may well behave as a promising anticancer agent. In this chapter, we summarized the current understanding on the effect that Se compounds have on different aspects of CSC biology, including self-renewal, differentiation, and migration processes.

Although current approaches in cancer therapy are grounded in the need to kill the majority of cancer cells, the emerging role of CSCs in cancer, along with the importance of redox-sensitive pathways in this peculiar subpopulation of stem cells, suggest new and stimulating strategies to improve treatment protocols. In this area of research, Se compounds have been demonstrated to interfere with main signaling pathways of CSCs that, above the others, include Wnt/β-catenin and PPAR-gamma via its response to arachidonic acid-derived bioactive metabolites, which are discussed in this chapter.

In this context, it has to be said that our knowledge of the therapeutic potential of Se compounds as agents to target CSCs remains very limited;

however, available evidence on specific redox properties and response to oxidants in these cells suggests the possibility of developing selective and CSC-targeted treatments. Of note, such a selective approach was already validated at the preclinical level in some tumors, such as CML, and the possibility to improve its efficacy is under investigation by the development of CSC-specific drug delivery formulations or strategies. As an example of this, studies on photoproducts of merocyanine analogues containing Se, such as selenomerocyanine dyes, showed potent cytotoxic activity against leukemia and other solid tumors, while being well tolerated by normal CD34-positive hematopoietic stem and progenitor cells (Sieber et al., 2005). This effect was reported to be the consequence of protein conjugate formation with serum albumin and lipoproteins by the elemental selenium generated during selenomerocyanine photobleaching, a response that was recently confirmed investigating other merocyanine analogues (Daziano et al., 2012). These conjugates are selectively cytotoxic to cancer cells in that these cells avidly take up serum albumin and lipoproteins, which thereby act as Trojan horses to deliver elemental selenium at cytotoxic levels to tumor cells.

Generalized toxicity and poor specificity of action remain an issue for the application of Se compounds in therapy, and in this respect, whether or not CSCs might represent a real target for Se or other anticancer therapies resembles the choice of Odysseus between Scylla and Charybdis, a current idiom deriving from Greek mythology meaning "having to choose between two evils." Scylla and Charybdis were mythical sea monsters noted by Homer: the former was rationalized as a rock shoal (described as a six-headed sea monster), while the latter was a whirlpool off the coast of Sicily. They were regarded as a sea hazard located close enough to each other that they posed an "inescapable threat" to passing sailors. According to Homer, Odysseus was forced to choose which monster to confront while passing through the strait. He opted to pass by Scylla and lose only a few sailors, rather than risk the loss of his entire ship in the whirlpool. Was this choice the right one? In that case, definitively yes. As far as selenium is concerned, we need to live and learn more.

ACKNOWLEDGMENTS

F.G.'s lab and M.P. are supported by the grant program of the Italian Ministry of University, National Technology Agrifood Cluster, Health and Nutrition program, PROS·IT project (CTN01_00230_413096).

D.B. is a fellow of research at F.G.'s lab supported by the FIRC/AIRC young investigator grant program.

DISCLOSURE STATEMENT

The authors have nothing to disclose.

REFERENCES

Ambrosone, C. (2000). Oxidants and antioxidants in breast cancer. *Antioxidants & Redox Signaling*, *2*, 903–917.

Barrett, C. W., Ning, W., Chen, X., Smith, J. J., Washington, M. K., Hill, K. E., et al. (2013). Tumor suppressor function of the plasma glutathione peroxidase gpx3 in colitis-associated carcinoma. *Cancer Research*, *73*, 1245–1255.

Barrett, C. W., Reddy, V. K., Short, S. P., Motley, A. K., Lintel, M. K., Bradley, A. M., et al. (2015). Selenoprotein P influences colitis-induced tumorigenesis by mediating stemness and oxidative damage. *The Journal of Clinical Investigation*, *125*, 2646–2660.

Bartolini, D., Ciffolilli, S., Piroddi, M., Murdolo, G., Tortoioli, C., & Galli, F. (2014). Biochemistry and nutrition of selenium: From inorganic forms to endogenous proteins (Chapter 9). In C. Santi (Ed.), *Organoselenium chemistry between synthesis and biochemistry (E-Book)* (pp. 268–327). United Arab Emirates: Bentham Science Publishers.

Bartolini, D., & Galli, F. (2016). The functional interactome of GSTP: A regulatory biomolecular network at the interface with the Nrf2 adaption response to oxidative stress. *Journal of Chromatography. B, Analytical Technologies in the Biomedical and Life Sciences*, *1019*, 29–44.

Bartolini, D., Piroddi, M., Tidei, C., Giovagnoli, S., Pietrella, D., Manevich, Y., et al. (2015). Reaction kinetics and targeting to cellular glutathione S-transferase of the glutathione peroxidase mimetic PhSeZnCl and its d,l-polylactide microparticle formulation. *Free Radical Biology & Medicine*, *78*, 56–65.

Birt, D. F. (1989). Effects of the intake of selected vitamins and minerals on cancer prevention. *Magnesium*, *8*, 17–30.

Blum, B., Bar-Nur, O., Golan-Lev, T., & Benvenisty, N. (2009). The anti-apoptotic gene survivin contributes to teratoma formation by human embryonic stem cells. *Nature Biotechnology*, *27*, 281–287.

Bomken, S., Fiser, K., Heidenreich, O., & Vormoor, J. (2010). Understanding the cancer stem cell. *British Journal of Cancer*, *103*, 439–445.

Bonnet, D., & Dick, J. (1997). Human acute myeloid leukemia is organized as a hierarchy that originates from a primitive hematopoietic cell. *Nature Medicine*, *3*, 730–737.

Boudreau, R. T., Conrad, D. M., & Hoskin, D. W. (2007). Differential involvement of reactive oxygen species in apoptosis caused by the inhibition of protein phosphatase 2A in Jurkat and CCRF-CEM human T-leukemia cells. *Experimental and Molecular Pathology*, *83*, 347–356.

Bronzetti, G., Cini, M., Andreoli, E., Caltavuturo, L., Panunzio, M., & Croce, C. D. (2001). Protective effects of vitamins and selenium compounds in yeast. *Mutation Research*, *496*, 105–115.

Cabrera, M., Hollingsworth, R., & Hurt, E. (2015). Cancer stem cell plasticity and tumor hierarchy. *World Journal of Stem Cells*, *7*, 27–36.

Cai, C., & Zhu, X. H. (2012). The Wnt/beta-catenin pathway regulates self-renewal of cancer stem-like cells in human gastric cancer. *Molecular Medicine Reports*, *5*, 1191–1196.

Calabrese, C., Poppleton, H., Kocak, M., Hogg, T. L., Fuller, C., Hamner, B., et al. (2007). A perivascular niche for brain tumor stem cells. *Cancer Cell*, *11*, 69–82.

Carroll, D., Howard, D., Zhu, H. N., Paumi, C. M., Vore, M., Bondada, S., et al. (2016). Simultaneous quantitation of oxidized and reduced glutathione via LC-MS/MS: An insight into the redox state of hematopoietic stem cells. *Free Radical Biology & Medicine*, *97*, 85–94.

Chen, Y. C., Prabhu, K. S., & Mastro, A. M. (2013). Is selenium a potential treatment for cancer metastasis? *Nutrients, 5*, 1149–1168.
Cheng, Y., Cheung, A. K. L., Ko, J. M. Y., Phoon, Y. P., Chiu, P. M., Lo, P. H. Y., et al. (2013). Physiological beta-catenin signaling controls self-renewal networks and generation of stem-like cells from nasopharyngeal carcinoma. *BMC Cell Biology, 14.*
Clark, L. C., Combs, G. F., Jr., Turnbull, B. W., Slate, E. H., Chalker, D. K., Chow, J., et al. (1996). Effects of selenium supplementation for cancer prevention in patients with carcinoma of the skin. A randomized controlled trial. Nutritional Prevention of Cancer Study Group. *JAMA, 276*, 1957–1963.
Clarke, M., Dick, J., Dirks, P., et al. (2006). Cancer stem cells perspectives on current status and future directions: AACR workshop on cancer stem cells. *Cancer Research, 66*, 9339–9344.
Clevers, H. (2006). Wnt/beta-catenin signaling in development and disease. *Cell, 127*, 469–480.
Council, N.R. (2000). *Dietary reference intakes for vitamin C, vitamin E, selenium, and carotenoids.* Washington, DC: The National Academies Press.
Daziano, J. P., Gunther, W. H. H., Krieg, M., Tsujino, I., Miyagi, K., Anderson, G. S., et al. (2012). Photochemically generated elemental selenium forms conjugates with serum proteins that are preferentially cytotoxic to leukemia and selected solid tumor cells. *Photochemistry and Photobiology, 88*, 448–460.
Dick, J. (2008). Stem cell concepts renew cancer research. *Blood, 112*, 4793–4807.
Dick, J. (2009). Looking ahead in cancer stem cell research. *Nature Biotechnology, 27*, 44–46.
Diehn, M., Cho, R. W., Lobo, N. A., Kalisky, T., Dorie, M. J., Kulp, A. N., et al. (2009). Association of reactive oxygen species levels and radioresistance in cancer stem cells. *Nature, 458*, 780–783.
Ding, S., Li, C., Cheng, N., Cui, X., Xu, X., & Zhou, G. (2015). Redox regulation in cancer stem cells. *Oxidative Medicine and Cellular Longevity, 2015*, 1–11.
Du, R., Wu, S. S., Lv, X. N., Fang, H. H., Wu, S. D., & Kang, J. B. (2014). Overexpression of brachyury contributes to tumor metastasis by inducing epithelial-mesenchymal transition in hepatocellular carcinoma. *Journal of Experimental & Clinical Cancer Research, 33*, 105.
Duffield-Lillico, A. J., Reid, M. E., Turnbull, B. W., et al. (2002). Baseline characteristics and the effect of selenium supplementation on cancer incidence in a randomized clinical trial: A summary report of the Nutritional Prevention of Cancer Trial. *Cancer Epidemiology, Biomarkers & Prevention, 11*, 630–639.
Eaves, C. (2008). Cancer stem cells: Here, there, everywhere? *Nature, 456*, 581–582.
Emmink, B. L., Laoukili, J., Kipp, A. P., Koster, J., Govaert, K. M., Fatrai, S., et al. (2014). GPx2 suppression of H2O2 stress links the formation of differentiated tumor mass to metastatic capacity in colorectal cancer. *Cancer Research, 74*, 6717–6730.
Engman, L., Cotgreave, I., Angulo, M., Taylor, C. W., Paine-Murrieta, G. D., & Powis, G. (1997). Diaryl chalcogenides as selective inhibitors of thioredoxin reductase and potential antitumor agents. *Anticancer Research, 17*(6D), 4599–4605.
Fang, W. F., Han, A. J., Bi, X. L., Xiong, B., & Yang, W. C. (2010). Tumor inhibition by sodium selenite is associated with activation of c-Jun NH2-terminal kinase 1 and suppression of beta-catenin signaling. *International Journal of Cancer, 127*, 32–42.
Ferlay, J., Soerjomataram, I., Dikshit, R., Eser, S., Mathers, C., Rebelo, M., et al. (2015). Cancer incidence and mortality worldwide: Sources, methods and major patterns in GLOBOCAN 2012. *International Journal of Cancer, 136*, E359–E386.
Finch, E. R., Tukaramrao, D. B., Goodfield, L. L., Quickel, M. D., Paulson, R. F., & Prabhu, K. S. (2017). Activation of PPARγ by endogenous prostaglandin J2 mediates the antileukemic effect of selenium in murine leukemia. *Blood, 129*(13), 1802–1810. http://dx.doi.org/10.1182/blood-2016-08-736405.

Gandhi, U., Kaushal, N., Hegde, S., Finch, E., Kudva, A., Kennett, M., et al. (2014). Selenium suppresses leukemia through the action of endogenous eicosanoids. *Cancer Research, 74*, 3890–3901.

Gao, M. J., Kong, Y. Y., Yang, G., Gao, L., & Shi, J. M. (2016). Multiple myeloma cancer stem cells. *Oncotarget, 7*, 35466–35477.

Glodkowska-Mrowka, E., Manda-Handzlik, A., Stelmaszczyk-Emmel, A., Seferynska, I., Stoklosa, T., Przybylski, J., et al. (2016). PPARγ ligands increase antileukemic activity of second- and third-generation tyrosine kinase inhibitors in chronic myeloid leukemia cells. *Blood Cancer Journal, 6*, e377.

Grumolato, L., Liu, G. Z., Haremaki, T., Mungamuri, S. K., Mong, P., Akiri, G., et al. (2013). Beta-catenin-independent activation of TCF1/LEF1 in human hematopoietic tumor cells through interaction with ATF2 transcription factors. *PLoS Genetics, 9*, e1003603.

Halliwell, B. (2007). Oxidative stress and cancer: Have we moved forward? *The Biochemical Journal, 401*, 1–11.

Heppner, G., & Miller, B. (1983). Tumor heterogeneity: Biological implications and therapeutic consequences. *Cancer Metastasis Reviews, 2*, 5–23.

Hurst, R., Elliott, R. M., Goldson, A. J., & Fairweather-Tait, S. J. (2008). Se-methylselenocysteine alters collagen gene and protein expression in human prostate cells. *Cancer Letters, 269*, 117–126.

Issa, M. E., Cretton, S., & Cuendet, M. (2017). Targeting multiple myeloma cancer stem cells with natural products—Lessons from other hematological malignancies. *Planta Medica, 83*, 752–760.

Jia, Y., Chen, J., Zhu, H., Jia, Z. H., & Cui, M. H. (2015). Aberrantly elevated redox sensing factor Nrf2 promotes cancer stem cell survival via enhanced transcriptional regulation of ABCG2 and Bcl-2/Bmi-1 genes. *Oncology Reports, 34*, 2296–2304.

Jia, S. Q., Qu, T. T., Wang, X. H., Feng, M. M., Yang, Y., Feng, X. M., et al. (2017). KIAA1199 promotes migration and invasion by Wnt/beta-catenin pathway and MMPs mediated EMT progression and serves as a poor prognosis marker in gastric cancer. *PLoS One, 12*, e0175058.

Jiang, C., Ganther, H., & Lu, J. (2000). Monomethyl selenium-specific inhibition of MMP-2 and VEGF expression: Implications for angiogenic switch regulation. *Molecular Carcinogenesis, 29*, 236–250.

Kipp, A., Banning, A., van Schothorst, E. M., Meplan, C., Schomburg, L., Evelo, C., et al. (2009). Four selenoproteins, protein biosynthesis, and Wnt signalling are particularly sensitive to limited selenium intake in mouse colon. *Molecular Nutrition & Food Research, 53*, 1561–1572.

Korbut, E., Ptak-Belowska, A., & Brzozowski, T. (2012). Mechanisms promoting physiological cells progression into tumorigenesis. *Journal of Physiology and Pharmacology, 63*, 565–570.

Li, Y., Wicha, M., Schwartz, S., & Sun, D. (2011). Implications of cancer stem cell theory for cancer chemoprevention by natural dietary compounds. *The Journal of Nutritional Biochemistry, 22*, 799–806.

Liu, S., Dontu, G., & Wicha, M. (2005). Mammary stem cells, self-renewal pathways, and carcinogenesis. *Breast Cancer Research, 7*.

Luchetti, F., Canonico, B., Bartolini, D., Arcangeletti, M., Ciffolilli, S., Murdolo, G., et al. (2014). Melatonin regulates mesenchymal stem cell differentiation: A review. *Journal of Pineal Research, 56*, 382–397.

Mao, J., Fan, S., Ma, W., Fan, P., Wang, B., Zhang, J., et al. (2014). Roles of Wnt/beta-catenin signaling in the gastric cancer stem cells proliferation and salinomycin treatment. *Cell Death & Disease, 5*, e1039.

Moon, R. T., Kohn, A. D., De Ferrari, G. V., & Kaykas, A. (2004). WNT and beta-catenin signalling: Diseases and therapies. *Nature Reviews. Genetics*, *5*, 689–699.

Moore, N., & Lyle, S. (2011). Quiescent, slow-cycling stem cell populations in cancer: A review of the evidence and discussion of significance. *Journal of Oncology*, *2011*, 11, Article ID 396076.

Murdolo, G., Bartolini, D., Tortoioli, C., Piroddi, M., Iuliano, L., & Galli, F. (2013). Lipokines and oxysterols: Novel adipose-derived lipid hormones linking adipose dysfunction and insulin resistance. *Free Radical Biology & Medicine*, *65*, 811–820.

Murdolo, G., Piroddi, M., Tortoioli, C., Bartolini, D., Schmelz, M., Luchetti, F., et al. (2016). Free radical-derived oxysterols: novel adipokines modulating adipogenic differentiation of adipose precursor cells. *The Journal of Clinical Endocrinology & Metabolism*, *101*, 4974–4983.

Narayanan, B. A., Narayanan, N. K., Desai, D., Pittman, B., & Reddy, B. S. (2004). Effects of a combination of docosahexaenoic acid and 1,4-phenylene bis(methylene) selenocyanate on cyclooxygenase 2, inducible nitric oxide synthase and beta-catenin pathways in colon cancer cells. *Carcinogenesis*, *25*, 2443–2449.

Navarro-Alarcon, M., & Cabrera-Vique, C. (2008). Selenium in food and the human body: A review. *Science of the Total Environment*, *400*, 115–141.

Oh, J., Hlatky, L., Jeong, Y., & Kim, D. (2016). Therapeutic effectiveness of anticancer phytochemicals on cancer stem cells. *Toxins*, *8*, 1–11.

Park, J. M., Kim, A., Oh, J., & Chung, A. (2007). Methylseleninic acid inhibits PMA-stimulated pro-MMP-2 activation mediated by Mt1-MMP expression and further tumor invasion through suppression of NF-kappaB activation. *Carcinogenesis*, *28*, 837–847.

Pérez, S., Talens-Visconti, R., Rius-Perez, S., Finamor, I., & Sastre, J. (2017). Redox signaling in the gastrointestinal tract. *Free Radical Biology & Medicine*, *104*, 75–103.

Peters, U., & Takata, Y. (2008). Selenium and the prevention of prostate and colorectal cancer. *Molecular Nutrition & Food Research*, *52*, 1261–1272.

Prost, S., Relouzat, F., Spentchian, M., Ouzegdouh, Y., Saliba, J., Massonnet, G., et al. (2015). Erosion of the chronic myeloid leukaemia stem cell pool by PPARγ agonists. *Nature*, *525*(7569), 380–383.

Rao, C. V., Cooma, I., Rodriguez, J. G. R., Simi, B., El-Bayoumy, K., & Reddy, B. S. (2000). Chemoprevention of familial adenomatous polyposis development in the APC(min) mouse model by 1,4-phenylene bis (methylene)selenocyanate. *Carcinogenesis*, *21*, 617–621.

Rayman, M. (2005). Selenium in cancer prevention: A review of the evidence and mechanism of action. *The Proceedings of the Nutrition Society*, *64*, 527–542.

Rayman, M. (2008). Food-chain selenium and human health: Emphasis on intake. *The British Journal of Nutrition*, *100*, 254–268.

Rayman, M. (2009). Selenoproteins and human health: Insights from epidemiological data. *Biochimica et Biophysica Acta*, *1790*, 1533–1540.

Rayman, M. P. (2012). Selenium and human health. *Lancet*, *379*, 1256–1268.

Redman, C., Scott, J., Baines, A., et al. (1998). Inhibitory effect of selenomethionine on the growth of three selected human tumor cell lines. *Cancer Letters*, *125*, 103–110.

Reya, T., & Clevers, H. (2005). Wnt signalling in stem cells and cancer. *Nature*, *434*, 843–850.

Reya, T., Morrison, S., Clarke, M., & Weissman, I. (2001). Stem cells, cancer, and cancer stem cells. *Nature*, *414*, 105–111.

Rodman, S. N., Spence, J. M., Ronnfeldt, T. J., Zhu, Y., Solst, S. R., O'Neill, R. A., et al. (2016). Enhancement of radiation response in breast cancer stem cells by inhibition of thioredoxin- and glutathione-dependent metabolism. *Radiation Research*, *186*, 385–395.

Ryoo, I. G., Lee, S. H., & Kwak, M. K. (2016). Redox modulating NRF2: A potential mediator of cancer stem cell resistance. *Oxidative Medicine and Cellular Longevity, 2016*, 14, Article ID 2428153.

Saifo, M. S., Rempinski, D. R., Rustum, Y. M., & Azrak, R. G. (2010). Targeting the oncogenic protein beta-catenin to enhance chemotherapy outcome against solid human cancers. *Molecular Cancer, 9*, 310.

Sanmartin, C., Plano, D., Sharma, A., & Palop, J. (2012). Selenium compounds, apoptosis and other types of cell death: An overview for cancer therapy. *International Journal of Molecular Sciences, 13*, 9649–9672.

Saretzki, G., Walter, T., Atkinson, S., Passos, J. F., Bareth, B., Keith, W. N., et al. (2008). Downregulation of multiple stress defense mechanisms during differentiation of human embryonic stem cells. *Stem Cells, 26*, 455–464.

Semenza, G. L. (2017). Hypoxia-inducible factors: Coupling glucose metabolism and redox regulation with induction of the breast cancer stem cell phenotype. *The EMBO Journal, 36*, 252–259.

Shi, X., Zhang, Y., Zheng, J., & Pan, J. (2012). Reactive oxygen species in cancer stem cells. *Antioxidants and Redox Signaling, 16*, 1215–1228.

Sieber, F., Daziano, J. P., Gunther, W. H. H., Krieg, M., Miyagi, K., Sampson, R. W., et al. (2005). Elemental selenium generated by the photobleaching of seleno-merocyanine photosensitizers forms conjugates with serum macro-molecules that are toxic to tumor cells. *Phosphorus, Sulfur, and Silicon and the Related Elements, 180*, 647–657.

Sies, H. (2017). Hydrogen peroxide as a central redox signaling molecule in physiological oxidative stress: Oxidative eustress. *Redox Biology, 11*, 613–619.

Smirnov, A., Panatta, E., Lena, A., Castiglia, D., Di Daniele, N., Melino, G., et al. (2016). FOXM1 regulates proliferation, senescence and oxidative stress in keratinocytes and cancer cells. *Aging, 8*, 1384–1397.

Speckmann, B., Bidmon, H. J., Pinto, A., Anlauf, M., Sies, H., & Steinbrenner, H. (2011). Induction of glutathione peroxidase 4 expression during enterocytic cell differentiation. *The Journal of Biological Chemistry, 286*, 10764–10772.

Suzuki, S., Pitchakarn, P., Ogawa, K., Naiki-Ito, A., Chewonarin, T., Punfa, W., et al. (2013). Expression of glutathione peroxidase 2 is associated with not only early hepatocarcinogenesis but also late stage metastasis. *Toxicology, 311*, 115–123.

Szatrowski, T., & Nathan, C. (1991). Production of large amounts of hydrogen peroxide by human tumor cells. *Cancer Research, 51*, 794–798.

Takahashi-Yanaga, F., & Kahn, M. (2010). Targeting Wnt signaling: Can we safely eradicate cancer stem cells? *Clinical Cancer Research, 16*, 3153–3162.

Tanaka, G., Inoue, K., Shimizu, T., Akimoto, K., & Kubota, K. (2016). Dual pharmacological inhibition of glutathione and thioredoxin systems synergizes to kill colorectal carcinoma stem cells. *Cancer Medicine, 5*, 2544–2557.

Trachootham, D., Alexandre, J., & Huang, P. (2009). Targeting cancer cells by ROS-mediated mechanisms: A radical therapeutic approach? *Nature Reviews Drug Discovery, 8*, 579–591.

Vinceti, M., Dennert, G., Crespi, C. M., Zwahlen, M., Brinkman, M., Zeegers, M. P., et al. (2014). Selenium for preventing cancer. *The Cochrane Database of Systematic Reviews, 3*, CD005195.

Vinceti, M., Filippini, T., Cilloni, S., Bargellini, A., Vergoni, A., Tsatsakis, A., et al. (2017). Health risk assessment of environmental selenium: Emerging evidence and challenges. *Molecular Medicine Reports, 15*, 3323–3335.

Visvader, J., & Lindeman, G. (2008). Cancer stem cells in solid tumours: Accumulating evidence and unresolved questions. *Nature Reviews. Cancer, 8*, 755–768.

Visvader, J., & Lindeman, G. (2012). Cancer stem cells: Current status and evolving complexities. *Cell Stem Cell, 10*, 717–728.

Vunta, H., Davis, F., Palempalli, U. D., Bhat, D., Arner, R. J., Thompson, J. T., et al. (2007). The anti-inflammatory effects of selenium are mediated through 15-deoxy-Delta12,14-prostaglandin J2 in macrophages. *The Journal of Biological Chemistry*, *282*, 17964–17973.
Whanger, P. (2004). Selenium and its relationship to cancer: An update. *The British Journal of Nutrition*, *91*, 11–28.
Yoo, M., & Hatfield, D. (2008). The cancer stem cell theory: Is it correct? *Molecules and Cells*, *26*, 514–516.
Yoshida, G. J. (2016). Emerging role of epithelial-mesenchymal transition in hepatic cancer. *Journal of Experimental & Clinical Cancer Research*, *35*, 141.
Yoshida, G. J., & Saya, H. (2016). Therapeutic strategies targeting cancer stem cells. *Cancer Science*, *107*, 5–11.
Zaccarin, M., Bosello-Travain, V., Di Paolo, M. L., Falda, M., Maiorino, M., Miotto, G., et al. (2017). Redox status in a model of cancer stem cells. *Archives of Biochemistry and Biophysics*, *617*, 120–128.
Zeng, H. W., Briske-Anderson, M., Idso, J. P., & Hunt, C. D. (2006). The selenium metabolite methylselenol inhibits the migration and invasion potential of HT1080 tumor cells. *The Journal of Nutrition*, *136*, 1528–1532.
Zhang, Y., Zeng, S., Ma, J., Deng, G. L., Qu, Y. L., Guo, C., et al. (2016). Nestin overexpression in hepatocellular carcinoma associates with epithelial-mesenchymal transition and chemoresistance. *Journal of Experimental & Clinical Cancer Research*, *35*, 111.
Zhao, H., Li, J., Li, X., Han, C., Zhang, Y., Zheng, L., et al. (2015). Silencing GPX3 expression promotes tumor metastasis in human thyroid cancer. *Current Protein & Peptide Science*, *16*, 316–321.

CHAPTER TEN

Selenocompounds in Cancer Therapy: An Overview

Desirée Bartolini*[,2], Luca Sancineto*[,2], Andreza Fabro de Bem[†,‡], Kenneth D. Tew[§], Claudio Santi*, Rafael Radi[¶], Pierangelo Toquato*, Francesco Galli*[,1]

*University of Perugia, Perugia, Italy
[†]Center of Biological Sciences (CCB), Federal University of Santa Catarina (UFSC), Florianópolis, SC, Brazil
[‡]Institute of Biological Sciences, University of Brasilia, Brasilia, Brazil
[§]Medical University of South Carolina, Charleston, SC, United States
[¶]Center for Free Radical and Biomedical Research (CEINBIO), Universidad de la República, Montevideo, Uruguay
[1]Corresponding author: e-mail address: francesco.galli@unipg.it

Contents

[2] These authors equally contributed to prepare this manuscript.

Advances in Cancer Research, Volume 136
ISSN 0065-230X
http://dx.doi.org/10.1016/bs.acr.2017.07.007

Abstract

In vitro and in vivo experimental models clearly demonstrate the efficacy of Se compounds as anticancer agents, contingent upon chemical structures and concentrations of test molecules, as well as on the experimental model under investigation that together influence cellular availability of compounds, their molecular dynamics and mechanism of action. The latter includes direct and indirect redox effects on cellular targets by the activation and altered compartmentalization of molecular oxygen, and the interaction with protein thiols and Se proteins. As such, Se compounds interfere with the redox homeostasis and signaling of cancer cells to produce anticancer effects that include alterations in key regulatory elements of energy metabolism and cell cycle checkpoints that ultimately influence differentiation, proliferation, senescence, and death pathways.

Cys-containing proteins and Se proteins involved in the response to Se compounds as sensors and transducers of anticancer signals, i.e., the pharmacoproteome of Se compounds, are described and include critical elements in the different phases of cancer onset and progression from initiation and escape of immune surveillance to tumor growth, angiogenesis, and metastasis. The efficacy and mode of action on these compounds vary depending on the inorganic and organic form of Se used as either supplement or pharmacological agent. In this regard, differences in experimental/clinical protocols provide options for either chemoprevention or therapy in different human cancers.

1. INTRODUCTION

Selenium (Se) is an essential micronutrient with the proposed physiological role of redox gatekeeper, detoxification, and cytoprotection factor (reviewed in Bartolini et al., 2014; Papp, Lu, Holmgren, & Khanna, 2007). Such essentials depend on its incorporation into selenoproteins, which on the one hand drives Se toward its biological roles and in the other avoids nonspecific effects of cytotoxicity.

Overall the biological properties of Se are important since this element, once organized in selenoproteins, has the capability to react with molecular oxygen and thiols to promote a series of fundamental homeostatic processes in cells and tissues. Foremost is protection against a range of cellular and extracellular prooxidants (antioxidant/detoxification function) and then the tuning of signaling proteins and transcription factors that contain redox-sensitive Cys residues (redox buffering and signaling function). These Se-proteins include two main families of enzymes, the glutathione peroxidases (GPx) and thioredoxin reductases (TrxRd). Other selenoproteins of the 25 identified in the human genome are important in regulating cellular calcium metabolism, the estrogen receptor stress response, and iodine and thyroid hormone metabolism.

Preserving the functioning and redox tone of tissues these selenoproteins are believed to prevent oncogene activation and cancer cell differentiation perhaps through direct scavenging of DNA-damaging species or even causing the expression of anticancer genes that include DNA repair genes, oncogene suppressors, and genes responsible for the earliest phases of immune surveillance of tumor cells. Selenoproteins may also interfere with the tumor microenvironment influencing cancer progression through the activation of inflammatory and immune escape pathways, growth and vascularization of the tumor mass, and then of metastasis (described elsewhere in this volume). Selenoproteins mediate such effects through a series of molecular mechanisms many of which still remain poorly understood. These include scavenging of ROS, tumor-promoting eicosanoids (Bellinger, Raman, Reeves, & Berry, 2009; Rayman, 2008) and other inflammatory mediators that support cancer development and progression (Brigelius-Flohe, 2008; Squires & Berry, 2006).

Most selenoprotein functions are influenced by the cellular availability of Se, which highlights the importance of reaching an optimal intake of this element with the diet. Failing such nutritional goal may increase the risk of developing chronic and degenerative diseases (Rayman, 2008).

Different forms of Se have been investigated in the chemotherapy of human cancers, but the promising findings obtained in a series of preclinical studies and pilot interventions have not been confirmed in clinical trials recently evaluated in meta-analysis studies and presented elsewhere in this volume. Rather, some of these trials have reported adverse effects that include an increased risk of diabetes and also of some forms of cancer. These findings clearly demonstrate the limits of the forms of Se investigated so far as chemotherapy agents; these include both inorganic and organic molecules with Se amino acids and sodium selenite as the most studied.

In the last few years, efforts have been made to design and develop a number of newer Se organic molecules (recently reviewed in Fernandes & Gandin, 2015; Santi, Tidei, Scalera, Piroddi, & Galli, 2013), hence expanding the opportunity to identify more efficient and cancer-specific chemoprevention and chemotherapy compounds that could help to mitigate and even overcome negative outcomes of past trials.

These aspects will be the focus of the material in this chapter.

2. SELENIUMO-CONTAINING MOLECULES IN CANCER THERAPY

2.1 Inorganic Selenium Species

Most studied inorganic Se compounds for cancer include, selenium dioxide (**1**), selenites in the form of sodium selenite (**2**), and sodium selenate (**3**) (Fig. 1).

In the first two compounds (**1, 2**) the selenium atom is in the oxidation state +IV, while in compound **3** it is +VI.

These forms of Se were demonstrated to possess anticancer activity in a series of in vitro studies (Fernandes & Gandin, 2015; Freitas, Alves, Sarmento-Ribeiro, & Mota-Pinto, 2011; Rudolf, Radocha, Cervinka, & Cerman, 2004). Compound **2** was studied alone or in combination with commercial anticancer agents. A synergistic effect with the drugs camptothecin and docetaxel was demonstrated in cervical and prostate cancer cell lines, respectively. Compound **2** was also investigated in vivo confirming its efficacy in solid tumors (Luo et al., 2013) and lymphoproliferative cancer models (An et al., 2013; Shi et al., 2013). Unfortunately, the clinical exploitability of sodium selenite is limited by its genotoxicity demonstrated in yeast (Anjaria & Madhvanath, 1988).

2.2 Selenium-Containing Amino Acids

Selenocysteine (SeCys), selenocystine (**4**), selenomethionine (SeMet) (**5**), and methylselenocysteine (**6**) are naturally occurring selenoaminoacids (Fig. 2). From a biochemical point of view, their synthesis is considerably more complex and energy demanding than that of the corresponding sulfur amino acids. This aspect suggests some evolutionary advantage in positioning Se in the place of S.

SeCys is considered the 21st amino acid; it is present in the 25 selenoproteins so far identified in human tissues with homeostatic effects on cellular redox and stress response pathways. Glutathione peroxidases

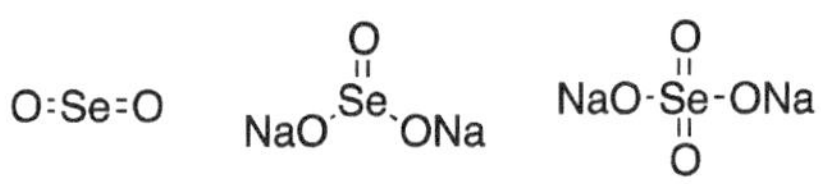

1 2 3

Fig. 1 Structures of compounds **1–3**.

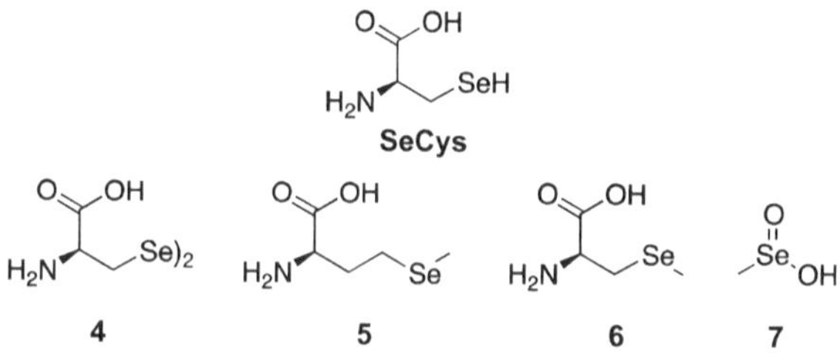

Fig. 2 Structures of compounds **4–7**.

are among the most studied selenoproteins with isoforms that specifically scavenge cellular peroxides. As with most of the selenol species, the corresponding form of SeCys is highly unstable and difficult to isolate in high purity for biochemical evaluation. On the other hand, Chen and Wong have studied the most stable dimeric form **4** in vivo in human melanoma implanted as a subcutaneous xenograft in mice (Chen & Wong, 2008). They demonstrated the efficacy of **4** in lowering tumor growth without effects on body weight. Other in vitro studies revealed that selenocystine (**4**) is endowed with antiproliferative activity in melanoma and cervical cancer cells (Chen & Wong, 2009a, 2009b; Wallenberg et al., 2014). Selectivity studies demonstrated that this effect of compound **4** on cellular proliferation discriminates between transformed and normal cells (Chen & Wong, 2009a, 2009b).

Selenomethionine (**5**) and methylselenocysteine (MeSeCys, **6**) tested in human breast cancer cells demonstrated antiproliferative activity (Redman et al., 1998; Sinha & Medina, 1997). In vivo studies in human colorectal and breast carcinoma as well as in head and neck squamous carcinoma, revealed the antiangiogenic properties of compound **6** by its influence on the tumor microenvironment. This effect on tumor vascularization was found to synergize with antineoplastic drugs such as irinotecan, cisplatin, and tamoxifen (Bhattacharya, 2011; Chintala et al., 2010).

2.3 Methylseleninic Acid

Methylseleninic acid (**7**) (Fig. 2) originates from the oxidative decomposition of methylselenocysteine (**6**); this is considered among the simplest organic selenium compounds with pharmacological properties. It has been tested in cellular models of prostate (Li, Hu, Jiang, Schuster, & Lu, 2007), breast (Jariwalla, Gangapurkar, & Nakamura, 2009), and lung cancer (Tarrado-Castellarnau et al., 2015); in the latter study, A549 human lung carcinoma cells responded to compound **7** that lowered intracellular

ROS levels in a dose-dependent manner thus leading to growth inhibition cell cycle arrest in G1 phase (Tarrado-Castellarnau et al., 2015), a condition that predisposes the tumor cell to senescence and even death pathway activation (reviewed in Pierpaoli et al., 2013).

Compound **7** was tested in vivo in mouse xenograft models of prostate (Lee et al., 2006; Wang et al., 2009) and colon cancer (Zeng & Wu, 2015), and in all these studies it reduced tumor growth without inducing animal weight loss or other signs of systemic toxicity or genotoxicity. Proapoptotic effects by caspase 3 activation and modulation of proinflammatory cytokines were identified as underlying events in the antiproliferative activity of this compound in the colon cancer study (Zeng & Wu, 2015). In prostate cancer compound **7** (similar to selenite) may act as an inhibitor of androgen receptor (AR) signaling. However, the AR suppression mechanism of methylseleninic acid differs from that of selenite (Husbeck, Bhattacharyya, Feldman, & Knox, 2006). Selenoprotein expression, namely Sp1, appears to play a role in this receptor suppression effect. Compound **7** has also been found to inhibit the activity of the alpha form of ER, a receptor isoform with broad-reaching signaling relevance (Shah, Kaul, Dong, Ip, & Rowan, 2005).

Chemopreventive effects of compound **7** were observed in vitro in esophageal squamous cell carcinoma (ESCC), by the interference with kelch-like ECH-associated protein 1 (Keap1), a redox-sensitive inhibitor of nuclear factor E2-related factor 2 (Nrf2) that consequently undergoes its stabilization as free form in the cytosol and nuclear translocation to promote the expression of cytoprotection and redox-signaling genes (Liu, Hu, et al., 2015). The chemopreventive role of **7** was then confirmed in vivo in a mouse model of prostate cancer in which tumor weight significantly decreased during the 25-week administration protocol, in association with decreased Akt phosphorylation and p53-mediated activation of cancer cell senescence (Wang et al., 2016).

2.4 Diselenides

Diselenides are organoseleno compounds characterized by the presence of a Se–Se bond largely used in organic synthesis because of their accessibility through a number of different synthetic protocols and relative stability.

Diselenides can be considered as convenient precursors for electrophilic, nucleophilic, and radical selenium reagents. In addition, aromatic diselenides are used as precatalysts in oxidation reactions being converted in situ into seleninic and perseleninic acid, which represent the actual oxygen transfer species. Given the availability of a number of diselenide molecules, a wealth of studies on anticancer properties of these compounds have recently appeared in the literature (reviewed in Fernandes & Gandin, 2015). For that reason, only some examples selected among the most relevant and recent ones will be reported and discussed in this chapter. The antiproliferative activity of a series of bis(aryl and aralkyl) diselenides has been tested in MCF-7 cells, a cellular model of breast cancer, and among the whole sets of derivatives, 1,2-bis(chloropyridazinyl) diselenide (compound **8**, Fig. 3) emerged as the strongest cell growth inhibitor with significant activity obtained at low micromolar concentrations (Kim, Lee, & Park, 2015).

Wessjohann and coworkers tested a series of diselenides synthesized in their laboratories (Shaaban, Negm, Sobh, & Wessjohann, 2015). Besides antimicrobial and antioxidant properties, the authors investigated the cytotoxic effects of these compounds in these cancer cells identifying the diselenide compound **9** as the most potent form with activity in the low micromolar range. The free radical-scavenging and GPx-like (thiol peroxidase) activity of such molecules were comparable to the reporter compounds ascorbic acid and Ebselen (discussed later in this chapter), respectively.

Worthy of note, Park and coworkers recently included the diselenide moiety of this compound as a functional and redox-sensitive epitope in a polymer used to formulate micelles that selectively target and release the anticancer drug doxorubicin (DOX) in cancer tissue (Deepagan et al., 2016; Fig. 4). The diselenide cross link responsible for the hydrophobic character of this polymeric material constitutes a physical barrier for the drug contained in the micellar core in nontumoral tissues; at the same time, it represents the cancer-targeting moiety of this formulation. In fact, the high flux of ROS by the tumor metabolism selectively promotes the oxidative

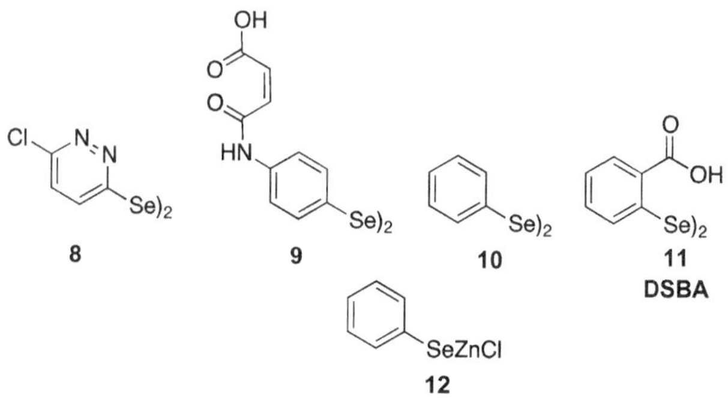

Fig. 3 Structures of diselenide derivatives (**8–12**).

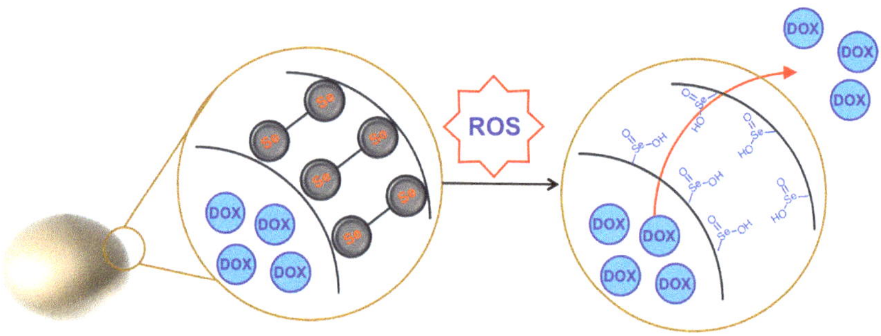

Fig. 4 Drug delivery system based on a micellar polymer containing the diselenide moiety of compound **9**. This redox-sensitive drug delivery system has been used to selectively target and release the anticancer drug doxorubicin in the prooxidant environment of cancer cells (Deepagan et al., 2016).

cleavage of the selenium–selenium bond in the hydrophobic polymer to form the corresponding hydrophilic derivative selenic acid, then releasing DOX in the tumor microenvironment. This example of a diselenide-mediated strategy of drug delivery allows the drug to be released and perform its pharmacological action locally in the tumor masses, provided that these have a sustained flux of oxidants, then reducing systemic and toxicological drawbacks of chemotherapy.

A small series of diselenides were recently designed and synthesized by some of us introducing modifications in the aryl moiety of the lead compound diphenyldiselenide (compound **10**, Fig. 3) with the intent of exploring their influence on the redox cycling activity of the Se–Se functional group (Fig. 5; Bartolini, Commodi, et al., 2015). Among this series of diselenides, DSBA (**11**) was observed to produce the lowest redox cycling (peroxidase) effect on cellular thiols and then very low levels of cytotoxicity; this molecule was then suitable to explore application as a selenohormetic agent. Selenohormetines are chemoprevention compounds endowed with the capability to train cells to reach higher levels of resistance against stressors by the stimulation of a "positive hormesis" effect that occurs by the (tolerable) stimulation of stress response pathways and induction of detoxification and antioxidant genes; these include the entire group of Nrf2-dependent genes responsible for the metabolism and redox function of GSH and Trx (Bartolini & Galli, 2016). The effect of diselenides with different degrees of reactivity with cellular thiols on this pathway has been investigated in our laboratories (Bartolini, Commodi, et al., 2015; Bartolini, Piroddi, et al., 2015). Highly reactive and cytotoxic compounds, such as compound

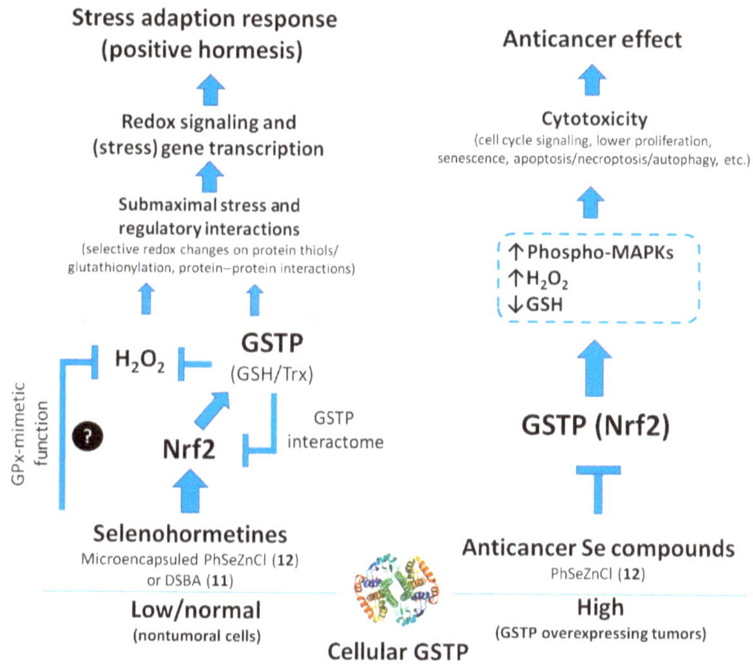

Fig. 5 Proposed effects of highly and mildly reactive diselenide formulations used to target GSTP in chemotherapy and chemoprevention protocols. *Modified from Bartolini, D., Piroddi, M., Tidei, C., Giovagnoli, S., Pietrella, D., Manevich, Y., et al. (2015). Reaction kinetics and targeting to cellular glutathione S-transferase of the glutathione peroxidase mimetic PhSeZnCl and its* D,L-*polylactide microparticle formulation.* Free Radical Biology & Medicine, *78, 56–65.*

10 $(PhSe)_2$ and **12** (PhSeZnCl), not only modify the GSH/GSSG ratio of cancer cells, but markedly depleted these cells of total glutathione (GSH plus GSSG), which suggests a leakage or export mechanism for this tripeptide or a lowered capability to maintain its de novo biosynthesis by transcriptional/functional control of glutamate cysteine ligase, formerly referred as to γ-glutamyl cysteine synthetase, and cysteine–cystine transport and cellular redox (further discussed in Section 3). These effects of highly reactive Se compounds also influenced protein Cys redox leading to a lowered PSH/PSSP ratio and increased glutathionylation (Bartolini, D., & Galli, F., unpublished observation); the latter is a posttranslational modification of protein Cys and SeCys residues with emerging role in the pathobiology of cancer (Grek, Zhang, Manevich, Townsend, & Tew, 2013; Tew et al., 2011). Glutathione S-transferases (GST) were among the first Cys-containing enzyme proteins identified to undergo functional inhibition by the interaction with the prototypal Se organic compound Ebselen (compound **31**, discussed later in this section). The isoform P, extensively investigated in our laboratories, has recently been demonstrated to represent an elective target of diselenide compounds (Bartolini, Commodi, et al., 2015; Bartolini, Piroddi, et al., 2015), and its essential role in selectively promoting glutathionylation of cellular proteins (Grek et al., 2013; Ye, Zhang, et al., 2017) provides a specific mechanism of action for these Se organic molecules which is worth further investigation to develop more efficient and selective applications of Se in chemoprevention protocols and cancer therapy.

Besides modifying the structure of diselenide compounds, the thiol peroxidase function of very reactive components in this group of molecules, such as PhSeZnCl (compound **12**, Fig. 3), can be finely tuned by means of formulation strategies that include, for example, microencapsulation in polylactide polymers (Bartolini, Piroddi, et al., 2015). With this strategy we demonstrated how the same Se compound can be used to selectively target GSTP protein of cancer cells to alternatively activate a positive (Se hormetic) stress response or apoptotic-promoting anticancer effects (Fig. 5). The possibility of using **11** as a Selenohormetine and chemoprevention agent is now under preclinical evaluation by some of us (KT's Lab).

As further discussed in Bartolini and Galli (2016) and in Section 4, targeting GSTP with these Se compounds may influence the cancer process at different levels. In fact, multiple regulatory protein–protein interactions are produced by this GST isoform (the GSTP interactome) and may justify a role as: (1) redox-sensitive regulation node and chaperone protein of Nrf2 that could intervene in the resolution and negative feedback of the stress adaption response (Bartolini, Commodi, et al., 2015), (2) glutathionylation effector of cellular proteins such as the 1-Cys protein with membrane peroxidase activity Prxd6 and several ER proteins with proven role in the ER stress response (Ye, Zhang, et al., 2017), as well as (3) physical inhibitor of stress-activated protein kinases such as TRAF2/ASK1 and JNK kinases and transcription factors, such as STAT3, with roles in the inflammatory and angiogenic responses of the tumor microenvironment (reviewed in Bartolini & Galli, 2016).

2.5 Selenocyanates and Isoselenocyanates

Starting in 2001, when the first compound of this class was tested in human oral carcinomas (Ghose, Fleming, El-Bayoumy, & Harrison, 2001), several selenocyanate and isoselenocyanate derivatives have been reported as

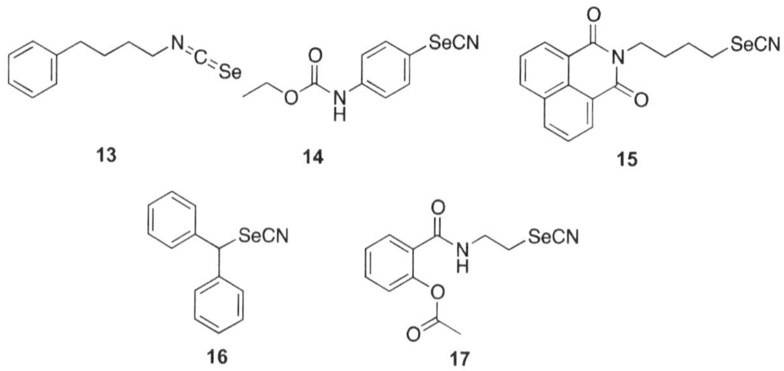

Fig. 6 Structure of isoselenocyanate (13) and selenocyanates (14–17).

antiproliferative agents. In 2008, Sharma et al. (2008) investigated a small series of arylalkyl isoselenocyanate-containing compounds on melanoma, prostate, breast, glioblastoma, sarcoma, and colon cancer cell lines, and in vivo experiments clearly established that the isoselenocyanate compound **13** (Fig. 6) effectively inhibited melanoma tumor growth in mice following intraperitoneal administration. Noteworthy, the corresponding sulfur analogue showed weaker antiproliferative activity.

A series of diselenides and selenocyanates were synthesized and investigated in vitro for cytotoxic and radical-scavenging properties (Romano, Plano, Encio, Palop, & Sanmartin, 2015). These compounds were tested in a panel of human cancerous cell lines including lymphoblastic leukaemia (CCRF-CEM), colon carcinoma (HT-29), lung carcinoma (HTB-54), prostate carcinoma (PC-3), breast adenocarcinoma (MCF-7), and others. Compound **14** (Fig. 6) provided the highest efficacy in growth inhibition with GI_{50} (the concentration that reduces of 50% the growth of treated cells) of 8.9 μM. In addition, the compound showed a total growth inhibition (the concentration that completely inhibits cell growth), and a lethal concentration LC_{50} (the concentration that kills 50% of the cells) of 36.7 and 67.8 μM, respectively.

Selenocyanate compounds **15** and **16** (Fig. 6) were investigated by Bhattacharya and coworkers (Chakraborty, Roy, Basu, & Bhattacharya, 2016; Ghosh, Bhattacharjee, Basu, Singha Roy, & Bhattacharya, 2015). The naphthalene derivative **15** decreased the pulmonary toxicity of cyclophosphamide when administered intraperitoneally to albino mice and was proven to improve cisplatin activity concomitantly constraining its ROS generation effects; this was believed to explain the observed in vivo restoration effect on reduced glutathione levels (Ghosh, Singha Roy, Basu, Bhattacharjee, & Bhattacharya, 2015). Compound **16** was tested in vitro alone and in combination with cyclophosphamide in a cellular model of mammary adenocarcinoma. A strong antiproliferative activity was shown, and an increased ROS generation was hypothesized as the causal event underlying the cytotoxic effects of this Se-containing molecule in cancer cells (Chakraborty et al., 2016).

A series of derivatives have been obtained by the introduction of a selenocyanate moiety into various antiinflammatory drugs (Plano et al., 2016). Among these, the acetylsalicylic-derived selenocyanate **17** (Fig. 6) was the most effective compound in reducing the viability of different cancer cell lines and particularly of colorectal cancer cells. That effect was 10-fold and more potent than the effect of 5-fluorouracil (5-FU), the drug currently used in the therapy of that type of human cancer, and was demonstrated to depend on cell growth inhibition by the p21-dependent blockade of cell cycle in G1 and G2/M.

2.6 Selenoureas

Among all the selenium-containing organic compounds selenoureas were poorly explored from a pharmacological point of view and only recently the interest on these molecules grew by the observation of interesting biological properties in comparison with corresponding sulfur analogues (Hussain, Badshah, Tahir, Tamoorul, & Bano, 2014).

In this context, Sharma and coworkers (Alcolea et al., 2016) produced a series of selenourea derivatives together with the corresponding thiourea analogues, thus assessing their antiproliferative properties in a panel of human cancer cell lines. This study demonstrated higher potency of selenium-containing derivatives with compound **18** (Fig. 7) endowed with IC_{50} values below 10 μM in all cell lines tested, and then it was selected for further biological evaluation in different colon cancer cell models. Such antiproliferative activity appears to depend on proapoptotic

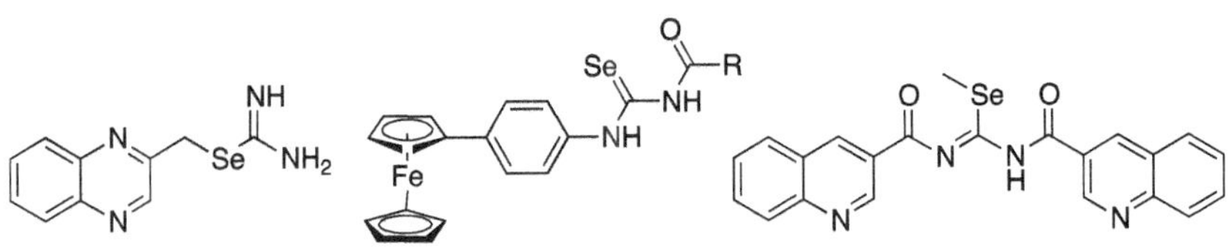

18 **19** **20**

Fig. 7 Structure of selenoureas (**19–20**).

effects of **18** by the activation of caspase-dependent pathways and inhibition of antiapoptotic proteins.

Seventeen ferrocenylselenoureas (among them compound **19**, Fig. 7) have been prepared and studied for their DNA binding properties, antioxidant activity, and cytotoxicity in neuroblastoma (MYCN2 and SK-N-SH cells), liver hepatocarcinoma (Hepa1c1c7), and breast cancer (MCF-7) cells. The majority of compounds killed up to 90% of cancer cells when used at low micromolar concentrations (Hussain et al., 2015).

Compound **20** (Fig. 7) is the first, and to the best of our knowledge the only, selenourea derivative to be tested in vivo. In a prostate cancer mouse xenograft model this compound suppressed tumor growth by 80% without causing untoward effects (Ibanez et al., 2012).

2.7 Selenocarbonyl Derivatives

The chemical transformation of a carbonyl-containing compound into the corresponding selone analogue was facilitated after the introduction of the Woollins reagent in the late 1990s; from this milestone in organic chemistry, several selone derivatives were prepared and their biological activities tested in different models. Examples of compounds bearing such functional groups (**21**–**23**) are reported in Fig. 8.

The selone analogue of chrysin (compound **21**) and of tetramethylquercetin (compound **22**) has recently been synthesized (Martins et al., 2015). When compared to their oxo and thio analogues, these compounds showed improved antioxidant activity and cytotoxicity. In particular, the IC_{50} of compound **21** in MCF-7 cells was lowered by 18 and 3 folds compared to chrysin and cisplatin, respectively. Additionally, these selenoderivatives were able to overcome cisplatin and multidrug resistance of cancer cells.

Selenohydantoin (**23**) was evaluated in vitro against a panel of cancer cell lines showing excellent (micromolar) antiproliferative activities coupled with good selectivity indexes (Ivanenkov et al., 2016); such antiproliferative

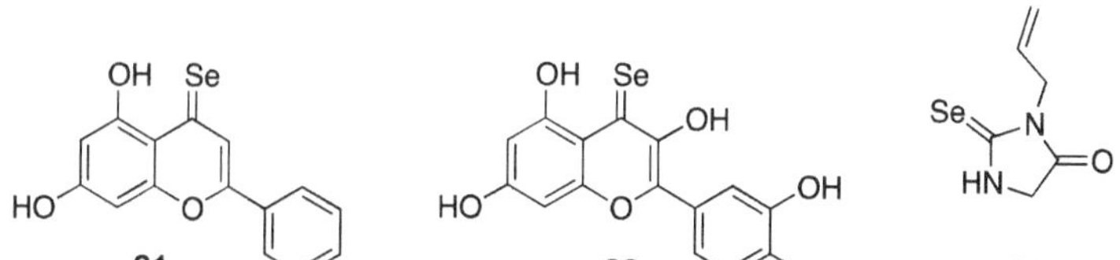

21 **22** **23**

Fig. 8 Selenocarbonyl derivatives (**21–23**).

effects were higher than the corresponding thiohydantoin analogues highlighting the importance of the selenium atom for the biological activity of this compound.

2.8 Selenoxides

Selenoxides are useful reagents in the synthesis of alkenes from alkanes because of the facility with which the functional group undergoes elimination reactions. When substituted with two aryl or etheroaryl groups the selenoxide is stable enough to allow biochemical evaluation. As an example, Zhang and coworkers reported the development of selenoxides through a scaffold-hopping approach starting from combretastatin A4 (compounds **24** in Fig. 9). Most of the compounds obtained from this synthetic expedient showed significant antiproliferative activity, some of which showed

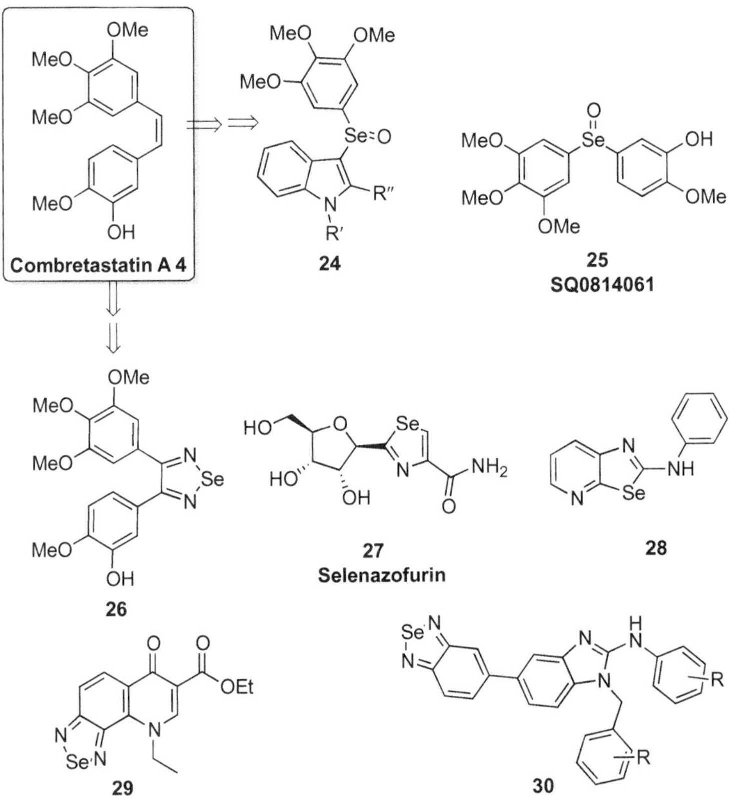

Fig. 9 Structure of selenoxides and selenazoles (**24–30**).

nanomolar IC_{50} values. Docking experiments further elucidated the action mechanism of these compounds that, similar to combretastatin A4, could bind to colchicine binding site on tubulin thus interfering with microtubule assembly (Wen et al., 2015).

The same group developed a selenodiazole derivative of combretastatin A4 (compound **26**, Fig. 9) that produced dose-dependent inhibition effects on cell growth in both HeLa and human fibrosarcoma HT-1080 cells (Zuo et al., 2015).

Microtubule assembly was reported to represent the target for another selenoxide derivative (compound **25**) studied by Wu and collaborators (Xu et al., 2016). Formerly known as SQ0814061, compound **25** promotes microtubule depolymerization in association with a strong effect of cell growth inhibition that was demonstrated in MCF-7 and MDA-MB-231 breast cancer cells by the arrest of cell cycle progression in G2/M phase and subsequent activation of apoptotic death pathways.

2.9 Selenazoles and Selenadiazoles

Selenazoles are five-membered rings characterized by the presence of selenium spaced one methylene from a nitrogen. The most studied compound belonging to this class of molecules is selenazofurin (compound **27**, Fig. 9) which was developed in 1983 and then studied in vivo as an anticancer agent in prostate carcinoma (Srivastava, Revankar, & Robins, 1984). Due to its toxicity, the drug development process was discontinued soon after. Nevertheless, that remains an important step in the field of anticancer organoselenium compounds and a stimulus for the development and investigation of new selenium-containing heterocycles.

Recently, Chen and coworkers synthetized a series of selenazole derivatives to be assessed as potential anticancer agents (Zhou et al., 2015). Among these, 2-(phenylamino)-selenazole[5,4-*b*]-pyridine (compound **28**, Fig. 9) showed the higher growth inhibition and proapoptotic activity in MCF-7 human breast carcinoma cells.

In HeLa cells an increased ROS generation was assessed as a mechanism behind the antiproliferative activity observed for the quinolone derivative **29** (Fig. 9; Jantova et al., 2016).

Compound **30** (Fig. 9) is interesting in this context since it was found to improve DOX activity in a hepatocellular carcinoma model mimicking diabetes-cancer comorbidity (Liu, Luo, Li, Zheng, & Chen, 2015).

2.10 Isoselenazolones

Among all selenoorganic compounds, Ebselen (compound **31**, also known as PZ 51 or SP1005, Fig. 10) has been the most investigated and continues to represent a reference molecule widely used to comparatively assess the pharmacological properties of other Se-compounds. Synthesized by Lesser & Weiss (1924) and considered pharmacologically irrelevant for 60 years, it developed new life when Sies and colleagues in 1984 highlighted the capability of this molecule to mimic the activity of glutathione peroxidase enzyme (Muller, Cadenas, Graf, & Sies, 1984). From there, a plethora of studies investigated the pharmacological and biochemical properties of Ebselen (reviewed in Azad & Tomar, 2014) suggesting applications that among the most recently reported include those of antimalarial (Harris et al., 2013), antitubercular (Favrot et al., 2013), and neuroprotection (Koizumi, Fujisawa, Suehiro, Shirao, & Suzuki, 2011) agent. The different

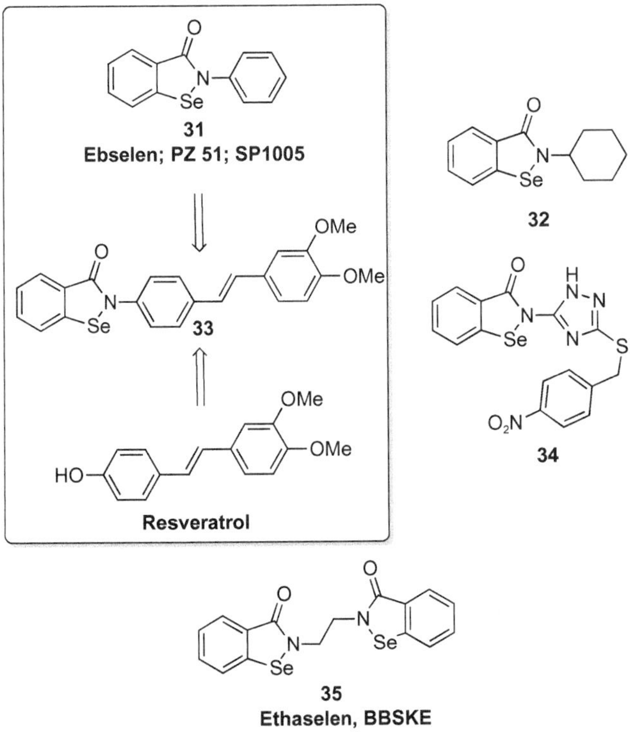

Fig. 10 Structures of Ebselen and Ebselen-inspired structures (**31–35**).

therapeutic indications of Ebselen (**31**) have been investigated in clinical trials that failed to conclusively demonstrate its efficacy.

Ebselen was also tested in vivo in animal models of human breast carcinoma (Engman et al., 1997). In these studies Ebselen was found to be a competitive inhibitor of human TrxRd (K_i 2.8 μM), thereby leading to develop a defective NADPH-dependent reduction of Trx. Thioredoxin is a redox-active protein overexpressed and actively involved in the abnormal redox that characterizes the development of several human cancers. In the studies of Engman et al. Ebselen was used as a reference compound to assess the activity of a series of organotellurium compounds that were demonstrated to behave as noncompetitive inhibitors (K_i from 2.3 to 35.2 μM).

High-throughput screening experiments carried out by Lake and coworkers, identified Ebselen as an inhibitor of quiescin sulfhydryl oxidase 1 (QSOX1), which is a redox enzyme overexpressed in several tumors. In pancreatic and renal cancer cell lines Ebselen stalled cell growth, and its oral administration to mice bearing human pancreatic tumor xenografts resulted in a 58% reduction in tumor growth (Hanavan et al., 2015). Ebselen was found to be a covalent inhibitor of QSOX1, perhaps to be expected since it reacts with cysteine residues of that enzyme protein. The former Cys-containing enzyme to be identified in cellular tests and in vivo as a target of Ebselen inhibitory activity was GST (Nikawa, Schuch, Wagner, & Sies, 1994a, 1994b), which is now confirmed to represent a preferential redox-sensitive node for mediating the pharmacological action of this and other Se organic compounds with high redox cycling activity on cellular thiols, such as compounds **10** and **12** (Bartolini, Piroddi, et al., 2015).

Chemical structure of Ebselen (**31**) was, and continues to be, a source of inspiration among the chemists who practice with selenoorganic compounds for biological purposes. For example, Schianowski and coworkers synthesized a series of alkyl derivatives and among these the cyclohexyl-containing compound **32** (Fig. 10) displayed superior redox properties also reducing cell growth in both MCF-7 and DU145 cancer cell lines with IC_{50}values that were in the low micromolar range (Pacula et al., 2017).

Other examples of this class of compounds include the combination of the pharmacophores of resveratrol and **31** (Fig. 10) designed to develop a series of multitarget redox-active agents, such as compound **33**. Resveratrol and **31** are antiproliferative agents, and this appears to represent an important anticancer property for the hybrid derivatives. Indeed, biological evaluations indicated that all the developed test compounds behaved as antiproliferative agents in the human cancer cell lines Bel-7402, A549, HeLa, and MCF-7. The test compounds were also demonstrated to be TrxRd inhibitors. Moreover, mechanistic studies clearly revealed that compound **33** increases intracellular ROS implicating this as a cause of apoptosis in cancer cells (Yan et al., 2015).

Chen and colleagues incorporated the scaffold of compound **31** in a series of 1,2,4-triazole derivatives that were assessed in human cancer cell lines SMMC-7721, HeLa, and A549. Among the compounds, derivative **34** is worthy of mention since it exhibited higher antiproliferative capacity than Ethaselen (compound **35**) and 5-FU (Li et al., 2016). Among all of the selenoorganic compounds, **35** is the most studied as an anticancer agent and currently (march 2017) is in phase I clinical trial for the treatment of TrxRd-overexpressing advanced nonsmall cell lung cancer (NSLCLC) (Wang et al., 2012). TrxRd is a component of several redox-sensitive pathways that mediate important biological processes, such as cell survival, growth, migration, and inhibition of apoptosis, though its overexpression is linked to various tumors via an unknown mechanism (Saccoccia et al., 2014). The mechanism by which compound **35** inhibits TrxRd and how this inhibition translates into an antiproliferative effect was studied by Zeng and coworkers using a NSLCLC xenograft mouse model (Ye, Li, Ji, Zeng, & Lu, 2017). In vitro experiments proved that this compound can reverse the malignant phenotype of MCF-7 cells (Dong et al., 2016). Synergistic anticancer effects were also observed in colorectal cancer cell models when **35** was coadministrated with the tyrosine kinase inhibitor Sunitinib (Sutent) (Zheng et al., 2016).

3. SELENIUM COMPOUNDS' MECHANISM OF ACTION IN TUMOR CELLS

3.1 The Selective Uptake of Se in Cancer Cells

A selective uptake of Se is observed in cancer cells and this could provide a tumor-specific mechanism of action for at least some Se compounds. The first evidence of a specific uptake dates back to the 1960s when ^{75}Se sodiumselenite and ^{75}Se-SeMet were used as tracers in cancer diagnostics; ^{75}Se was employed as a tumor radiotracer since it guarantees a high precision in localizing intracranial, thoracic, and abdominal neoplasms (Cavalieri & Scott, 1968; Cavalieri, Scott, & Sairenji, 1966). However, Se uptake in cancer is not completely understood and differs between Se compounds. Selenide has been proposed to be transported by ATPases (Ganyc & Self, 2008), whereas selenite uptake can be mediated by anion transporters (Galanter, Hakimian, & Labotka, 1993). The entrance of selenite into various cell lines is favored by the presence of reduced thiols in the extracellular medium (Olm et al., 2009), indicating that the redox state of Se is important in this context and possibly that the formation of Se–S adducts can facilitate its entrance into the cell. Se–S adducts may simulate cystine and mixed disulfides in their uptake mechanisms. The cystine/glutamate antiport system (xCT or X_c^-) that cooperates with TrxRd to control redox of both the extracellular and intracellular environment (Mandal et al., 2010) may have a major role in this context. Se accumulation in the tumor could be partly explained by the fact that in many cancers xCT is overexpressed (Lewerenz et al., 2013). This system helps produce a more reducing extracellular environment so promoting the absorption of Se in the (reduced) form of selenide (Olm et al., 2009). Alternatively, similar to mixed disulfides (reviewed in Banjac et al., 2008), the formation of extracellular S–Se adducts could directly promote Se uptake via the transport system for bulky amino acids and then releasing cysteine and selenite by the reducing conditions of the intracellular environment. This latter mechanism remains to be experimentally verified.

The fact that the cellular availability of Se and its capability to react with intracellular thiols depend on Cys availability and then on the cysteine/cystine redox cycle itself (Olm et al., 2009), highlights the important interaction between Se and GSH/Cys metabolism of cells. Cys uptake is the (substrate)-limiting step in GSH biosynthesis and as such it influences the cellular redox and the activity of redox-sensitive signali transduction pathways that regulate the proliferation rate and apoptotic susceptibility of cancer cells; other key aspects of tumor biology can be affected with the same mechanism, such as the cross talk between cancer cells and immune and vascular cells in the tumor microenvironment (reviewed in Banjac et al., 2008; Mandal et al., 2010). Drug resistance could also be under the direct influence of Cys availability cancers that overexpress GSH and its detoxification and antioxidant enzymes (Tew, 2016). Therefore, Cys and GSH metabolism may represent, at the same time, the Trojan horse and the final target for Se to enter and selectively interfere with tumor metabolism.

Interspecies analysis of selenite-treated lung cancer cells revealed that a significant proportion of Se bound to protein, while the remainder remains unchanged (Lunoe et al., 2011). Although the nature of the Se bonding within the protein fraction was not explored, a preferential reaction with protein Cys residues conceivably occurs at least at pharmacological doses. Experiments in rat liver and cell models demonstrate that the glutathione/TrxRd/glutaredoxin system can reduce inorganic selenite to selenodiglutathione (GSSeSG) that is further reduced via a selenopersulfide (GSSeH) to hydrogen selenide (HSe^-) by the same systems (reviewed in Weekley et al., 2013), and therefore, this cellular reduction system may help prevent protein Cys to react with inorganic Se under normal (nonpharmacological) conditions and may hinder Se and other prooxidant agents from producing selective effects.

HSe^- also is a product of Se amino acid metabolism; it is formed during SeCys metabolism by the activity of β-lyase on SeMet (**5**; Fig. 2) and may occur after conversion to SeCys via the transsulfuration pathway. Alternatively, **5** can substitute Met into proteins, and be metabolized by γ-lyase to $MeSe^-$, but the meaning of this route in cancer cells remains uncertain. In fact, compound **5** is poorly metabolized and does not react with cytosolic proteins when administered to cancer cells, such as PC-3, HT-29, and Jurkat E6-1 cells (Lunoe et al., 2011). On the contrary, compound **7** (Fig. 2) was found to be completely transformed to other Se metabolites by cellular incubation that also included the production of the volatile species dimethyl diselenide. Opposite to **5**, compound **7** showed noticeable protein binding, a response that was also observed for other compounds, such as selenite (**2**; Fig. 1) and Se methylselenocysteine (**6**; Fig. 2); the latter is another Se amino acid with metabolic pathway distinct from that of SeMet. Compound **7** is often used in place of MeSeCys in cell culture experiments. In fact, in cancer cells besides its protein interaction properties, it was found to generate compound **6**, γ-glutamyl-MeSeCys and DMeDSe, a $MeSe^-$ metabolite derivative, or eventually to be converted to SeMet (**5**) by a currently unknown metabolic pathway (Infante et al., 2007).

Although, a correlation between the levels of cell death and transformation to Se metabolites (along with elemental Se these include, R–Se–R, R–Se–Se–R, and S–Se–S components) or protein binding was not demonstrated for these Se compounds (reviewed elsewhere Weekley et al., 2013), it appears probable that metabolite formation and protein interactions are critical aspects in assessing the anticancer activity of such forms of Se, the most commonly used in dietary interventions (namely, **2**, **6**, and **5**). In this respect, further efforts in developing more efficient metabolite and proteomic strategies are needed to identify and interpret mechanistic and pharmacological aspects of Se therapy. This might beneficially be extended to in vivo models and possibly to clinical trials that clearly suffer from the lack of biomarkers of appropriateness and efficacy of therapy protocols. Blood cell and fingernail Se levels can be assessed as indicators of Se status along with selenoprotein synthesis and the excretion of $TMSe^+$ and selenosugars in urine (Tsuji, Suzuki, Suzuki, & Ogra, 2009), but these indicators do not provide exhaustive information on Se metabolism, pharmacodynamics, and mechanisms of action. In this context, it is important to highlight that animal studies demonstrated how the accumulation and metabolism of these compounds in some organs, such as liver, kidney and testis, can represent limiting factors in their in vivo bioavailability and efficacy in cancer therapy (Tsuji et al., 2009).

Protein binding of Se compounds also occurs at the extracellular level, and this may further help in developing selective drug delivery strategies for some tumors, such as leukaemia. Studies on photoproducts of merocyanine analogues containing Se, such as selenomerocyanine dyes, showed potent cytotoxic activity against leukemia and different solid tumors, while being well tolerated by normal CD34-positive hematopoietic stem and progenitor cells (Sieber et al., 2005); this effect was reported to be the consequence of protein conjugate formation by elemental selenium generated during selenomerocyanine photobleaching with serum albumin and lipoproteins, a response that was recently confirmed for other merocyanine analogues (Daziano et al., 2012). These conjugates are selectively cytotoxic to cancer cells because of selective uptake, which thereby act as Trojan horses serving to deliver elemental selenium at cytotoxic levels.

3.2 The Redox-Active Character of Organoselenium Compounds Relevant to Their Anticancer Properties

Cancer cells are characterized by increased aerobic glycolysis (termed the Warburg effect) and usually exhibit high basal levels of reactive species (which includes molecular oxygen, nitric oxide, Cl-derived species, and other reactive metabolites) balanced by a concomitant upregulation of antioxidant defense systems. Although such compensatory responses in cancer cells favor malignancy, conversely its existence suggests a limited capacity to adapt to oxidative stress or alternatively provide a plausible window for therapeutic intervention with anticancer strategies selectively interfering with reactive species flux (Gorrini, Harris, & Mak, 2013; Misra, Boylan, Selvam, Spallholz, & Bjornstedt, 2015). This redox-active property is essential for the application of organoselenium compounds to cancer therapeutics. Many organoselenium compounds, including the prototype Ebselen

(**31**; Fig. 10), are highly redox reactive and generate RS upon reaction with and oxidation of thiols (Nogueira & Rocha, 2011). The basic molecular mechanism(s) involved in organoselenium-mediated cytotoxicity are still not completely understood; however, their interaction with thiols seems to play a central role in their prooxidant effects. In fact, to be activated the organoselenium compounds react with thiols and, consequently, consume thiols. In this way, the reaction with thiols represents an important and preliminary event in both the cytotoxic and cytoprotective effects of the different members in this class of compounds. These opposite, but redox-related, pharmacological responses have been previously described in the case of diselenide compounds (described in Section 2.4) by Nogueira, Zeni, and Rocha (2004). The reactive cycle, presented in Fig. 11, depicts a cytotoxic/oxidant or a cytoprotective/antioxidant pathway of diselenide-mediated reaction with cellular thiols. In both pathways, the generation of selenol intermediates (RSe^-) is a prerequisite to their pharmacological effects. This selenol formation can occur either chemically by the oxidation of any accessible parent compounds with thiol-containing molecules, e.g., GSH or thiol-containing proteins, or enzymatically generated by the action of the selenoenzyme TrxR, which catalyzes the transfer of reducing equivalents from NADPH to organoselenium compounds (Nascimento et al., 2014; Wallenberg, Olm, Hebert, Bjornstedt, & Fernandes, 2010).

The schematic cytotoxic and prooxidant pathway of diselenides (Fig. 11, left) is initiated by a one-electron transfer from RSe^- to dioxygen, yielding superoxide and the putative selenyl radical $RSe^\bullet$, which decays very rapidly to RSe–SeR. Part of the $RSe^\bullet$ would be reversibly trapped by excess GSH yielding the glutathione radical $GS^\bullet$, which would decay to GSSG. Following, RSe^- can reduce superoxide to hydrogen peroxide. The reactive intermediates generated in this cycle efficiently react with oxygen and thiols leading to a nonstoichiometric consumption of thiols and NADPH, huge oxidative stress, and finally cell death (Misra et al., 2015; Nogueira et al., 2004).

Otherwise, in the cytoprotective pathway (Fig. 11, right), the selenol intermediate (RSe^-) can promote the thiol peroxidase-like cycle, reacting and neutralizing peroxides (Nogueira & Rocha, 2011).

It is important to highlight that the cytotoxic pathway promoted by these promising anticancer agents is based upon their chemical form, the dose, and the time of exposure (Misra et al., 2015). Moreover, the absence of reactive specificity of organochalcogens with different types of thiol-containing molecules represents a barrier that must be overcome in the pharmacological development of organoselenium compounds as anticancer agents.

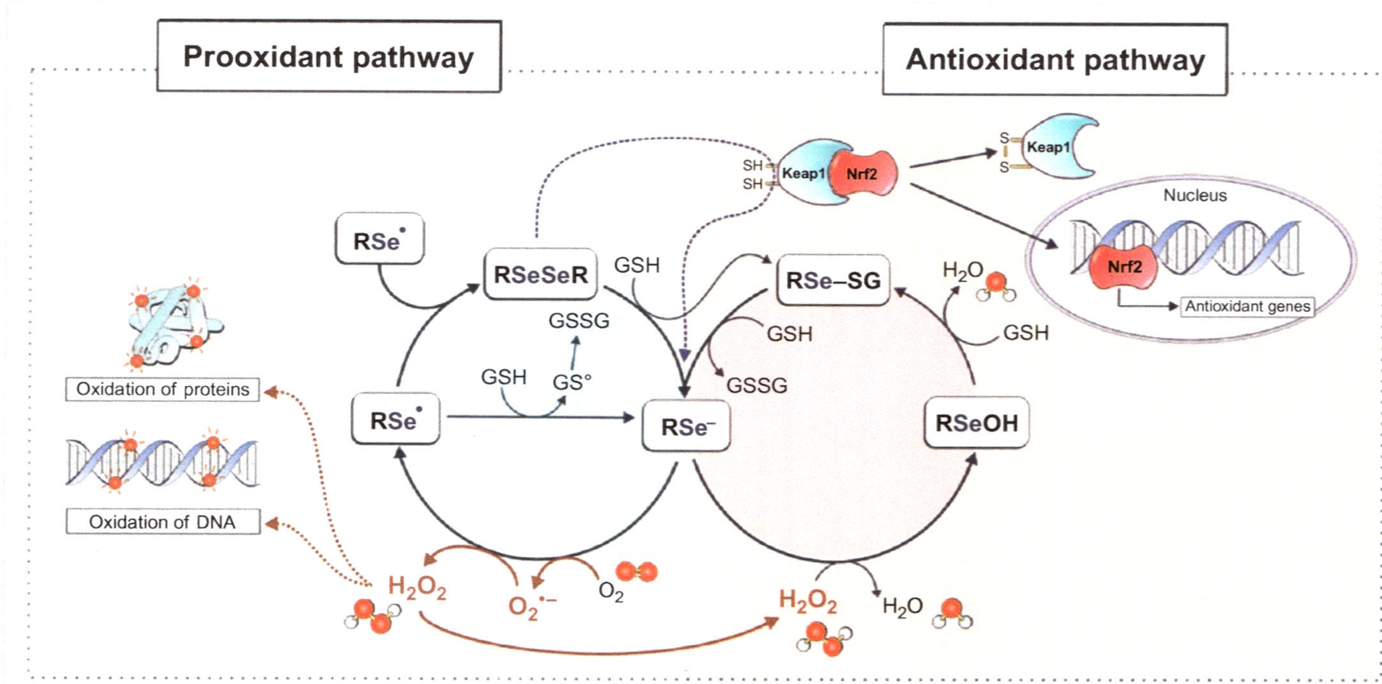

Fig. 11 Redox cycling of organoselenium compounds: pro- or antioxidant pathways. The ability of diselenides to be reduced by GSH, as well as other thiols, promotes the generation of two equivalents of selenolate ion (RSe^-), a redox-active intermediate. The antioxidant pathway of diselenides can be triggered by the direct reaction of RSe^- with hydrogen peroxide (and other peroxides, such as peroxynitrite) resulting in the production of selenenic acid (RSeOH), which is trapped by excess GSH to yield a selenylsulfide intermediate (RSe–SG). Diselenides can activate Nrf2 signaling through oxidation of Keap1 cysteine thiols, promoting its nuclear translocation and the consequent transcription of antioxidants genes. On the other hand, the cytotoxic and anticancer effect of organoselenium compounds may be attributable to its ability to generate reactive species in a prooxidant redox cycle. The first step is the one-electron transfer from RSe^- to dioxygen, yielding superoxide ($O_2^{\bullet -}$) and the selenyl radical intermediate ($RSe^{\bullet}$) that evolves rapidly to RSe–SeR. $RSe^{\bullet}$, which can react with and oxidize GSH promoting GSSG generation. Finally, RSe^- can reduce $O_2^{\bullet -}$ to H_2O_2 and also evolve to $RSe^{\bullet}$ ultimately yielding RSe–SeR. *Adapted from Nogueira, C. W., & Rocha, J. B. (2011). Toxicology and pharmacology of selenium: Emphasis on synthetic organoselenium compounds.* Archives of Toxicology, *85(11), 1313–1359.*

4. CELL SIGNALING PATHWAYS INFLUENCED BY Se COMPOUNDS

Se compounds stimulate a series of cellular responses relevant to cancer therapy such as mitochondrial damage and ER stress, the arrest of cell cycle progression with reduced proliferation, activation of cellular senescence and then stimulation of death pathways or autophagy as final events. How these cellular responses develop and which signals promote their activation is still under investigation, but it is now universally accepted that these are the consequence of the interaction of Se compounds with the cellular redox and then with the oxidation status of critical Cys resides of cellular proteins (see Section 3.2). Protein thiol oxidation may lead to development of conformational changes of structural proteins and a modified activity of enzymes and signaling proteins, including protein kinases, phosphatases, and transcription factors that include for instance the nuclear factor kappa-B (NF-κB) (reviewed in Pantano, Reynaert, van der Vliet, & Janssen-Heininger, 2006; further discussed in this section) and other redox-sensitive elements such as AP-1, Jun, APE-1/Ref-1, and Sp1, the oncogene protein p53 with a role in DNA repair and other kinases with regulatory effects on cell cycle progression, some members of the caspase family of enzymes, the stress-activated protein kinase JNK, and the Trx-regulated kinase ASK1 (Flohe, Brigelius-Flohe, Saliou, Traber, & Packer, 1997; Husbeck et al., 2006; Ranawat & Bansal, 2008; Zou et al., 2008). The Grx and Trx systems appear to control thiol modifications of these Se sensitive proteins stimulating compensatory responses (Arner & Holmgren, 2006; Matthews, Wakasugi, Virelizier, Yodoi, & Hay, 1992). Alterations of critical thiols in proteins may affect signaling pathways and gene response through other and so far poorly understood mechanisms. These include an impaired biogenesis in Fe–Zn clusters (Lill et al., 2012) and changes in iron and calcium homeostasis (Muhlenhoff et al., 2010; Sag, Wagner, & Maier, 2013; Sheftel, Stehling, & Lill, 2010). Selenoorganic compounds are able to interact with metal proteins containing zinc–thiolate coordination sites (Chen & Maret, 2001; Jacob, Maret, & Vallee, 1998a, 1998b), and in an environment rich in GSH, Se compounds are able to promote the release of zinc, which itself can influence signal transduction pathways (Blessing, Kraus, Heindl, Bal, & Hartwig, 2004; Larabee, Hocker, & Hanas, 2009; Larabee, Hocker, Hanas, Kahn, & Hanas, 2002).

4.1 PI3K/Akt and MAPK Pathways

Se compounds have been repeatedly reported to influence the expression and activity of PI3K/Akt and MAPK pathways. The efficacy of Se on such prominent cancer cell pathways with roles in tumor growth and apoptotic susceptibility varies depending on several aspects that include the form of Se and the experimental model used for the investigation. For example, in different types of cancer cells selenite was demonstrated to arrest the cell cycle in S phase leading to caspase-independent apoptosis, while methylated Se was reported to arrest the cycle in G1 phase leading to caspase-mediated apoptosis (Jiang, Wang, Ganther, & Lu, 2002). Again, focusing on a cellular model, namely the NB4 promyelocytic leukemia cell line, selenite was found to downregulate ERK along with kinase-regulated transcription factors, such as *c*-myc, *c*-fos, and *c*-jun, which plays a major role in regulating cancer cell metabolism and death pathways (Han et al., 2007); in the same cell model, selenite promotes heat shock protein Hsp90-mediated inactivation of NF-κB responsible for the switch from autophagy to apoptosis by becn1 transcriptional downregulation (Jiang et al., 2011). In this latter study the role of ERK was not investigated in detail, but it is known that ERK and other survival-promoting and stress-activated protein kinases modify their signal transduction activities to coordinate transcriptional and functional responses to the redox changes that the Se compound imposes on the cell system; NF-κB is among a series of transcription factors downstream of ERK and other protein kinases that coordinate this physiological mechanism of response to control different groups of (stress adaption response) genes with a role in inflammatory, detoxification, cytoprotection, and death pathways (Luchetti et al., 2010). At the same time, converging mechanistic aspects in these two studies on NB4 leukemia cells include the fact that ERK is now demonstrated to influence the role of becl1 as an autophagy pathway activator (Wang et al., 2017 and references therein).

ERK1/2 and its coordinated regulation with other MAPKs have been extensively investigated in a series of studies on Se compounds representing an ideal target to inhibit proliferation and stimulate cancer cell death. As the most investigated inorganic form, selenite was reported to inhibit this survival-promoting kinase in different cancer cell models (Jiang et al., 2002); this effect was associated with a decreased phosphorylation of Akt, upregulation of other protein kinases, such as p27kip1, p21cip1, and the stress-MAPK p38, which together interfere with cell cycle to arrest its progression and activate the apoptotic program; in the same studies, compound **7** (MeSeH) precursors were found to provoke the opposite effect.

A similar mechanism of response to selenite (**2**) was observed by Rudolf, Rudolf, and Cervinka (2008) showing increased levels of p21 and phosphorylated p53; along with p38 pathway activation, these cell cycle kinases stimulate the accumulation of Bax protein, a proapoptotic component of the mitochondrial (or intrinsic) pathway of apoptotic cell death.

The PI3K–Akt pathway represents a major node in the anticancer response to Se compounds upstream of MAPKs, particularly of ERK1/2, promoting cell survival and growth in response to hormones and various growth factors (Lee, Shin, Kang, Lee, & Bae, 2008). Se compounds like compound **7** show similar signal transduction patterns as selenite, with Akt dephosphorylation and modulation effects on PI3K, ERK1/2, and p38 pathways (Hu, Jiang, Li, & Lu, 2005; Wang, Jiang, Ganther, & Lu, 2001; Wu, Zu, Warren, Wallace, & Ip, 2006). In human oral squamous carcinoma cells, the thioselenide GSSeSG stimulates the stress-activated protein kinases JNK and p38 MAPK, also activating the survival kinase ERK1/2 and Akt (Ghose et al., 2001). Se methylselenocysteine (**6**) has been shown to decrease the activity of PI3K, resulting in Akt and p38 dephosphorylation, and leading to inhibitory effects on the Raf/MEK/ERK signaling pathways (Unni, Koul, Yung, & Sinha, 2005), which is compatible with the downstream activation of apoptotic cell death effectors (Bartolini, Piroddi, et al., 2015). Similarly, methylselenol inhibits the activation of ERK1/2 and *c*-myc expression (Zeng, Briske-Anderson, Wu, & Moyer, 2012; Zeng, Wu, & Botnen, 2009). Remarkably, this compound exhibits a strong inhibition effect on cell signaling of colon cancer cells (Zeng et al., 2012). In prostate cancer cells, compound **7** decreases the levels of phospho-Akt and phospho-ERK1/2, and these effects appear to be independent of p38 and JNK1/2 stress kinase expression and activity (Jiang et al., 2002).

The selenite compound (**2**) inhibits Akt activity, an activity proposed to influence the activity of the Wnt/β-catenin pathway (Fang, Han, Bi, Xiong, & Yang, 2010; Luo et al., 2012; Park et al., 2000; Rudolf et al., 2008). The PI3K-mediated inhibition of Akt by this Se compound promoted FoxO3a nuclear accumulation and thus the transcriptional upregulation of the targeted genes Bim and PTEN in colorectal cancer (Luo et al., 2013). According to previous studies on SeMet (compound **5** in Fig. 2), PTEN tumor suppressor protein may have a role in Akt inactivation (Lee et al., 2008). Treatments with this Se amino acid stimulated a kinase-mediated transient activation of Akt in which PTEN intervenes in the inactivation step of this protein kinase leading to its degradation through both the caspase and proteasome pathways.

The effect of Se compounds on these cellular kinases is reported to involve upstream signal transduction elements of growth factor and G protein-coupled receptors, such as protein kinase C (PKC). With its different isoforms, this kinase influences the activity of different cancer cell pathways, such as MEK/ERK, GSK3β, and NF-κB (Mackay & Twelves, 2008). For example, compound **7**, **2**, GSSeSG, and **4** are reported to oxidize critical Cys residues in the PKC active site and as such these compounds behave as potent modulators of downstream protein kinases important in cancer (Gopalakrishna, Chen, & Gundimeda, 1997; Gopalakrishna, Gundimeda, & Chen, 1997; Gundimeda et al., 2008). At the same time, only GSSeSG and **4** appear to inhibit protein kinase A, which differs from PKC in terms of critical Cys pattern and redox sensitivity.

4.2 ER Stress Response to Protein Misfolding

Se-induced redox modifications of thiols/disulfides in proteins may cause protein unfolding; also an incorrect incorporation of SeCys in the place of Cys can occur leading to misfolded protein formation with altered structure, biological activity, and function (Muller, Heider, & Bock, 1997; Zorn, Ihling, Golbik, Sawers, & Sinz, 2013). When cell stress occurs, the endoplasmic reticulum (ER) becomes a major signaling hub to direct a process for cell survival known as UPR (*unfolded protein response*) (Wu, Zhang, Dong, Park, & Ip, 2005; Zu et al., 2006). The process is coordinated through three signal transduction pathways: phospho-ERK, ATF alpha, and XBP1. The ER stress markers CHOP and phospho-ERK are altered by compound **7**. Also SeCys treatment causes ER stress with effects on CHOP, Bim, ERdj5, and Bip proteins and induces a multitargeted cell death process (Wallenberg et al., 2014).

These inorganic forms of Se have been reported to influence mTOR protein kinase activity, another signal transduction hub in the autophagy process downstream of the PI3K/Akt pathway. Moreover, the mTOR signal transduction function operates at the interface with other protein kinases that control cellular proliferation in response to different membrane receptor agonists and drugs (Ciuffreda, Di Sanza, Incani, & Milella, 2010; Fleming, Noda, Yoshimori, & Rubinsztein, 2011). Compound **3** suppresses the mTOR pathway in colon cancer cells acting in both an Akt-dependent and -independent manner (Lee et al., 2010). Compound **7** also affects mTOR signaling under conditions of hypoxia in prostate cancer cells, and this effect appears to involve the induction of REDD1 and Akt expression (Sinha, Allen, Pinto, & Sinha, 2014).

The identification of GSTP as a selective target of anticancer Se organic compounds (Bartolini, Piroddi, et al., 2015) and the recently identified role of this transferase in ER protein glutathionylation (Ye, Zhang, et al., 2017) indicate a new GSTP-dependent mechanism for these Se-based anticancer drugs that warrant further exploration (further discussed in Section 4.4).

4.3 Angiogenesis and Cancer Progression

Some of the selenium compounds described earlier influence the vascularization process interfering with different components of the tumor microenvironment (Jiang et al., 2002). Rats supplemented with high levels of selenite showed a decreased microvascular density in mammary tumors (Jiang, Jiang, Ip, Ganther, & Lu, 1999). Methylated Se was proposed to be more effective than selenite in interfering with the tumor vasculature (Li et al., 2008) and promoted cell cycle arrest of microvascular endothelial cells leading to a reduced microvessel density in prostate cancer xenografts (Wang et al., 2008). Therapeutic synergies with other anticancer drugs have been proposed for compound **6**. This Se compound was demonstrated to alter vascular maturation and increased doxorubicin uptake in tumor cells (Bhattacharya et al., 2008).

Selenite is by far the most investigated antiangiogenic Se compound. The treatment with this molecule in cultured biopsy-derived glioma cells lowered the mRNA levels of matrix tissue inhibitors of metalloproteinases (TIMPs) and epidermal growth factor receptor (Rooprai et al., 2007). The same results were obtained in selenite-treated colon cancer cells (Li, Meng, Xu, Qin, & Zhou, 2013). Furthermore, this Se compound was consistently shown to inhibit the expression of vascular endothelial growth factor (VEGF), a key protein in angiogenesis (Jiang et al., 1999). In LPS-stimulated PC-3 human prostate carcinoma cells, selenite inhibited TGF-beta1, IL-6, and VEGF expression also producing an inhibitory effect on NF-κB (Pei, Li, Guo, Jin, & Lin, 2010). Lowered levels of VEGF were also demonstrated in melanoma cell lines treated with selenite, along with decreased levels of hypoxia-inducible factor-1α (HIF-1α) and IL-18 (Song et al., 2011), perhaps relevant to selenite inhibition of tumor metastasis (Song et al., 2009).

VEGF expression and activity are also sensitive to compound **7** and this correlated with decreased tumor growth in a nude mouse model of metastatic breast cancer (Wu et al., 2012). VEGF secretion decreased together with HIF-1α expression in lymphoma and in invasive prostate cancer after treatment with this Se compound (Sinha et al., 2012). This effect was confirmed in rats with metastatic prostate cancer in which GLUT1 levels and microvascular density were also decreased by the treatment (Sinha et al., 2012). Moreover, compound **7** was found to inhibit PMA-stimulated pro-MMP-2 activation, which is mediated by MT1-MMP, and lowered the secretion and expression of MMP-2 and TIMP-1 and tumor invasiveness that was associated with NF-κB suppression (Conley, Bruhn, Morgan, & Stamer, 2004; Jiang, Ganther, & Lu, 2000; Park, Kim, Oh, & Chung, 2007). Dietary supplementation with **7**, but not **5**, reduced spontaneous metastasis of Lewis lung carcinoma in C57BL/6 mice (Yan & DeMars, 2012).

In line with compound **2** and compound **7** activity, compound **6** inhibited HIF-1α and HIF-2α in renal cell carcinoma leading to in vivo tumor growth inhibition (Chintala et al., 2012). Angiogenic chemomodulation by **6** was effective in limiting colon cancer growth through a reduction of microvessel density (Bhattacharya et al., 2009). HIF-1α and VEGF modulation by **6** was confirmed in other tumor models with resulting effects on microvascular density and vascular maturation (Bhattacharya et al., 2008).

The migration and invasion potential of HT-1080 cells were both inhibited by methylselenol, decreasing the active form of MMP-2 and increasing TIMP-1 and -2 protein levels (Zeng, Briske-Anderson, Idso, & Hunt, 2006). When human umbilical vein endothelial cells are treated with methylselenol they undergo caspase-dependent apoptotic death via p38 MAPK activation (Jiang, Kim, Wang, & Lu, 2004).

Dietary administration of **2** or **5** constrained the growth of secondary cancers in in vivo mice models (Yan, Yee, Li, McGuire, & Graef, 1999; Yan, Yee, McGuire, & Graef, 1997) perhaps through changes in collagen gene expression. This type of gene response is also caused by methylated Se in human prostate cells (Hurst, Elliott, Goldson, & Fairweather-Tait, 2008). Moreover, **5** decreased ROS generation, Akt-dependent secretion of matrix metalloproteinases, and cancer cell invasiveness (Kim et al., 2008).

4.4 GSTP as Target of Se Organic Anticancer Agents

Among the inorganic and organic forms of Se discussed so far, a commonality in the mechanism of action is the reaction with free and protein-bound thiols (thiol-peroxidase activity) and with selenoproteins such as TrxR (see, e.g., the activity of Ebselen (**31**) described in Section 2). Redox activity of Se compounds markedly varies within the different subgroups of molecules and in a cell-specific manner. Moreover, such activity is contingent upon threshold dose effects in cancer cells, which suggests the possibility to identify specific cellular targets associated with this type of response. Different types of tumors show marked variations in the levels of cellular thiols and this obviously influences the dose–response relationship of the different selenocompounds assessed as anticancer agents. At the same time, not all cellular thiols are the same level of importance in making a cancer cell susceptible of the cytotoxic activity of Se compounds. Several laboratories have assessed the GPx-mimetic (thiol peroxidase) activity of Se compounds and in some studies the correlation of this activity with the impact on cellular thiols and the apoptotic signaling of different cancer cell models has been demonstrated (Bartolini, Piroddi, et al., 2015; Tidei, Piroddi, Galli, & Santi, 2012; references therein). In our recent studies we also explored the role of GST, one of the main protein thiols of several cancer cells, in the response to highly reactive Se-compounds. The fact that GSTs are an elective protein target for Se organic molecules, particularly compound **31** , was known since the early 1990s of the last century (Nikawa, Schuch, Wagner, & Sies, 1994a, 1994b). What we found in our studies is that the extent of GST expression influences the antiproliferative and proapoptotic response to compounds **10** and **12**, two forms with very high thiol-peroxidase and pro-apoptotic activity (Bartolini, Commodi, et al., 2015; Bartolini, Piroddi, et al., 2015; Fig. 5). Accordingly, manipulating GSTP gene expression resulted in a modified apoptotic signaling of cells during the exposure to those Se compounds. These effects were associated with changes in the levels and redox balance of cellular thiols, and increased H_2O_2 generation and phosphorylation of survival and stress-activated protein kinases, ERK and JNK, respectively. Assessing the response to a microencapsulated form of **12** or to compound **11** (a Se hormetine; the least reactive of the corresponding diselenide compounds produced and assessed in our laboratories), GSTP expression was shown to be a prerequisite to preventing H_2O_2-induced oxidative stress. Intriguingly, the allelic form of GSTP with higher affinity with Prxd6 produced the higher level of protection against hydrogen peroxide in MCF7 cells transfected with the different allelic forms of human GSTP. At the same time, the studies on the Se hormetine **11** provided evidence of other protein–protein interactions of GSTP that could be relevant to both the hormetic and anticancer activity of Se compounds (Fig. 5); in fact, a physical interaction between GSTP and Nrf2 was demonstrated in human hepatocarcinoma cells and MEF during the treatment with compound **11**. As expected such interaction was absent in MEF obtained from GSTP knockout mice thus leading to an altered expression and transcriptional control of Nrf2; that defect resulted in a lowered resistance to

H_2O_2-induced oxidative stress. The overexpression of GSTP and Nrf2 of some tumors such as ovarian, NSCLC, breast, colon, pancreas, and lymphomas and in a wide range of drug-resistant cell lines and tumors (Tew, 1994) provides the rationale for GSTP-targeted intervention with Se organic compounds (Fig. 5) and perhaps their use in combination chemotherapy. Besides Nrf2 interaction, the GSTP interactome may itself provide a platform for Se organic molecules to simultaneously targeting different cancer-related pathways; these include 1-Cys Prxd6, ER stress-related proteins, stress-activated protein kinases such as TRAF2/ASK1 and JNK, and STAT3 transcription factor (reviewed in Bartolini & Galli, 2016; Ye, Zhang, et al., 2017).

GSTP is also a key player in protein glutathionylation, a posttranslational modification of Cys residues in cellular proteins with emerging role in cancer cell signal transduction (Tew et al., 2011). Other players include other Nrf2-dependent genes such as the family of glutaredoxins and TrxRD. This further highlights the importance of developing agents that target GSTP and its multiple functions in the cellular interactome with included its effects on this process. Accordingly, recent studies revealed how glutathionylation may effect the chemical species-specific anticancer response to Se compounds (Wallenberg et al., 2014). Selenodiglutathione causes glutathionylation of free protein thiols in HeLa cells. This might provide an explanation for early differences in cytotoxic effects induced by this compound in comparison with selenite, as well as the different mode of death program execution for these two Se compounds, i.e., necroptosis-like cell death for selenite, and apoptosis-like for selenodiglutathione. In contrast, seleno-DL-cystine activates two distinct cell death types, one with apoptosis-like morphological phenotype, the other reminiscent of paraptosis-like cell death, characterized by the occurrence of large cytoplasmic vacuoles and induction of UPR and ER stress, also involving GSTP-dependent glutathionylation in the process (Ye, Zhang, et al., 2017). Collectively, these recent findings suggest that GSTP and its role in protein glutathionylation may intervene in the cytotoxic mechanism and variable potential of redox-active Se compounds to mediate between death and survival responses, which holds some potential to developing further targeted and selective therapies.

4.5 Epigenetics of Se Compounds

Se compounds may interfere with cancer gene expression through epigenetic mechanisms that are described elsewhere in this volume. Here, we summarize some emerging evidence on this mechanism that characterizes the activity of some Se compounds. These include α-keto-γ-methylselenobutyrate (KMSB) and β-methylselenopyruvate (MSP), two products of **5** and **6** transamination reactions with the structure of short-chain fatty acids; these metabolites exert cytotoxic effects through inhibition of histone deacetylases (HDACs), consequently influence histone protein function at the posttranslational level, and are promising anticancer targets since they are upregulated in many tumors. KMSB and MSP may act as HDAC competitive inhibitors (Lee et al., 2009; Nian, Bisson, Dashwood, Pinto, & Dashwood, 2009), and **7** has been shown to inhibit HDAC activity in diffuse large B-cell lymphoma cell lines (Kassam et al., 2011) and in ESCC (Hu et al., 2015). In the latter tumor an induction of H3 histone acetylation has been observed. Selenite (**2**) (as well as **7**) increases acetylation levels while decreasing methylation levels on Lys 9 of this histone protein (Xiang, Zhao, Song, & Zhong, 2008). The levels of HDAC and DNA methylation decreased during exposure to such Se compounds. Compound **7** decreased the heterochromatin-associated histone marker H3K9me3 and increased acetylated histone H4 at lysine 16 (H4K16ac), while only compound **2** reduced acetylation in breast cancer (de Miranda et al., 2014). In addition, class I HDACs may represent a target for the inhibitory activity of Se compounds since their activity is influenced by the oxidation of Cys residues (Doyle & Fitzpatrick, 2010; Kassam et al., 2011).

5. CONCLUSION

An impressive number of in vitro and in vivo (animal) studies have demonstrated the anticancer potential of selenium compounds in the last two decades. Their activity as antiproliferative and proapoptotic agents, however, was not confirmed in clinical trials, and effects on normal tissues are uncertain (discussed elsewhere in this volume). Toxicity has been a main issue of clinical trials carried out so far with selenite, Se amino acids, and other forms of Se used in human therapy. This and the primarily negative results of previous trials have led to a breakdown in clinical investigation even if the development of a series of innovative Se organic compounds clearly suggests the possibility of more cancer-specific effects and tolerability. However, apart from Ebselen (**31**) discussed in Section 2, in recent years only a few Se compounds have entered clinical trials for cancer therapy, including Ethaselen (**35**) now phase I in China in combination with cisplatin (Tan et al., 2010).

Additional clinical trials are needed and it is likely that the full potential of selenium compounds as anticancer agents in both solid and hematological cancers will only be realized once novel and more selective (cancer-specific) selenium compounds will be developed. In this respect, expanding the library of synthetic compounds and understanding their underlying mechanisms of action are essential steps of research that surely warrant further investments in the future.

ACKNOWLEDGMENTS

Co-authors (PT and FG) of this review paper have been financially supported by the grant program of the Italian Ministry of University and Research (MIUR), National Technology Agrifood Cluster, Health and Nutrition area, PROS.IT project (CTN01_00230_413096) and Bartolini Desirée was supported by the Fondazione Italiana per la Ricerca sul Cancro (FIRC-AIRC). This manuscript is part of the activity of the international scientific network "SeS Redox and Catalysis." Prof. R. Radi and Prof. A.F. de Bem were funded by grants from Universidad de la República (CSIC and EI), Uruguay and Conselho Nacional de Desenvolvimento Científico e Tecnológico, CNPq-(PVE 401065/2014-6), Brazil

REFERENCES

Alcolea, V., Plano, D., Karelia, D. N., Palop, J. A., Amin, S., Sanmartin, C., et al. (2016). Novel seleno- and thio-urea derivatives with potent in vitro activities against several cancer cell lines. *European Journal of Medicinal Chemistry*, *113*, 134–144.

An, J. J., Shi, K. J., Wei, W., Hua, F. Y., Ci, Y. L., Jiang, Q., et al. (2013). The ROS/JNK/ATF2 pathway mediates selenite-induced leukemia NB4 cell cycle arrest and apoptosis in vitro and in vivo. *Cell Death & Disease*, *4*, e973.

Anjaria, K. B., & Madhvanath, U. (1988). Genotoxicity of selenite in diploid yeast. *Mutation Research*, *204*(4), 605–614.

Arner, E. S., & Holmgren, A. (2006). The thioredoxin system in cancer. *Seminars in Cancer Biology*, *16*(6), 420–426.

Azad, G. K., & Tomar, R. S. (2014). Ebselen, a promising antioxidant drug: Mechanisms of action and targets of biological pathways. *Molecular Biology Reports*, *41*(8), 4865–4879.

Banjac, A., Perisic, T., Sato, H., Seiler, A., Bannai, S., Weiss, N., et al. (2008). The cystine/cysteine cycle: A redox cycle regulating susceptibility versus resistance to cell death. *Oncogene*, *27*(11), 1618–1628.

Bartolini, D., Ciffolilli, S., Piroddi, M., Murdolo, G., Tortoioli, C., & Galli, F. (2014). Biochemistry and nutrition of selenium: From inorganic forms to endogenous proteins. In C. Santi (Ed.), *Organoselenium chemistry—Between synthesis and biochemistry* (pp. 268–327). Bentham e-books, United Arab Emirates.

Bartolini, D., Commodi, J., Piroddi, M., Incipini, L., Sancineto, L., Santi, C., et al. (2015). Glutathione S-transferase pi expression regulates the Nrf2-dependent response to hormetic diselenides. *Free Radical Biology & Medicine*, *88*(Pt. B), 466–480.

Bartolini, D., & Galli, F. (2016). The functional interactome of GSTP: A regulatory biomolecular network at the interface with the Nrf2 adaption response to oxidative stress. *Journal of Chromatography. B, Analytical Technologies in the Biomedical and Life Sciences, 1019*, 29–44.

Bartolini, D., Piroddi, M., Tidei, C., Giovagnoli, S., Pietrella, D., Manevich, Y., et al. (2015). Reaction kinetics and targeting to cellular glutathione S-transferase of the glutathione peroxidase mimetic PhSeZnCl and its D,L-polylactide microparticle formulation. *Free Radical Biology & Medicine, 78*, 56–65.

Bellinger, F. P., Raman, A. V., Reeves, M. A., & Berry, M. J. (2009). Regulation and function of selenoproteins in human disease. *The Biochemical Journal, 422*(1), 11–22.

Bhattacharya, A. (2011). Methylselenocysteine: A promising antiangiogenic agent for overcoming drug delivery barriers in solid malignancies for therapeutic synergy with anticancer drugs. *Expert Opinion on Drug Delivery, 8*(6), 749–763.

Bhattacharya, A., Seshadri, M., Oven, S. D., Toth, K., Vaughan, M. M., & Rustum, Y. M. (2008). Tumor vascular maturation and improved drug delivery induced by methylselenocysteine leads to therapeutic synergy with anticancer drugs. *Clinical Cancer Research, 14*(12), 3926–3932.

Bhattacharya, A., Toth, K., Sen, A., Seshadri, M., Cao, S., Durrani, F. A., et al. (2009). Inhibition of colon cancer growth by methylselenocysteine-induced angiogenic chemomodulation is influenced by histologic characteristics of the tumor. *Clinical Colorectal Cancer, 8*(3), 155–162.

Blessing, H., Kraus, S., Heindl, P., Bal, W., & Hartwig, A. (2004). Interaction of selenium compounds with zinc finger proteins involved in DNA repair. *European Journal of Biochemistry, 271*(15), 3190–3199.

Brigelius-Flohe, R. (2008). Selenium compounds and selenoproteins in cancer. *Chemistry & Biodiversity, 5*(3), 389–395.

Cavalieri, R. R., & Scott, K. G. (1968). Sodium selenite Se 75. A more specific agent for scanning tumors. *JAMA, 206*(3), 591–595.

Cavalieri, R. R., Scott, K. G., & Sairenji, E. (1966). Selenite (75Se) as a tumor-localizing agent in man. *Journal of Nuclear Medicine*, 7(3), 197–208.

Chakraborty, P., Roy, S. S., Basu, A., & Bhattacharya, S. (2016). Sensitization of cancer cells to cyclophosphamide therapy by an organoselenium compound through ROS-mediated apoptosis. *Biomedicine & Pharmacotherapy, 84*, 1992–1999.

Chen, Y., & Maret, W. (2001). Catalytic oxidation of zinc/sulfur coordination sites in proteins by selenium compounds. *Antioxidants & Redox Signaling, 3*(4), 651–656.

Chen, T., & Wong, Y. S. (2008). Selenocystine induces apoptosis of A375 human melanoma cells by activating ROS-mediated mitochondrial pathway and p53 phosphorylation. *Cellular and Molecular Life Sciences, 65*(17), 2763–2775.

Chen, T., & Wong, Y. S. (2009a). Selenocystine induces caspase-independent apoptosis in MCF-7 human breast carcinoma cells with involvement of p53 phosphorylation and reactive oxygen species generation. *The International Journal of Biochemistry & Cell Biology, 41*(3), 666–676.

Chen, T., & Wong, Y. S. (2009b). Selenocystine induces reactive oxygen species-mediated apoptosis in human cancer cells. *Biomedicine & Pharmacotherapy, 63*(2), 105–113.

Chintala, S., Najrana, T., Toth, K., Cao, S., Durrani, F. A., Pili, R., et al. (2012). Prolyl hydroxylase 2 dependent and Von-Hippel-Lindau independent degradation of Hypoxia-inducible factor 1 and 2 alpha by selenium in clear cell renal cell carcinoma leads to tumor growth inhibition. *BMC Cancer, 12*, 293.

Chintala, S., Toth, K., Cao, S., Durrani, F. A., Vaughan, M. M., Jensen, R. L., et al. (2010). Se-methylselenocysteine sensitizes hypoxic tumor cells to irinotecan by targeting hypoxia-inducible factor 1alpha. *Cancer Chemotherapy and Pharmacology, 66*(5), 899–911.

Ciuffreda, L., Di Sanza, C., Incani, U. C., & Milella, M. (2010). The mTOR pathway: A new target in cancer therapy. *Current Cancer Drug Targets*, *10*(5), 484–495.

Conley, S. M., Bruhn, R. L., Morgan, P. V., & Stamer, W. D. (2004). Selenium's effects on MMP-2 and TIMP-1 secretion by human trabecular meshwork cells. *Investigative Ophthalmology & Visual Science*, *45*(2), 473–479.

Daziano, J. P., Gunther, W. H. H., Krieg, M., Tsujino, I., Miyagi, K., Anderson, G. S., et al. (2012). Photochemically generated elemental selenium forms conjugates with serum proteins that are preferentially cytotoxic to leukemia and selected solid tumor cells. *Photochemistry and Photobiology*, *88*(2), 448–460.

Deepagan, V. G., Kwon, S., You, D. G., Nguyen, V. Q., Um, W., Ko, H., et al. (2016). In situ diselenide-crosslinked polymeric micelles for ROS-mediated anticancer drug delivery. *Biomaterials*, *103*, 56–66.

de Miranda, J. X., Andrade Fde, O., Conti, A., Dagli, M. L., Moreno, F. S., & Ong, T. P. (2014). Effects of selenium compounds on proliferation and epigenetic marks of breast cancer cells. *Journal of Trace Elements in Medicine and Biology*, *28*(4), 486–491.

Dong, C., Zhang, L., Sun, R., Liu, J., Yin, H., Li, X., et al. (2016). Role of thioredoxin reductase 1 in dysplastic transformation of human breast epithelial cells triggered by chronic oxidative stress. *Scientific Reports*, *6*, 36860.

Doyle, K., & Fitzpatrick, F. A. (2010). Redox signaling, alkylation (carbonylation) of conserved cysteines inactivates class I histone deacetylases 1, 2, and 3 and antagonizes their transcriptional repressor function. *The Journal of Biological Chemistry*, *285*(23), 17417–17424.

Engman, L., Cotgreave, I., Angulo, M., Taylor, C. W., Paine-Murrieta, G. D., & Powis, G. (1997). Diaryl chalcogenides as selective inhibitors of thioredoxin reductase and potential antitumor agents. *Anticancer Research*, *17*(6D), 4599–4605.

Fang, W., Han, A., Bi, X., Xiong, B., & Yang, W. (2010). Tumor inhibition by sodium selenite is associated with activation of c-Jun NH2-terminal kinase 1 and suppression of beta-catenin signaling. *International Journal of Cancer*, *127*(1), 32–42.

Favrot, L., Grzegorzewicz, A. E., Lajiness, D. H., Marvin, R. K., Boucau, J., Isailovic, D., et al. (2013). Mechanism of inhibition of Mycobacterium tuberculosis antigen 85 by ebselen. *Nature Communications*, *4*, 2748.

Fernandes, A. P., & Gandin, V. (2015). Selenium compounds as therapeutic agents in cancer. *Biochimica et Biophysica Acta*, *1850*(8), 1642–1660.

Fleming, A., Noda, T., Yoshimori, T., & Rubinsztein, D. C. (2011). Chemical modulators of autophagy as biological probes and potential therapeutics. *Nature Chemical Biology*, *7*(1), 9–17.

Flohe, L., Brigelius-Flohe, R., Saliou, C., Traber, M. G., & Packer, L. (1997). Redox regulation of NF-kappa B activation. *Free Radical Biology & Medicine*, *22*(6), 1115–1126.

Freitas, M., Alves, V., Sarmento-Ribeiro, A. B., & Mota-Pinto, A. (2011). Combined effect of sodium selenite and docetaxel on PC3 metastatic prostate cancer cell line. *Biochemical and Biophysical Research Communications*, *408*(4), 713–719.

Galanter, W. L., Hakimian, M., & Labotka, R. J. (1993). Structural determinants of substrate specificity of the erythrocyte anion transporter. *The American Journal of Physiology*, *265*(4 Pt. 1), C918–926.

Ganyc, D., & Self, W. T. (2008). High affinity selenium uptake in a keratinocyte model. *FEBS Letters*, *582*(2), 299–304.

Ghose, A., Fleming, J., El-Bayoumy, K., & Harrison, P. R. (2001). Enhanced sensitivity of human oral carcinomas to induction of apoptosis by selenium compounds: Involvement of mitogen-activated protein kinase and Fas pathways. *Cancer Research*, *61*(20), 7479–7487.

Ghosh, P., Bhattacharjee, A., Basu, A., Singha Roy, S., & Bhattacharya, S. (2015). Attenuation of cyclophosphamide-induced pulmonary toxicity in Swiss albino mice by naphthalimide-based organoselenium compound 2-(5-selenocyanatopentyl)-benzo[de] isoquinoline 1,3-dione. *Pharmaceutical Biology, 53*(4), 524–532.

Ghosh, P., Singha Roy, S., Basu, A., Bhattacharjee, A., & Bhattacharya, S. (2015). Sensitization of cisplatin therapy by a naphthalimide based organoselenium compound through modulation of antioxidant enzymes and p53 mediated apoptosis. *Free Radical Research, 49*(4), 453–471.

Gopalakrishna, R., Chen, Z. H., & Gundimeda, U. (1997). Selenocompounds induce a redox modulation of protein kinase C in the cell, compartmentally independent from cytosolic glutathione: Its role in inhibition of tumor promotion. *Archives of Biochemistry and Biophysics, 348*(1), 37–48.

Gopalakrishna, R., Gundimeda, U., & Chen, Z. H. (1997). Cancer-preventive selenocompounds induce a specific redox modification of cysteine-rich regions in Ca(2+)-dependent isoenzymes of protein kinase C. *Archives of Biochemistry and Biophysics, 348*(1), 25–36.

Gorrini, C., Harris, I. S., & Mak, T. W. (2013). Modulation of oxidative stress as an anticancer strategy. *Nature Reviews. Drug Discovery, 12*(12), 931–947.

Grek, C. L., Zhang, J., Manevich, Y., Townsend, D. M., & Tew, K. D. (2013). Causes and consequences of cysteine S-glutathionylation. *The Journal of Biological Chemistry, 288*(37), 26497–26504.

Gundimeda, U., Schiffman, J. E., Chhabra, D., Wong, J., Wu, A., & Gopalakrishna, R. (2008). Locally generated methylseleninic acid induces specific inactivation of protein kinase C isoenzymes: Relevance to selenium-induced apoptosis in prostate cancer cells. *The Journal of Biological Chemistry, 283*(50), 34519–34531.

Han, B., Wei, W., Hua, F., Cao, T., Dong, H., Yang, T., et al. (2007). Requirement for ERK activity in sodium selenite-induced apoptosis of acute promyelocytic leukemia-derived NB4 cells. *Journal of Biochemistry and Molecular Biology, 40*(2), 196–204.

Hanavan, P. D., Borges, C. R., Katchman, B. A., Faigel, D. O., Ho, T. H., Ma, C. T., et al. (2015). Ebselen inhibits QSOX1 enzymatic activity and suppresses invasion of pancreatic and renal cancer cell lines. *Oncotarget, 6*(21), 18418–18428.

Harris, M. T., Walker, D. M., Drew, M. E., Mitchell, W. G., Dao, K., Schroeder, C. E., et al. (2013). Interrogating a hexokinase-selected small-molecule library for inhibitors of Plasmodium falciparum hexokinase. *Antimicrobial Agents and Chemotherapy, 57*(8), 3731–3737.

Hu, H., Jiang, C., Li, G., & Lu, J. (2005). PKB/AKT and ERK regulation of caspase-mediated apoptosis by methylseleninic acid in LNCaP prostate cancer cells. *Carcinogenesis, 26*(8), 1374–1381.

Hu, C., Liu, M., Zhang, W., Xu, Q., Ma, K., Chen, L., et al. (2015). Upregulation of KLF4 by methylseleninic acid in human esophageal squamous cell carcinoma cells: Modification of histone H3 acetylation through HAT/HDAC interplay. *Molecular Carcinogenesis, 54*(10), 1051–1059.

Hurst, R., Elliott, R. M., Goldson, A. J., & Fairweather-Tait, S. J. (2008). Se-methylselenocysteine alters collagen gene and protein expression in human prostate cells. *Cancer Letters, 269*(1), 117–126.

Husbeck, B., Bhattacharyya, R. S., Feldman, D., & Knox, S. J. (2006). Inhibition of androgen receptor signaling by selenite and methylseleninic acid in prostate cancer cells: Two distinct mechanisms of action. *Molecular Cancer Therapeutics, 5*(8), 2078–2085.

Hussain, R. A., Badshah, A., Pezzuto, J. M., Ahmed, N., Kondratyuk, T. P., & Park, E. J. (2015). Ferrocene incorporated selenoureas as anticancer agents. *Journal of Photochemistry and Photobiology. B, Biology, 148*, 197–208.

Hussain, R. A., Badshah, A., Tahir, M. N., Tamoorul, H., & Bano, A. (2014). Synthesis, chemical characterization, DNA binding, antioxidant, antibacterial, and antifungal activities of ferrocence incorporated selenoureas. *Journal of Biochemical and Molecular Toxicology, 28*(2), 60–68.

Ibanez, E., Agliano, A., Prior, C., Nguewa, P., Redrado, M., Gonzalez-Zubeldia, I., et al. (2012). The quinoline imidoselenocarbamate EI201 blocks the AKT/mTOR pathway and targets cancer stem cells leading to a strong antitumor activity. *Current Medicinal Chemistry, 19*(18), 3031–3043.

Infante, H. G., Joel, S. P., Warburton, E., Hopley, C., Hearn, R., & Juliger, S. (2007). Investigation of the selenium species distribution in a human B-cell lymphoma line by HPLC- and GC-ICP-MS in combination with HPLCESIMS/MS and GC-TOFMS after incubation with methylseleninic acid. *Journal of Analytical Atomic Spectrometry, 22*(8), 888–896.

Ivanenkov, Y. A., Veselov, M. S., Rezekin, I. G., Skvortsov, D. A., Sandulenko, Y. B., Polyakova, M. V., et al. (2016). Synthesis, isomerization and biological activity of novel 2-selenohydantoin derivatives. *Bioorganic & Medicinal Chemistry, 24*(4), 802–811.

Jacob, C., Maret, W., & Vallee, B. L. (1998a). Control of zinc transfer between thionein, metallothionein, and zinc proteins. *Proceedings of the National Academy of Sciences of the United States of America, 95*(7), 3489–3494.

Jacob, C., Maret, W., & Vallee, B. L. (1998b). Ebselen, a selenium-containing redox drug, releases zinc from metallothionein. *Biochemical and Biophysical Research Communications, 248*(3), 569–573.

Jantova, S., Mrvova, N., Hudec, R., Sedlak, J., Panik, M., & Milata, V. (2016). Pro-apoptotic effect of new quinolone 7- ethyl 9-ethyl-6-oxo-6,9-dihydro[1,2,5] selenadiazolo [3,4-h]quinoline-7-carboxylate on cervical cancer cell line HeLa alone/with UVA irradiation. *Toxicology In Vitro: An International Journal Published in Association with BIBRA, 33*, 35–44.

Jariwalla, R. J., Gangapurkar, B., & Nakamura, D. (2009). Differential sensitivity of various human tumour-derived cell types to apoptosis by organic derivatives of selenium. *The British Journal of Nutrition, 101*(2), 182–189.

Jiang, C., Ganther, H., & Lu, J. (2000). Monomethyl selenium—Specific inhibition of MMP-2 and VEGF expression: Implications for angiogenic switch regulation. *Molecular Carcinogenesis, 29*(4), 236–250.

Jiang, C., Jiang, W., Ip, C., Ganther, H., & Lu, J. (1999). Selenium-induced inhibition of angiogenesis in mammary cancer at chemopreventive levels of intake. *Molecular Carcinogenesis, 26*(4), 213–225.

Jiang, C., Kim, K. H., Wang, Z., & Lu, J. (2004). Methyl selenium-induced vascular endothelial apoptosis is executed by caspases and principally mediated by p38 MAPK pathway. *Nutrition and Cancer, 49*(2), 174–183.

Jiang, C., Wang, Z., Ganther, H., & Lu, J. (2002). Distinct effects of methylseleninic acid versus selenite on apoptosis, cell cycle, and protein kinase pathways in DU145 human prostate cancer cells. *Molecular Cancer Therapeutics, 1*(12), 1059–1066.

Jiang, Q., Wang, Y., Li, T., Shi, K., Li, Z., Ma, Y., et al. (2011). Heat shock protein 90-mediated inactivation of nuclear factor-kappaB switches autophagy to apoptosis through becn1 transcriptional inhibition in selenite-induced NB4 cells. *Molecular Biology of the Cell, 22*(8), 1167–1180.

Kassam, S., Goenaga-Infante, H., Maharaj, L., Hiley, C. T., Juliger, S., & Joel, S. P. (2011). Methylseleninic acid inhibits HDAC activity in diffuse large B-cell lymphoma cell lines. *Cancer Chemotherapy and Pharmacology, 68*(3), 815–821.

Kim, Y., Lee, Y. S., Choe, J., Lee, H., Kim, Y. M., & Jeoung, D. (2008). CD44-epidermal growth factor receptor interaction mediates hyaluronic acid-promoted cell motility by activating protein kinase C signaling involving Akt, Rac1, Phox, reactive oxygen species,

focal adhesion kinase, and MMP-2. *The Journal of Biological Chemistry*, *283*(33), 22513–22528.
Kim, C., Lee, J., & Park, M. S. (2015). Synthesis of new diorganodiselenides from organic halides: Their antiproliferative effects against human breast cancer MCF-7 cells. *Archives of Pharmacal Research*, *38*(5), 659–665.
Koizumi, H., Fujisawa, H., Suehiro, E., Shirao, S., & Suzuki, M. (2011). Neuroprotective effects of ebselen following forebrain ischemia: Involvement of glutamate and nitric oxide. *Neurologia Medico-Chirurgica (Tokyo)*, *51*(5), 337–343.
Larabee, J. L., Hocker, J. R., & Hanas, J. S. (2009). Mechanisms of inhibition of zinc-finger transcription factors by selenium compounds ebselen and selenite. *Journal of Inorganic Biochemistry*, *103*(3), 419–426.
Larabee, J. L., Hocker, J. R., Hanas, R. J., Kahn, F. M., & Hanas, J. S. (2002). Inhibition of zinc finger protein-DNA interactions by sodium selenite. *Biochemical Pharmacology*, *64*(12), 1757–1765.
Lee, J. I., Nian, H., Cooper, A. J., Sinha, R., Dai, J., Bisson, W. H., et al. (2009). Alpha-keto acid metabolites of naturally occurring organoselenium compounds as inhibitors of histone deacetylase in human prostate cancer cells. *Cancer Prevention Research (Philadelphia, Pa.)*, *2*(7), 683–693.
Lee, Y. K., Park, S. Y., Kim, Y. M., Kim, D. C., Lee, W. S., Surh, Y. J., et al. (2010). Suppression of mTOR via Akt-dependent and -independent mechanisms in selenium-treated colon cancer cells: Involvement of AMPKalpha1. *Carcinogenesis*, *31*(6), 1092–1099.
Lee, J. H., Shin, S. H., Kang, S., Lee, Y. S., & Bae, S. (2008). A novel activation-induced suicidal degradation mechanism for Akt by selenium. *International Journal of Molecular Medicine*, *21*(1), 91–97.
Lee, S. O., Yeon Chun, J., Nadiminty, N., Trump, D. L., Ip, C., Dong, Y., et al. (2006). Monomethylated selenium inhibits growth of LNCaP human prostate cancer xenograft accompanied by a decrease in the expression of androgen receptor and prostate-specific antigen (PSA). *The Prostate*, *66*(10), 1070–1075.
Lesser, R., & Weiss, R. (1924). Uber selenhaltige aromatische Verbindungen (VI). *Chemische Berichte*, *57B*, 1077–1082.
Lewerenz, J., Hewett, S. J., Huang, Y., Lambros, M., Gout, P. W., Kalivas, P. W., et al. (2013). The cystine/glutamate antiporter system x(c)(-) in health and disease: From molecular mechanisms to novel therapeutic opportunities. *Antioxidants & Redox Signaling*, *18*(5), 522–555.
Li, G. X., Hu, H., Jiang, C., Schuster, T., & Lu, J. (2007). Differential involvement of reactive oxygen species in apoptosis induced by two classes of selenium compounds in human prostate cancer cells. *International Journal of Cancer*, *120*(9), 2034–2043.
Li, G. X., Lee, H. J., Wang, Z., Hu, H., Liao, J. D., Watts, J. C., et al. (2008). Superior in vivo inhibitory efficacy of methylseleninic acid against human prostate cancer over selenomethionine or selenite. *Carcinogenesis*, *29*(5), 1005–1012.
Li, B. L., Li, B., Zhang, R. L., Zhao, J. J., Wang, X. F., Liu, Y. M., et al. (2016). Synthesis and antiproliferative evaluation of novel 1,2,4-triazole derivatives incorporating benzisoselenazolone scaffold. *Bioorganic & Medicinal Chemistry Letters*, *26*(4), 1279–1281.
Li, Z., Meng, J., Xu, T. J., Qin, X. Y., & Zhou, X. D. (2013). Sodium selenite induces apoptosis in colon cancer cells via Bax-dependent mitochondrial pathway. *European Review for Medical and Pharmacological Sciences*, *17*(16), 2166–2171.
Lill, R., Hoffmann, B., Molik, S., Pierik, A. J., Rietzschel, N., Stehling, O., et al. (2012). The role of mitochondria in cellular iron-sulfur protein biogenesis and iron metabolism. *Biochimica et Biophysica Acta*, *1823*(9), 1491–1508.
Liu, M., Hu, C., Xu, Q., Chen, L., Ma, K., Xu, N., et al. (2015). Methylseleninic acid activates Keap1/Nrf2 pathway via up-regulating miR-200a in human oesophageal

squamous cell carcinoma cells. *Bioscience Reports, 35*(5), e00256. http://dx.doi.org/10.1042/BSR20150092.

Liu, Y., Luo, Y., Li, X., Zheng, W., & Chen, T. (2015). Rational design of selenadiazole derivatives to antagonize hyperglycemia-induced drug resistance in cancer cells. *Chemistry, An Asian Journal, 10*(3), 642–652.

Luchetti, F., Canonico, B., Betti, M., Arcangeletti, M., Pilolli, F., Piroddi, M., et al. (2010). Melatonin signaling and cell protection function. *The FASEB Journal, 24*(10), 3603–3624.

Lunoe, K., Gabel-Jensen, C., Sturup, S., Andresen, L., Skov, S., & Gammelgaard, B. (2011). Investigation of the selenium metabolism in cancer cell lines. *Metallomics: Integrated Biometal Science, 3*(2), 162–168.

Luo, H., Yang, Y., Duan, J., Wu, P., Jiang, Q., & Xu, C. (2013). PTEN-regulated AKT/FoxO3a/Bim signaling contributes to reactive oxygen species-mediated apoptosis in selenite-treated colorectal cancer cells. *Cell Death & Disease, 4*, e481.

Luo, H., Yang, Y., Huang, F., Li, F., Jiang, Q., Shi, K., et al. (2012). Selenite induces apoptosis in colorectal cancer cells via AKT-mediated inhibition of beta-catenin survival axis. *Cancer Letters, 315*(1), 78–85.

Mackay, H. J., & Twelves, C. J. (2008). Targeting the protein kinase C family: Are we there yet? *Nature Reviews Cancer, 8*(3), 255.

Mandal, P. K., Seiler, A., Perisic, T., Kolle, P., Canak, A. B., Forster, H., et al. (2010). System x(c)(-) and thioredoxin reductase 1 cooperatively rescue glutathione deficiency. *Journal of Biological Chemistry, 285*(29), 22244–22253.

Martins, I. L., Charneira, C., Gandin, V., Ferreira da Silva, J. L., Justino, G. C., Telo, J. P., et al. (2015). Selenium-containing chrysin and quercetin derivatives: Attractive scaffolds for cancer therapy. *Journal of Medicinal Chemistry, 58*(10), 4250–4265.

Matthews, J. R., Wakasugi, N., Virelizier, J. L., Yodoi, J., & Hay, R. T. (1992). Thioredoxin regulates the DNA binding activity of NF-kappa B by reduction of a disulphide bond involving cysteine 62. *Nucleic Acids Research, 20*(15), 3821–3830.

Misra, S., Boylan, M., Selvam, A., Spallholz, J. E., & Bjornstedt, M. (2015). Redox-active selenium compounds—From toxicity and cell death to cancer treatment. *Nutrients*, 7(5), 3536–3556.

Muhlenhoff, U., Molik, S., Godoy, J. R., Uzarska, M. A., Richter, N., Seubert, A., et al. (2010). Cytosolic monothiol glutaredoxins function in intracellular iron sensing and trafficking via their bound iron-sulfur cluster. *Cell Metabolism, 12*(4), 373–385.

Muller, A., Cadenas, E., Graf, P., & Sies, H. (1984). A novel biologically active seleno-organic compound—I. Glutathione peroxidase-like activity in vitro and antioxidant capacity of PZ 51 (Ebselen). *Biochemical Pharmacology, 33*(20), 3235–3239.

Muller, S., Heider, J., & Bock, A. (1997). The path of unspecific incorporation of selenium in Escherichia coli. *Archives of Microbiology, 168*(5), 421–427.

Nascimento, V., Ferreira, N. L., Canto, R. F., Schott, K. L., Waczuk, E. P., Sancineto, L., et al. (2014). Synthesis and biological evaluation of new nitrogen-containing diselenides. *European Journal of Medicinal Chemistry, 87*, 131–139.

Nian, H., Bisson, W. H., Dashwood, W. M., Pinto, J. T., & Dashwood, R. H. (2009). Alpha-keto acid metabolites of organoselenium compounds inhibit histone deacetylase activity in human colon cancer cells. *Carcinogenesis, 30*(8), 1416–1423.

Nikawa, T., Schuch, G., Wagner, G., & Sies, H. (1994a). Interaction of albumin-bound ebselen with rat-liver glutathione-S-transferase and microsomal proteins. *Biochemistry and Molecular Biology International, 32*(2), 291–298.

Nikawa, T., Schuch, G., Wagner, G., & Sies, H. (1994b). Interaction of ebselen with glutathione-S-transferase and papain in-vitro. *Biochemical Pharmacology, 47*(6), 1007–1012.

Nogueira, C. W., & Rocha, J. B. (2011). Toxicology and pharmacology of selenium: Emphasis on synthetic organoselenium compounds. *Archives of Toxicology, 85*(11), 1313–1359.

Nogueira, C. W., Zeni, G., & Rocha, J. B. (2004). Organoselenium and organotellurium compounds: Toxicology and pharmacology. *Chemical Reviews, 104*(12), 6255–6285.

Olm, E., Fernandes, A. P., Hebert, C., Rundlof, A. K., Larsen, E. H., Danielsson, O., et al. (2009). Extracellular thiol-assisted selenium uptake dependent on the x(c)- cystine transporter explains the cancer-specific cytotoxicity of selenite. *Proceedings of the National Academy of Sciences of the United States of America, 106*(27), 11400–11405.

Pacula, A. J., Kaczor, K. B., Wojtowicz, A., Antosiewicz, J., Janecka, A., Dlugosz, A., et al. (2017). New glutathione peroxidase mimetics-insights into antioxidant and cytotoxic activity. *Bioorganic & Medicinal Chemistry, 25*(1), 126–131.

Pantano, C., Reynaert, N. L., van der Vliet, A., & Janssen-Heininger, Y. M. (2006). Redox-sensitive kinases of the nuclear factor-kappaB signaling pathway. *Antioxidants & Redox Signaling, 8*(9–10), 1791–1806.

Papp, L. V., Lu, J., Holmgren, A., & Khanna, K. K. (2007). From selenium to selenoproteins: Synthesis, identity, and their role in human health. *Antioxidants & Redox Signaling, 9*(7), 775–806.

Park, J. M., Kim, A., Oh, J. H., & Chung, A. S. (2007). Methylseleninic acid inhibits PMA-stimulated pro-MMP-2 activation mediated by MT1-MMP expression and further tumor invasion through suppression of NF-kappaB activation. *Carcinogenesis, 28*(4), 837–847.

Park, H. S., Park, E., Kim, M. S., Ahn, K., Kim, I. Y., & Choi, E. J. (2000). Selenite inhibits the c-Jun N-terminal kinase/stress-activated protein kinase (JNK/SAPK) through a thiol redox mechanism. *The Journal of Biological Chemistry, 275*(4), 2527–2531.

Pei, Z., Li, H., Guo, Y., Jin, Y., & Lin, D. (2010). Sodium selenite inhibits the expression of VEGF, TGFbeta(1) and IL-6 induced by LPS in human PC3 cells via TLR4-NF-(K)B signaling blockage. *International Immunopharmacology, 10*(1), 50–56.

Pierpaoli, E., Viola, V., Barucca, A., Orlando, F., Galli, F., & Provinciali, M. (2013). Effect of annatto-tocotrienols supplementation on the development of mammary tumors in HER-2/neu transgenic mice. *Carcinogenesis, 34*(6), 1352–1360.

Plano, D., Karelia, D. N., Pandey, M. K., Spallholz, J. E., Amin, S., & Sharma, A. K. (2016). Design, synthesis, and biological evaluation of novel selenium (Se-NSAID) molecules as anticancer agents. *Journal of Medicinal Chemistry, 59*(5), 1946–1959.

Ranawat, P., & Bansal, M. P. (2008). Decreased glutathione levels potentiate the apoptotic efficacy of selenium: Possible involvement of p38 and JNK MAPKs—In vitro studies. *Molecular and Cellular Biochemistry, 309*(1–2), 21–32.

Rayman, M. P. (2008). Food-chain selenium and human health: Emphasis on intake. *The British Journal of Nutrition, 100*(2), 254–268.

Redman, C., Scott, J. A., Baines, A. T., Basye, J. L., Clark, L. C., Calley, C., et al. (1998). Inhibitory effect of selenomethionine on the growth of three selected human tumor cell lines. *Cancer Letters, 125*(1–2), 103–110.

Romano, B., Plano, D., Encio, I., Palop, J. A., & Sanmartin, C. (2015). In vitro radical scavenging and cytotoxic activities of novel hybrid selenocarbamates. *Bioorganic & Medicinal Chemistry, 23*(8), 1716–1727.

Rooprai, H. K., Kyriazis, I., Nuttall, R. K., Edwards, D. R., Zicha, D., Aubyn, D., et al. (2007). Inhibition of invasion and induction of apoptosis by selenium in human malignant brain tumour cells in vitro. *International Journal of Oncology, 30*(5), 1263–1271.

Rudolf, E., Radocha, J., Cervinka, M., & Cerman, J. (2004). Combined effect of sodium selenite and campthotecin on cervical carcinoma cells. *Neoplasma, 51*(2), 127–135.

Rudolf, E., Rudolf, K., & Cervinka, M. (2008). Selenium activates p53 and p38 pathways and induces caspase-independent cell death in cervical cancer cells. *Cell Biology and Toxicology, 24*(2), 123–141.

Saccoccia, F., Angelucci, F., Boumis, G., Carotti, D., Desiato, G., Miele, A. E., et al. (2014). Thioredoxin reductase and its inhibitors. *Current Protein & Peptide Science, 15*(6), 621–646.

Sag, C. M., Wagner, S., & Maier, L. S. (2013). Role of oxidants on calcium and sodium movement in healthy and diseased cardiac myocytes. *Free Radical Biology & Medicine, 63*, 338–349.

Santi, C., Tidei, C., Scalera, C., Piroddi, M., & Galli, F. (2013). Selenium containing compounds from poison to drug candidates: A review on the GPx-like activity. *Current Chemical Biology*, 7(1), 25–36.

Shaaban, S., Negm, A., Sobh, M. A., & Wessjohann, L. A. (2015). Organoselenocyanates and symmetrical diselenides redox modulators: Design, synthesis and biological evaluation. *European Journal of Medicinal Chemistry, 97*, 190–201.

Shah, Y. M., Kaul, A., Dong, Y., Ip, C., & Rowan, B. G. (2005). Attenuation of estrogen receptor alpha (ERalpha) signaling by selenium in breast cancer cells via downregulation of ERalpha gene expression. *Breast Cancer Research and Treatment, 92*(3), 239–250.

Sharma, A. K., Sharma, A., Desai, D., Madhunapantula, S. V., Huh, S. J., Robertson, G. P., et al. (2008). Synthesis and anticancer activity comparison of phenylalkyl isoselenocyanates with corresponding naturally occurring and synthetic isothiocyanates. *Journal of Medicinal Chemistry, 51*(24), 7820–7826.

Sheftel, A., Stehling, O., & Lill, R. (2010). Iron-sulfur proteins in health and disease. *Trends in Endocrinology and Metabolism, 21*(5), 302–314.

Shi, K., Jiang, Q., Li, Z., Shan, L., Li, F., An, J., et al. (2013). Sodium selenite alters microtubule assembly and induces apoptosis in vitro and in vivo. *Journal of Hematology & Oncology, 6*, 7.

Sieber, F., Daziano, J. P., Gunther, W. H. H., Krieg, M., Miyagi, K., Sampson, R. W., et al. (2005). Elemental selenium generated by the photobleaching of seleno-merocyanine photosensitizers forms conjugates with serum macro-molecules that are toxic to tumor cells. *Phosphorus, Sulfur and Silicon and the Related Elements, 180*(3-4), 647–657.

Sinha, I., Allen, J. E., Pinto, J. T., & Sinha, R. (2014). Methylseleninic acid elevates REDD1 and inhibits prostate cancer cell growth despite AKT activation and mTOR dysregulation in hypoxia. *Cancer Medicine, 3*(2), 252–264.

Sinha, R., & Medina, D. (1997). Inhibition of cdk2 kinase activity by methylselenocysteine in synchronized mouse mammary epithelial tumor cells. *Carcinogenesis, 18*(8), 1541–1547.

Sinha, I., Null, K., Wolter, W., Suckow, M. A., King, T., Pinto, J. T., et al. (2012). Methylseleninic acid downregulates hypoxia-inducible factor-1alpha in invasive prostate cancer. *International Journal of Cancer, 130*(6), 1430–1439.

Song, H., Hur, I., Park, H. J., Nam, J., Park, G. B., Kong, K. H., et al. (2009). Selenium inhibits metastasis of murine melanoma cells through the induction of cell cycle arrest and cell death. *Immune Network, 9*(6), 236–242.

Song, H., Kim, J., Lee, H. K., Park, H. J., Nam, J., Park, G. B., et al. (2011). Selenium inhibits migration of murine melanoma cells via down-modulation of IL-18 expression. *International Immunopharmacology, 11*(12), 2208–2213.

Squires, J., & Berry, M. J. (2006). Selenium, selenoproteins, and cancer. *Hawaii Medical Journal, 65*(8), 239–240.

Srivastava, P. C., Revankar, G. R., & Robins, R. K. (1984). Synthesis and biological activity of nucleosides and nucleotides related to the antitumor agent 2-beta-D-ribofuranosylthiazole-4-carboxamide. *Journal of Medicinal Chemistry, 27*(3), 266–269.

Tan, Q., Li, J., Yin, H. W., Wang, L. H., Tang, W. C., Zhao, F., et al. (2010). Augmented antitumor effects of combination therapy of cisplatin with ethaselen as a novel thioredoxin reductase inhibitor on human A549 cell in vivo. *Investigational New Drugs, 28*(3), 205–215.

Tarrado-Castellarnau, M., Cortes, R., Zanuy, M., Tarrago-Celada, J., Polat, I. H., Hill, R., et al. (2015). Methylseleninic acid promotes antitumour effects via nuclear FOXO3a translocation through Akt inhibition. *Pharmacological Research*, *102*, 218–234.

Tew, K. D. (1994). Glutathione-associated enzymes in anticancer drug resistance. *Cancer Research*, *54*(16), 4313–4320.

Tew, K. D. (2016). Glutathione-associated enzymes in anticancer drug resistance. *Cancer Research*, *76*(1), 7–9.

Tew, K. D., Manevich, Y., Grek, C., Xiong, Y., Uys, J., & Townsend, D. M. (2011). The role of glutathione S-transferase P in signaling pathways and S-glutathionylation in cancer. *Free Radical Biology and Medicine*, *51*(2), 299–313.

Tidei, C., Piroddi, M., Galli, F., & Santi, C. (2012). Oxidation of thiols promoted by PhSeZnCl. *Tetrahedron Letters*, *53*(2), 232–234.

Tsuji, Y., Suzuki, N., Suzuki, K. T., & Ogra, Y. (2009). Selenium metabolism in rats with long-term ingestion of Se-methylselenocysteine using enriched stable isotopes. *Journal of Toxicological Sciences*, *34*(2), 191–200.

Unni, E., Koul, D., Yung, W. K., & Sinha, R. (2005). Se-methylselenocysteine inhibits phosphatidylinositol 3-kinase activity of mouse mammary epithelial tumor cells in vitro. *Breast Cancer Research*, 7(5), R699–707.

Wallenberg, M., Misra, S., Wasik, A. M., Marzano, C., Bjornstedt, M., Gandin, V., et al. (2014). Selenium induces a multi-targeted cell death process in addition to ROS formation. *Journal of Cellular and Molecular Medicine*, *18*(4), 671–684.

Wallenberg, M., Olm, E., Hebert, C., Bjornstedt, M., & Fernandes, A. P. (2010). Selenium compounds are substrates for glutaredoxins: A novel pathway for selenium metabolism and a potential mechanism for selenium-mediated cytotoxicity. *The Biochemical Journal*, *429*(1), 85–93.

Wang, L., Bonorden, M. J., Li, G. X., Lee, H. J., Hu, H., Zhang, Y., et al. (2009). Methylselenium compounds inhibit prostate carcinogenesis in the transgenic adenocarcinoma of mouse prostate model with survival benefit. *Cancer Prevention Research (Philadelphia, Pa.)*, *2*(5), 484–495.

Wang, L., Guo, X., Wang, J., Jiang, C., Bosland, M. C., Lu, J., et al. (2016). Methylseleninic acid superactivates p53-senescence cancer progression barrier in prostate lesions of Pten-knockout mouse. *Cancer Prevention Research (Philadelphia, Pa.)*, *9*(1), 35–42.

Wang, Z., Hu, H., Li, G., Lee, H. J., Jiang, C., Kim, S. H., et al. (2008). Methylseleninic acid inhibits microvascular endothelial G1 cell cycle progression and decreases tumor microvessel density. *International Journal of Cancer*, *122*(1), 15–24.

Wang, Z., Jiang, C., Ganther, H., & Lu, J. (2001). Antimitogenic and proapoptotic activities of methylseleninic acid in vascular endothelial cells and associated effects on PI3K-AKT, ERK, JNK and p38 MAPK signaling. *Cancer Research*, *61*(19), 7171–7178.

Wang, Y. F., Xu, Y. L., Tang, Z. H., Li, T., Zhang, L. L., Chen, X. P., et al. (2017). Baicalein induces beclin 1-and extracellular signal-regulated kinase-dependent autophagy in ovarian cancer cells. *American Journal of Chinese Medicine*, *45*(1), 123–136.

Wang, L., Yang, Z., Fu, J., Yin, H., Xiong, K., Tan, Q., et al. (2012). Ethaselen: A potent mammalian thioredoxin reductase 1 inhibitor and novel organoselenium anticancer agent. *Free Radical Biology & Medicine*, *52*(5), 898–908.

Weekley, C. M., Aitken, J. B., Finney, L., Vogt, S., Witting, P. K., & Harris, H. H. (2013). Selenium metabolism in cancer cells: The combined application of XAS and XFM techniques to the problem of selenium speciation in biological systems. *Nutrients*, *5*(5), 1734–1756.

Wen, Z., Xu, J., Wang, Z., Qi, H., Xu, Q., Bai, Z., et al. (2015). 3-(3,4,5-Trimethoxyphenylselenyl)-1H-indoles and their selenoxides as combretastatin A-4 analogs: Microwave-assisted synthesis and biological evaluation. *European Journal of Medicinal Chemistry*, *90*, 184–194.

Wu, Y., Zhang, H., Dong, Y., Park, Y. M., & Ip, C. (2005). Endoplasmic reticulum stress signal mediators are targets of selenium action. *Cancer Research*, *65*(19), 9073–9079.
Wu, X., Zhang, Y., Pei, Z., Chen, S., Yang, X., Chen, Y., et al. (2012). Methylseleninic acid restricts tumor growth in nude mice model of metastatic breast cancer probably via inhibiting angiopoietin-2. *BMC Cancer*, *12*, 192.
Wu, Y., Zu, K., Warren, M. A., Wallace, P. K., & Ip, C. (2006). Delineating the mechanism by which selenium deactivates Akt in prostate cancer cells. *Molecular Cancer Therapeutics*, *5*(2), 246–252.
Xiang, N., Zhao, R., Song, G., & Zhong, W. (2008). Selenite reactivates silenced genes by modifying DNA methylation and histones in prostate cancer cells. *Carcinogenesis*, *29*(11), 2175–2181.
Xu, J., Zuo, D., Qi, H., Shen, Q., Bai, Z., Han, M., et al. (2016). 2-Methoxy-5((3,4,5-trimethosyphenyl)seleninyl) phenol (SQ0814061), a novel microtubule inhibitor, evokes G2/M cell cycle arrest and apoptosis in human breast cancer cells. *Biomedicine & Pharmacotherapy*, *78*, 308–321.
Yan, L., & DeMars, L. C. (2012). Dietary supplementation with methylseleninic acid, but not selenomethionine, reduces spontaneous metastasis of Lewis lung carcinoma in mice. *International Journal of Cancer*, *131*(6), 1260–1266.
Yan, J., Guo, Y., Wang, Y., Mao, F., Huang, L., & Li, X. (2015). Design, synthesis, and biological evaluation of benzoselenazole-stilbene hybrids as multi-target-directed anticancer agents. *European Journal of Medicinal Chemistry*, *95*, 220–229.
Yan, L., Yee, J. A., Li, D., McGuire, M. H., & Graef, G. L. (1999). Dietary supplementation of selenomethionine reduces metastasis of melanoma cells in mice. *Anticancer Research*, *19*(2A), 1337–1342.
Yan, L., Yee, J. A., McGuire, M. H., & Graef, G. L. (1997). Effect of dietary supplementation of selenite on pulmonary metastasis of melanoma cells in mice. *Nutrition and Cancer*, *28*(2), 165–169.
Ye, S. F., Li, J., Ji, S. M., Zeng, H. H., & Lu, W. (2017). Dose-biomarker-response modeling of the anticancer effect of ethaselen in a human non-small cell lung cancer xenograft mouse model. *Acta Pharmacologica Sinica*, *38*(2), 223–232.
Ye, Z. W., Zhang, J., Ancrum, T., Manevich, Y., Townsend, D. M., & Tew, K. D. (2017). Glutathione S-transferase P-mediated protein S-glutathionylation of resident endoplasmic reticulum proteins influences sensitivity to drug-induced unfolded protein response. *Antioxidants & Redox Signaling*, *26*(6), 248–262.
Zeng, H., Briske-Anderson, M., Idso, J. P., & Hunt, C. D. (2006). The selenium metabolite methylselenol inhibits the migration and invasion potential of HT1080 tumor cells. *The Journal of Nutrition*, *136*(6), 1528–1532.
Zeng, H., Briske-Anderson, M., Wu, M., & Moyer, M. P. (2012). Methylselenol, a selenium metabolite, plays common and different roles in cancerous colon HCT116 cell and noncancerous NCM460 colon cell proliferation. *Nutrition and Cancer*, *64*(1), 128–135.
Zeng, H., & Wu, M. (2015). The inhibitory efficacy of methylseleninic acid against colon cancer xenografts in C57BL/6 mice. *Nutrition and Cancer*, *67*(5), 831–838.
Zeng, H., Wu, M., & Botnen, J. H. (2009). Methylselenol, a selenium metabolite, induces cell cycle arrest in G1 phase and apoptosis via the extracellular-regulated kinase 1/2 pathway and other cancer signaling genes. *The Journal of Nutrition*, *139*(9), 1613–1618.
Zheng, X., Zhang, Y., Zhang, L., Xu, W., Ma, W., Sun, R., et al. (2016). Synergistic inhibition of sunitinib and ethaselen against human colorectal cancer cells proliferation. *Biomedicine & Pharmacotherapy*, *83*, 212–220.
Zhou, M., Ji, S., Wu, Z., Li, Y., Zheng, W., Zhou, H., et al. (2015). Synthesis of selenazolopyridine derivatives with capability to induce apoptosis in human breast carcinoma MCF-7 cells through scavenge of intracellular ROS. *European Journal of Medicinal Chemistry*, *96*, 92–97.

Zorn, M., Ihling, C. H., Golbik, R., Sawers, R. G., & Sinz, A. (2013). Selective selC-independent selenocysteine incorporation into formate dehydrogenases. *PLoS One*, *8*(4), e61913.

Zou, Y., Niu, P., Yang, J., Yuan, J., Wu, T., & Chen, X. (2008). The JNK signaling pathway is involved in sodium-selenite-induced apoptosis mediated by reactive oxygen in HepG2 cells. *Cancer Biology & Therapy*, 7(5), 689–696.

Zu, K., Bihani, T., Lin, A., Park, Y. M., Mori, K., & Ip, C. (2006). Enhanced selenium effect on growth arrest by BiP/GRP78 knockdown in p53-null human prostate cancer cells. *Oncogene*, *25*(4), 546–554.

Zuo, D., Guo, D., Jiang, X., Guan, Q., Qi, H., Xu, J., et al. (2015). 3-(3-Hydroxy-4-methoxyphenyl)-4-(3,4,5-trimethoxyphenyl)-1,2,5-selenadiazole (G-1103), a novel combretastatin A-4 analog, induces G2/M arrest and apoptosis by disrupting tubulin polymerization in human cervical HeLa cells and fibrosarcoma HT-1080 cells. *Chemico-Biological Interactions*, *227*, 7–17.

Kolibri 360 GmbH
Ferdinand-Jühlke-Straße 7
99095 Erfurt, Deutschland
produktsicherheit@kolibri360.de